Practical Audio Electronics

Practical Audio Electronics is a comprehensive introduction to basic audio electronics and the fundamentals of sound circuit building, providing the reader with the necessary knowledge and skills to undertake projects from scratch.

Imparting a thorough foundation of theory alongside the practical skills needed to understand, build, modify, and test audio circuits, this book equips the reader with the tools to explore the sonic possibilities that emerge when electronics technology is applied innovatively to the making of music. Suitable for all levels of technical proficiency, this book encourages a deeper understanding through highlighted sections of advanced material and example projects including circuits to make, alter, and amplify audio, providing a snapshot of the wide range of possibilities of practical audio electronics.

An ideal resource for students, hobbyists, musicians, audio professionals, and those interested in exploring the possibilities of hardware-based sound and music creation.

Kevin Robinson is a degree-qualified electronics engineer who also holds diplomas in electronic music production and live sound engineering. He has taught electronics and audio technology on various diploma- and degree-level sound engineering courses, and also works as a live sound engineer.

Practical Audio Electronics

Kevin Robinson

LONDON AND NEW YORK

First published 2020
by Routledge
2 Park Square, Milton Park, Abingdon, Oxon OX14 4RN

and by Routledge
52 Vanderbilt Avenue, New York, NY 10017

Routledge is an imprint of the Taylor & Francis Group, an informa business

British Library Cataloguing-in-Publication Data
A catalogue record for this book is available from the British Library

Library of Congress Cataloging-in-Publication Data
Names: Robinson, Kevin (Electronics engineer), author.
Title: Practical audio electronics / Kevin Robinson.
Description: Abingdon, Oxon : Routledge, an imprint of the Taylor & Francis Group, 2020. | Includes bibliographical references and index.
Identifiers: LCCN 2019047479 (print) | LCCN 2019047480 (ebook) | ISBN 9780367359850 (paperback) | ISBN 9780367359867 (hardback) | ISBN 9780429343056 (ebook)
Subjects: LCSH: Sound–Recording and reproducing–Equipment and supplies. | Electronic circuits.
Classification: LCC TK7881.4 .R624 2020 (print) | LCC TK7881.4 (ebook) | DDC 621.389/3–dc23
LC record available at https://lccn.loc.gov/2019047479
LC ebook record available at https://lccn.loc.gov/2019047480

ISBN: 978-0-367-35986-7 (hbk)
ISBN: 978-0-367-35985-0 (pbk)
ISBN: 978-0-429-34305-6 (ebk)

Typeset in CharterBT

Publisher's Note
This book has been prepared from camera-ready copy provided by the author.

For Jill

and
Pigasus, *the flying pig*

The author welcomes feedback at: pigsmayfly@eircom.net

Contents

Appendices

Tables

Learning by Doing

Proofs and Derivations

Preface

The purpose of this book is to present basic electronics in an audio context. The vast majority of modern audio technology is electronic in nature. Furthermore, a growing interest in custom and circuit-bent musical instruments and equipment renders a basic level of electronics expertise ever more relevant and valuable to the modern musician, audio professional, and sonic experimenter.

Electronics can be approached at a variety of levels. As a highly academic engineering discipline it merits study in the context of a technically rigorous electronics degree, and a career of application, and deepening theoretical knowledge and comprehension.

This is not the electronics addressed here.

As a practical tool for technically minded musicians, music producers, and sonic artists, electronics can open up a world of experimentation and artistic possibilities, without the need for the depth and rigour of the professional electronics engineer. Creativity wedded to a modest but solid technical knowledge-base opens this rich world of experimentation and inspiration, expanding the boundaries to create new sounds in interesting and unusual ways.

Of the very many electronics textbooks available, few enough approach the subject from a more informal standpoint, better suited to application in less technical disciplines. For the most part they are aimed at electronics engineers, and rely on detailed technical analysis and much mathematics. Fewer still diverge from the general approach to the subject, and provide a treatment specifically focused on the audio electronics sub-specialty. This book aims to address these dual requirements, with an accessible, audio-centric presentation. It provides a relatively rigorous but easily digested treatment, with enough technical detail to allow the reader to expand their understanding, while avoiding the use of complex mathematics and analysis.

This book is aimed at the technically minded electronics non-specialist who wishes to explore circuit building and hardware hacking, without getting bogged down in electronics theory and advanced mathematics. It allows students, musicians, hobbyists, and audio professionals to develop the skills and understanding needed to engage creatively in the rich and varied hacking subculture which exists where music and engineering intersect.

The book combines this moderate level of technicality with a firm focus on the practical. The aim is to allow the reader to become an electronics builder and experimenter, understanding more than is achieved by simply following step by step circuit building tutorials, but avoiding the limiting requirements of a full technical treatment.

KPR
October 2019

1 Introduction

This brief introductory chapter outlines the broad framework adopted for the material presented in the rest of the book, and highlights some of the important features of the text. These include a number of special 'optional asides' presenting more technical material apart from the main flow of the text. There are also numerous practical exercises providing the opportunity for a deepened understanding of the subject through guided practical work. The main body of the book is presented in three parts – Electrical Theory, Practical Electronics, and Component Reference – followed by a number of appendices, and closing with a bibliography and index.

Part I – Electrical Theory starts off by introducing the fundamental concepts and quantities in electricity, including voltage, current, resistance, and power. As semiconductor materials play such a key role in modern electronics, a brief introduction to these is provided, outlining the science behind their structure and operation. Later this knowledge is applied when looking at how various semiconductor devices operate. While for the most part the material in this book does not involve any work with mains power, it is well worth acquiring a basic understanding; a chapter on mains electricity and electrical safety covers the important aspects of working in this potentially hazardous environment. Part I is rounded off with a chapter on signal characteristics, providing some background to the key aspects of audio signals.

Part II – Practical Electronics looks at the most important tools of the trade and how to use them. All stages of designing, prototyping, building, and testing of audio circuits are covered. After a quick overview of common components, the drafting and interpretation of circuit diagrams is examined. Next, the primary tools and equipment used for developing and testing new circuits are described. Some simple analysis procedures are introduced to allow basic circuit behaviour to be understood and predicted, and then the hands-on activities involved in actually building circuits are addressed. Part II is concluded with a set of fully worked projects to build, including a simple battery powered amplifier, an interesting sound synthesis circuit and a useful audio effects processor – all the building blocks needed to start making some noise.

Part III – Component Reference is composed of a series of chapters detailing the most important facts and figures for all the different components likely to be encountered in audio electronic circuits, including numerous applications and circuit examples throughout. The focus is on that information which can prove most useful in understanding and working with these components, while the more obscure theory aspects are kept to a minimum.

Appendices round out the body of the book, presenting useful reference material as well as bringing together a catalogue of the most interesting circuits encountered throughout, complete with breadboard and stripboard layout designs for each, allowing them to be easily and quickly built, tested, and experimented with.

LEARNING BY DOING

The core guiding principle of this book is that the best way to learn electronics is to experiment with electronic concepts; to build and test electronic circuits, to see how they work first hand. Throughout the text, the reader will encounter highlighted sections such as this, and bearing a 'Learning by Doing' tag line as above. Each exercise aims to encourage the reader to expand and deepen their knowledge and understanding of a particular topic by engaging with it in a practical way.

As with the Proofs and Derivations above, a list of these Learning by Doing exercises is included in the front matter, see Learning by Doing, p. xv.

Proofs and Derivations

As this volume is intended to be first and foremost a practical book, maths has been kept to a minimum in the main body of the text. For the reader who wishes to take their understanding a little further, and perhaps gain a more in-depth appreciation of some of the theory underpinning the concepts encountered, a few simple mathematical proofs and derivations are presented throughout the text, indicated by their inclusion within a shaded box such as this.

The aim is to provide only the very basics of a more rigorous approach to the technical aspects of the subject matter and to highlight some of the relationships between different concepts and ideas presented in the main text. An illustration of the relationships and dependencies which exist between key details can assist greatly in developing a fuller understanding of the subject matter at hand. Such understanding is not essential to the core goals of this book, which are to provide the tools to enter and explore the possibilities of practical audio electronics building and experimentation, but it is hoped the interested reader will find these excursions both interesting and enlightening.

A full list of these proofs and derivations can be found in the front matter, see Proofs and Derivations, p. xvii.

Part I

Electrical Theory

2 | Electricity

Key Concepts in Electricity

This chapter introduces a number of key concepts. In so doing it provides a solid framework for the development of an understanding of the behaviour of electronic components and electrical circuits. In keeping with the stated aims of the book to avoid unnecessary levels of detail, the topics discussed here are presented in a relatively broad fashion, with the goal of facilitating a sufficient degree of understanding to guide the practical electronics to come. The technically inclined reader has at their disposal many more rigorous, academic textbooks to choose from if a deeper treatment is sought (see for example Horowitz and Hill, 2015; Sedra and Smith, 2014). The primary goal in this brief chapter is to provide a description of the importance of, and the relationships between, the following key concepts:

Electric Charge – positively and negatively charged particles are the basis of electricity
Static and Current Electricity – charge can accumulate (static) or it can flow (current)
Current and Circuits – a loop or circuit is generally needed in order for current to flow
Direct and Alternating Current – the flow of current can be one way or bidirectional
Voltage – the push that causes current to flow *(voltage up $\rightarrow$ current up)*
Resistance – the opposition to the flow of current *(resistance up $\rightarrow$ current down)*
Power – when current flows through a circuit work is done, dissipating power
Electricity and Magnetism – the two mutually dependant aspects of electromagnetism
Conductors and Insulators – materials through which current can and can not flow

Electricity deals with the accumulation and movement of charged particles. Two of the key constituents of all matter are positively charged particles called protons, and negatively charged particles called electrons. Overall the positive charge of the protons and the negative charge of the electrons usually cancel each other out. In certain materials (most notably metals, like copper) some of the electrons are not very tightly bound to any particular location, but rather tend to float about in what can be thought of as a cloud of quite mobile negative charge. There can therefore arise an accumulation of negative charge in one place due to an excess of electrons, and a corresponding accumulation of positive charge in another place due to a scarcity of electrons. There are a number of different ways in which this can happen but, however it comes about, such migrations of charged particles result in the creation of either a static electric charge or a flowing electrical current. Electricity is all about the accumulation and movement of charged particles. When the charged particles are moving the phenomenon is often referred to as current electricity, whereas a localised build-up of charge is called static electricity. All this leads to the three most fundamental and important quantities in basic electrical theory: current, voltage, and resistance.

An analogy is often used when explaining the basics of current electricity flowing in an electrical circuit in which there is a system of pipes and valves with water flowing through them (Figure 2.1). The pipes and valves are the wires and components making up the circuit, and the water is the electricity. When a valve is opened, water flows. How quickly the water flows depends on two things: the water pressure in the pipes and how far the valve is opened. In this analogy the water pressure equates to electrical voltage, and how much the valve has been opened equates to the electrical resistance in the circuit. These two factors combine to control the rate at which water flows in the pipes, one seeking to increase the rate of flow (the voltage or water pressure) while the other tries to limit it (the resistance or the valve). The flow of water itself equates to electrical current.

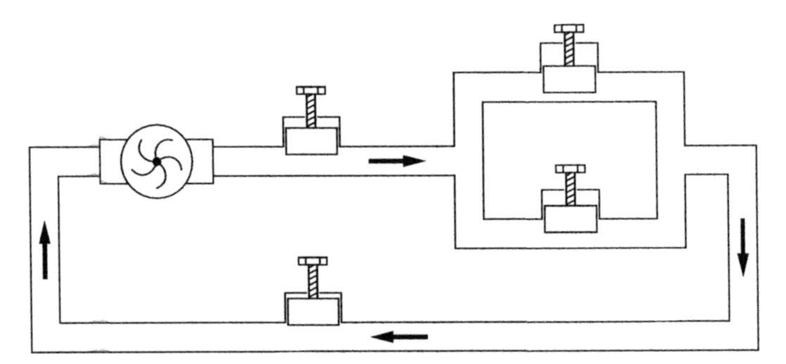

Figure 2.1 Water in pipes as an analogy for an electrical circuit. The speed of the pump corresponds to the voltage of the power supply, the valves represent resistance, and the water flowing is the electrical current.

Add to these three concepts the familiar idea of power, and the four primary quantities used in the analysis of electronics are assembled. Formally power is 'the rate of doing work'. In other words the amount of work done in a second. Ultimately the kind of work most often of interest here is going to be something like moving the cone of a loudspeaker in and out so as to produce sound. It can therefore be seen why the power of an amplifier is a commonly quoted figure. A one hundred watt guitar amplifier goes louder than a ten watt amp because the one hundred watt amplifier is capable of doing more work per second driving the loudspeaker, and thus a louder sound can be generated.

The terms alternating current and direct current (AC and DC) are also likely to be familiar. These are just two different ways in which current might flow, either bouncing back and forth, constantly changing the direction in which it is flowing (AC), or flowing steadily at a moderately constant rate always in the same direction (DC). Both types of current flow play crucial roles in different aspects of audio electronics.

The close relationship between electricity and magnetism is also explored. Indeed the two are so inextricably linked that they are often talked about in terms of a single compound concept, electromagnetism. The interactions between these two quantities are at the heart of some of the most important audio electrical devices, including the most common types of microphones and loudspeakers. While the details are unimportant in this context, a basic appreciation of their close relationship can help in understanding how audio electronic technology works.

And finally in this chapter on key concepts, conductors and insulators are addressed. How easily electricity flows through various materials is fundamental to controlling and utilising it to do useful things like driving loudspeakers and amplifying signals. Chapter 3 extends this discussion to take in the consideration of some of the most important materials in all of electronics, semiconductors, but for now the focus remains on the more fundamental concepts of simple conductors and insulators.

With a firm grasp on all of this terminology and all of these concepts, the material that follows in subsequent chapters of this book falls more easily into place. A good working knowledge and understanding of the theory of audio electronics can thus be developed, without the need to expand into the more involved theory of the subject. As such, the much more interesting and useful practical aspects of the discipline, which this book makes its primary focus, can be explored and applied to optimum effect.

Electric Charge

As has already been mentioned, charge comes in two flavours, positive and negative, and the movement and accumulation of this charge is at the root of all electricity. Some electrical components can store charge (e.g. capacitors, see Chapter 13) or generate it (e.g. batteries and power supplies), and often such components are referred to as being charged up and discharged. Electric charge is measured in coulombs (C) but for the purposes of this book that particular quantity is not needed.

Electric charge can also be specified in amp-hours, written Ah (1Ah = 3,600C). The place this quantity is most likely to be encountered is written on the side of a rechargeable battery. The capacity of such a battery is often quoted in milliamp-hours (mAh), as in Figure 2.2. (If the 'milli-' prefix is unfamiliar see Appendix A for a list of commonly encountered prefix multipliers.) The larger the number, the longer the battery should last on one charge (for batteries of the same rated voltage). Beyond the fact that bigger is generally better, the details of milliamp-hours need not be considered further.

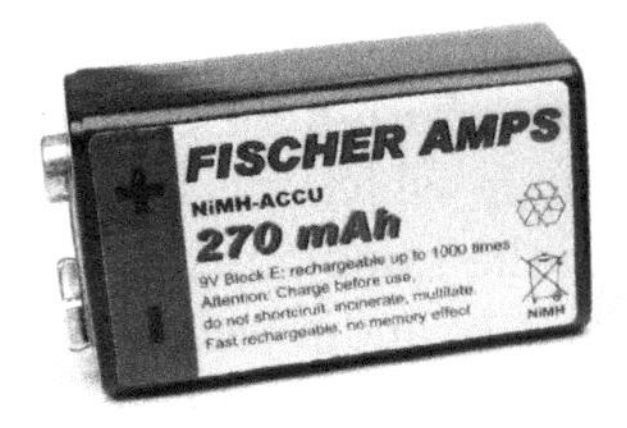

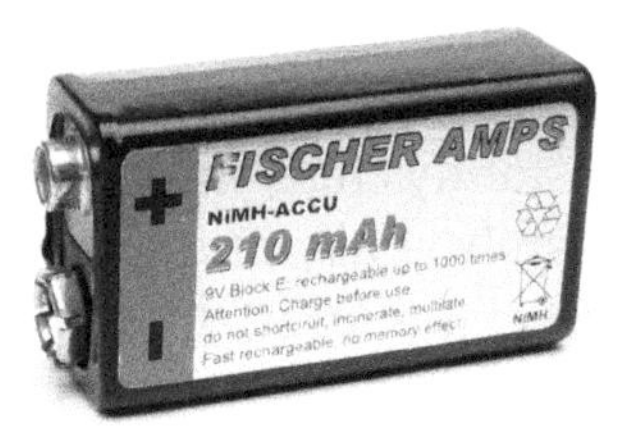

Figure 2.2 Two rechargeable 9V batteries showing their charge holding capacities of 270mAh and 210mAh respectively. A bigger number corresponds to a longer lasting battery.

Static and Current Electricity

Rub a plastic biro or pen through your hair and it can pick up small pieces of paper, or you can deflect the flow of water from a gently running tap. What is observed here is a build up of a little static charge on the plastic. The charge dissipates quickly enough but for a short while the biro becomes an electrically charged rod. Another familiar example of static electricity is the cling experienced when certain synthetic fabrics are rubbed together or pulled over one another. One surface develops a negative charge by accumulating excess electrons while the other becomes positively charged due to a decrease in electron numbers.

By and large static electricity is not of very much interest when talking about audio electronics, although there are a couple of places where it appears. Firstly, remembering what was just said about some fabrics building up a static charge, it is important to be aware that certain electronic components are very sensitive to electrostatic discharge. If such components are touched by someone carrying a static charge they can very easily be damaged. This is most common in certain types of integrated circuits (silicon chips) and transistors. It is important to avoid picking up such components while carrying a static charge, and in general handling them should be kept to a minimum.

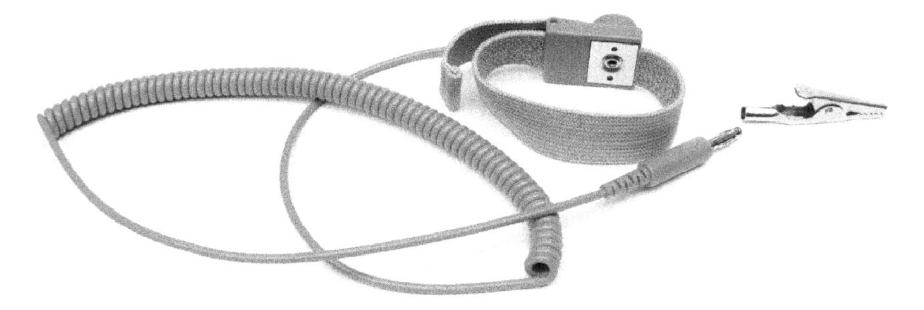

Figure 2.3 An antistatic wristband connected to earth can be worn in order to dissipate any build-up of static charge which might otherwise damage sensitive electronic components.

Learning by Doing 2.1

Static Electricity

Materials

- Plastic pen
- Paper scraps
- Water tap

Theory

Many familiar objects are capable of holding a static charge. In this experiment we use a plastic pen or biro, but other suitable objects can be substituted for a similar effect. A common party balloon is another thing often used in this kind of demonstration.

Rubbing the pen (or balloon) through your hair or against an item of clothing made from a suitable fabric results in a build up of static charge on the surface of the object in question. Typically, manmade fabrics such as polyester are good while natural fabrics such as cotton won't work.

Once the surface of the pen has been charged up with static electricity, the presence of the charge can be observed in a number of ways. The accumulated charge will tend to dissipate relatively quickly, but can be easily refreshed as you go along.

Practice

Rub the pen back and forth through your hair for a few seconds. This will build up a static charge on its surface. Repeat the charging process as needed.

Scatter a few small scraps of paper on the table, and hover over them with the pen, moving it closer until the paper scraps start to react. See how many pieces of paper you can get hanging from the pen. How large a piece of paper can the static charge lift? How long does the effect last before the paper scraps begin to fall away?

Set water flowing from a tap in a slow, steady stream (a gurgling tap is not much good, the water flow needs to be smooth for the best effect). Bring the charged pen in from the side, slowly approaching the stream of water. The flow will deflect towards the pen (be sure not to let the water and pen touch or the charge will quickly dissipate). How far can you get the stream of water to deflect? Is the effect easier to achieve high up and close to the tap, or lower down the stream of water? How strong can you make the flow before no significant deflection can be observed?

Does the top you are wearing serve to introduce a charge onto the pen's surface? Check the label to see what fabrics are used in the material. As noted above, manmade fabrics often work well for introducing static electricity while natural fibres tend not to. Try a few different fabrics.

When working with these components it is common to wear a special wrist band, as shown in Figure 2.3, connected to earth in order to avoid any chance of static build-up. Static can also be discharged by touching the metal chassis of an earthed piece of equipment. Some dedicated electronics work benches have an earthed strip along their front edge, while others will be covered in an antistatic mat. Both are designed to keep the user and work area free from any build-up of static charge. These are small but important points, which are worth remembering when it comes time to start building circuits or working inside electronic equipment.

There are also one or two places where static charge plays a key role in the operation of audio electronic components. There is a type of microphone capsule called an electret condenser microphone which depends for its operation on a small permanent static charge held by a piece of electret material within the microphone capsule. Electret material is just a particular type of substance capable of maintaining a static charge over a long period of time. It can be thought of as the electrical equivalent of a permanent magnet. Once an appropriate piece of iron has been magnetised it can stay that way indefinitely. Similarly once a piece of electret material has been charged up it can hold that charge for a very long time. The electret material inside an electret microphone can slowly lose its charge over time, and thus its ability to generate an audio signal, usually over the course of tens of years or more.

While static electricity does come into play in a small way in audio electronics, there is not much about it which needs to be known. For the most part what is of real interest is not static but rather current electricity, electricity flowing around a circuit, and in the process doing useful and interesting things. This is described next.

Current and Circuits

The first thing to remember about electricity in the context of electronic circuits is that in order for an electric current to flow, a closed circuit is needed. That is to say a loop such that the current can flow around and end up back where it started, tracing an unbroken path through wires and components. The other thing needed for current to flow is a source, something capable of actually generating the current flow. For a first, simple (and somewhat unlikely) electrical circuit imagine a battery (that's the source), and a piece of copper wire running from one terminal to the other. This is a valid electric circuit, although also not very useful and generally not a good idea. It would be referred to as a 'short circuit', and it allows as much current as the source can provide to flow through the wire. The battery is likely to get very hot and could be damaged or destroyed. Although there are exceptions, in the context of practical circuit building, short circuits such as this are generally to be avoided.

What is needed is to add a non-zero load to the circuit to limit the amount of current flowing (and hopefully do something useful in the process). In electrical terms anything attached between the terminals of a battery constitutes a load, but let's keep it simple and just add a little torch bulb (Figure 2.4). Now current flows out of the positive terminal of the battery, through a piece of wire, through the bulb, and then through a second piece of wire and back to the negative terminal of the battery. Assuming the battery is charged and the bulb is suitable and the filament intact, light will be produced. No matter how big and complex a circuit gets this basic principle holds. Current will flow out of the positive terminal of the source along the various paths in the circuit and back into the negative side of the source.

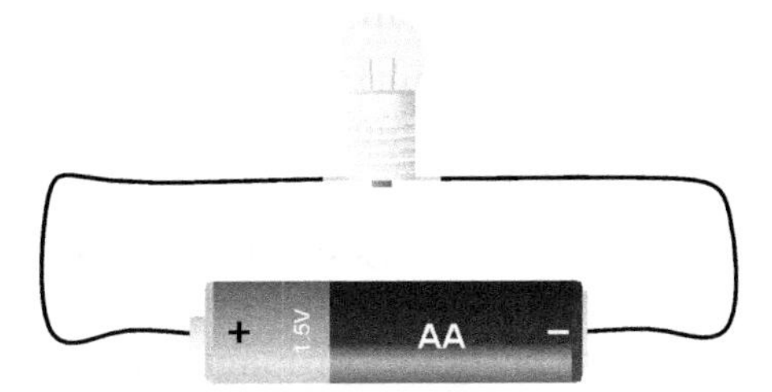

Figure 2.4 In order for any electronic device (no matter how simple or how complex) to work, a loop or circuit must exist providing an unbroken path from one side of the power source back to the other.

It is often useful to be able to say exactly how much current is flowing. Electrical current is measured in amperes or amps (A), and is an indication of how much electricity is flowing. Different circuits will allow different amounts of current to flow depending on the characteristics of the circuit and of the source. One ampere is quite a large current for the kind of battery powered circuits involved here, but it is a fairly modest current for many familiar electrical devices especially mains powered equipment: kettles, toasters, and of course larger audio power amplifiers.

Some mains power plugs (mainly the type used in the UK and Ireland, see Chapter 4, p. 41), contain a fuse. These fuses are typically rated at between about three and thirteen amps. A fuse is really just a tiny piece of wire designed to burn out if more than the rated current is allowed to flow through it. This is a crude but effective safety feature used to limit the chances of large, dangerous currents being allowed to flow unchecked due to a fault in wiring or equipment.

The smaller amounts of current which are usually involved in battery powered audio circuits can be measured in milliamps (mA). A milliamp is one thousandth of an amp. It can be a good idea to keep an eye on how much current a circuit draws. For battery powered devices this gives an idea of how long the circuit will be able to run before the batteries will need changing or recharging. And if the required current in a circuit starts rising too high it may be necessary to look at how much heat is being generated in the circuit, and if any components need to be replaced with ones that can better handle the higher levels of current flowing. These considerations are addressed as individual components and general questions of circuit design are explored later.

Electric Current and Electrons

Electric current is taken to flow from positive to negative within a circuit, while physical electrons actually propagate in the other direction. This positive to negative direction is called conventional current flow. It does not really matter which direction is used so long as it is applied consistently, and so some authors have taken to equating the direction of current flow with the direction of electron movement. This is never done in this book but it is worth being aware that this reversed convention for the direction of current flow might occasionally be encountered elsewhere. So long as it is applied consistently it does not represent a problem.

Sticking to the conventional positive to negative direction for current flow might be seen to have the added merit of highlighting that an electric current may be associated with the movement of charged particles (negative or positive) other than electrons.

The separation between electrons and electric current is further re-enforced by another important distinction. Electrical signals propagate around a circuit at a significant fraction of the speed of light (in other words very fast indeed) whereas the electrons (or other charged particles) that generate the electric current that constitutes such a signal typically move much, much more slowly.

Electric current is the propagation of an electric field, not the movement of the charged particles themselves. It is only the propagation of the field which is of interest here, i.e. the electrical signal not the moving particles that are responsible for generating it, but again reference may occasionally be encountered to this much slower movement, usually referred to as the drift velocity of the electrons, typically on the order of millimetres per second in electrical wires, and these two speeds of propagation are sometimes confused.

Direct and Alternating Current

A direct current means that a (moderately) constant amount of current (number of amps) flows always in the same direction. For an alternating current the direction in which the current flows keeps reversing, first one way and then the other, so the actual electrons never really go anywhere they just keep jiggling back and forth. The graphs in Figure 2.5 illustrate the difference between a DC and an AC signal.

Here the DC plot is a good way to think of what happens at the terminals of a battery and the AC signal, represented here as a simple sine wave, would be a very good visualisation of the signal present on the live conductor of a mains power socket. However an AC signal can also be a much more complex affair and this is of particular interest here, as this is how an acoustic sound signal is represented electrically, as for example in the short audio sample displayed in Figure 2.6, where the graph shows a very complex signal variation over time. This same graph could represent either the air pressure changes of an acoustic sound signal or the voltage variations of an electrical sound signal.

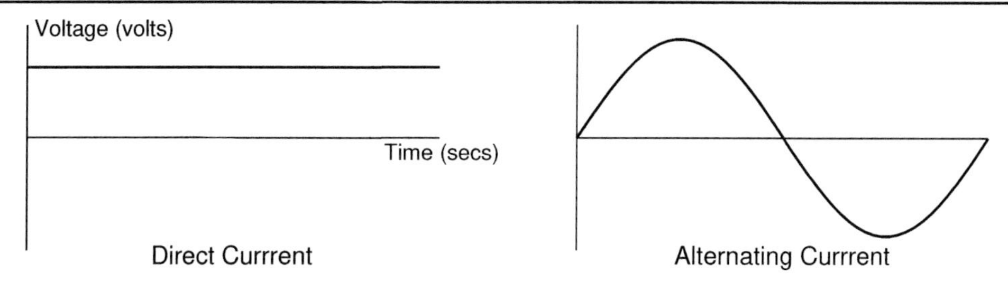

Figure 2.5 Direct current (DC) and alternating current (AC) electrical signals. With DC the level and polarity stays constant over time whereas with AC it continually swings between positive and negative.

As has been said, conventional current always flows in a circuit from points at a more positive voltage to points at a more negative voltage within the circuit (recall it flowed out the positive battery terminal and around the circuit, always heading back towards the negative terminal). In the case of this simple circuit the only voltage source present is the battery. Batteries only provide constant, steady voltages (the voltage will drop slowly as the battery discharges, but as far as any analysis is concerned a battery is a constant voltage source). So in the circuit in Figure 2.4 above only direct current (DC) is present. Compare this with a mains powered desk lamp, where a mains AC signal is used to run the light bulb. This circuit is effectively identical to the one shown in the figure, except for the nature of the source driving it. A standard incandescent light bulb can run equally well from a DC or an AC source, as all it needs to do in order to give off light is to get hot and glow. Most circuits are rather more particular as to how they are powered, and audio circuits usually require a DC power supply.

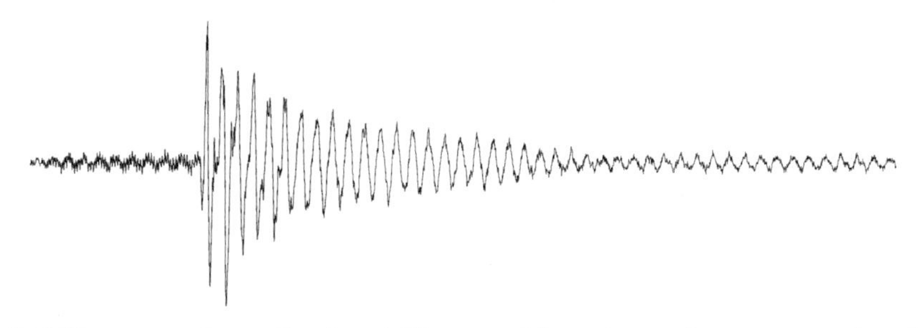

Figure 2.6 The example audio signal illustrated here is much more complex than the simple sine wave AC signal shown above in Figure 2.5 but both are AC signals.

Finally on the subject of DC and AC signals, another way in which audio signals are often encountered is as a combination of an AC component and a DC component. How this is usually described is as an AC signal with a DC offset or bias, see Figure 2.7. This idea of applying a constant offset or bias to an audio signal can be quite useful, as is seen when it comes time to examine various audio circuits later in this book.

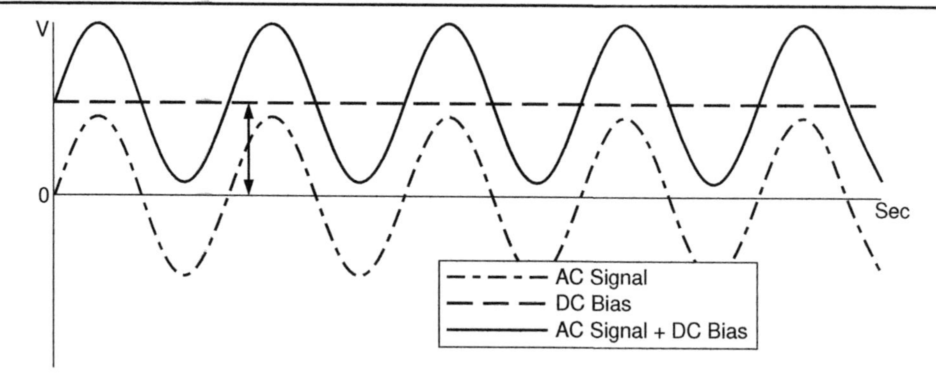

Figure 2.7 An AC signal with a DC offset or bias. The dashed sine wave oscillates equally either side of the zero line whereas with a DC bias added the mid point of the resulting sine wave (solid line) is located at a non-zero position.

VOLTAGE

Voltage, measured in volts (V), is the electrical push that attempts to cause current to flow. So for example the 9V battery in Figure 2.2 pushes harder than the AA or AAA batteries shown in Figure 2.8. Voltages are measured between two points in a circuit; when referring to voltages they are described as being between point A and point B, or across component X.

Figure 2.8 Fresh AA and AAA batteries typically generate a steady voltage or EMF of approximately 1.5V. This voltage will of course slowly fall off as the battery is used and the charge it holds is dissipated. It is also worth noting that, while historically 1.5V is the established output for such batteries, newer battery types, especially rechargeables, can vary from this level, with 1.2V often being quoted (see inset).

It is worth noting that a number of alternative terms can be encountered all meaning much the same thing, although sometimes applied in different contexts. They are all talking about voltages, but it is worth being aware of them so that they do not cause confusion if encountered. The voltage generated by a battery or other source will sometimes be referred to as an electromotive force, or EMF for short. So it might be said that the AA or AAA batteries in Figure 2.8 each generates an EMF of 1.2V.

When talking about a voltage across components or parts of a circuit other than a source, it is often called a potential difference or PD. So the bulb in Figure 2.4 has a PD of 1.5 volts across it since it is connected directly to the two terminals of the battery, while the battery itself might be quoted as having an EMF of 1.5 volts. The related terms 'potential' or 'electrical potential' are often used when talking about the voltage at a point in a circuit relative to a predefined reference point designated as being zero volts. For a simple battery powered circuit this zero reference point is often taken to be the negative terminal of the battery but this is not essential. All voltages are relative and it is just necessary to know where they are being measured relative to. These ideas are important and are encountered again when it comes to examining the operation of circuits throughout the rest of this book.

There is one more term which is sometimes used to refer to voltage, and that is tension, used especially where talking about very high voltages. So the big power lines that carry electricity across the countryside are often called high tension lines. This does not mean that they are stretched very tight (indeed they usually hang quite slack). It refers to the extremely high voltages used to transmit power over long distances. The acronym EHT meaning extra high tension is often encountered written on signs around power substations or other places where high voltages exist.

Resistance

Resistance is measured in ohms (Ω) and can be thought of as the opposition to the flow of electricity in a circuit. Resistance is a property of the components themselves and exists independent of the presence of any voltage or current. In other words a component's resistance is there all the time regardless of whether or not the component is in a circuit, and regardless of whether or not electricity has been applied to the circuit.

Another name which is often used almost interchangeably with resistance is impedance. Resistance and impedance are both measured in ohms and in general impedance is actually made up of two components called resistance and reactance (which is never much talked about on its own here). One useful way of thinking about the relationship between resistance and impedance is to remember that resistance is more to do with DC circuits and impedance is more to do with the more complex interactions present when considering AC circuits.

The places where impedance is most commonly encountered when working in audio in general are loudspeakers and amplifiers (remember audio signals are AC so usually relate to impedance rather than resistance). Figure 2.9 shows the back of a loudspeaker where the impedance is marked as 8Ω. Four, eight, and sixteen are common impedances for loudspeakers. Power amplifiers are typically designed to operate with a specific range of loudspeaker impedances.

Figure 2.9 Eight ohm (8Ω), half watt (0.5W), two and a half inch diameter loudspeakers. Speakers are most commonly found in impedances of 4Ω, 8Ω, and 16Ω.

Power

Electrical power is a measure of how much work per second is being done by the components or a circuit. Power is measured in watts (W), or for many of the smaller, low power audio circuits of interest here, in milliwatts (mW). In audio circuits (as in many other circuits) inefficient use of incoming power leading to the generation of unwanted heat can be a major problem, and various strategies are employed in order to cool circuits to prevent them from overheating and failing. Fans and heat sinks are designed to remove excess heat from a circuit. High power circuits such as amplifiers need careful attention paid to the questions of ventilation and cooling if they are to operate without problems over long periods of time. A heat sink is a piece of material with good thermal conduction and a large surface area so that it can radiate the heat it gathers into the surrounding air (or occasionally water or some other fluid). Heat sinks are attached firmly to any components that get especially hot, like for instance the power transistors in an amplifier. The job of a fan is to keep cool air moving over the surface of the heat sink so as to maximise the transfer of heat into the air and away from the hot circuitry, thus dissipating the maximum amount of power.

Amplifiers are very often described in terms of their power output capabilities (a one hundred watt amplifier etc.), and the second important characteristic for a loudspeaker along with the impedance mentioned before is its power handling capability, i.e. how much power an amplifier can drive into it before it starts distorting badly, and running the risk of being damaged.

Electricity and Magnetism

The finer details of the nature of and the relationship between electricity and magnetism are areas of ongoing advanced practical and theoretical research, and these questions are deeply entangled with the areas of quantum mechanics, relativity, and high energy physics. Fortunately these details need not be considered here. It is however useful to develop a passing acquaintance with some of the more basic but nonetheless important concepts involved. The first thing worth remembering is that the effects of an electrical signal flowing in a circuit are not limited to within the components and the wires that make up the circuit. An electromagnetic field extends out from the current carriers and can have both wanted and unwanted effects in the general vicinity.

Electric and magnetic fields can to a large extent be viewed as just two aspects of the greater single electromagnetic field. The idea of a permanent magnet should be familiar, with a field surrounding it which can have an effect on certain metals and other magnets close by, without any physical contact. One magnet can be made to push another across a tabletop without the two ever touching. A compass can be relied upon to find north because the compass needle is a magnet and its field interacts with the earth's magnetic field to line up showing north–south. But remember also that a compass can easily be fooled by stronger magnetic fields produced by electrical devices and other magnets in the vicinity. The freely rotating compass needle lines up with the strongest magnetic field it finds. In fact a compass is occasionally used in this fashion when working with electric guitar pickups as it provides an easy method of determining the orientation of the magnets in the pickup, which can be useful information when wiring multiple pickups into a guitar (see Learning by Doing 20.4).

Electromagnetic induction is the process used to generate the majority of electricity around the world. All that is needed in order to generate an electric current is to move an electrical conductor in close proximity to a magnet. This is what is found in the generators in an electric power plant, and it is also what happens in probably the commonest type of microphone, the ubiquitous dynamic mic. Fundamentally all that a dynamic microphone consists of is a very fine coil of wire attached to a diaphragm and suspended in a magnetic field. Sound travels through the air, hits the diaphragm and makes it vibrate back and forth in time with the sound. This moves the coil which is sitting in the magnetic field, and a tiny electrical signal is generated by electromagnetic induction. The generated signal voltage varies up and down exactly as the diaphragm is moved back and forth, and so an electrical signal is created that is an excellent representation of the original acoustic sound arriving at the diaphragm.

In fact electromagnetic induction is a two way process. The previous description showed that moving a conductor in a magnetic field produces an electric current, but it also works in the other direction. Placing a conductor in a magnetic field and sending an electric current through it produces a force on the conductor. This is the principle of the electric motor, as illustrated in Figure 2.10. Motors and generators have identical configurations at their core, and indeed if an electric motor for example from a CD player is spun by hand a voltage can be measured at its terminals – the motor is acting as a generator. So it can be seen that electromagnetic induction is widely used in both its directions of operation: motion used to generate a signal as in a dynamic microphone, and a signal used to generate motion as in a moving coil loudspeaker.

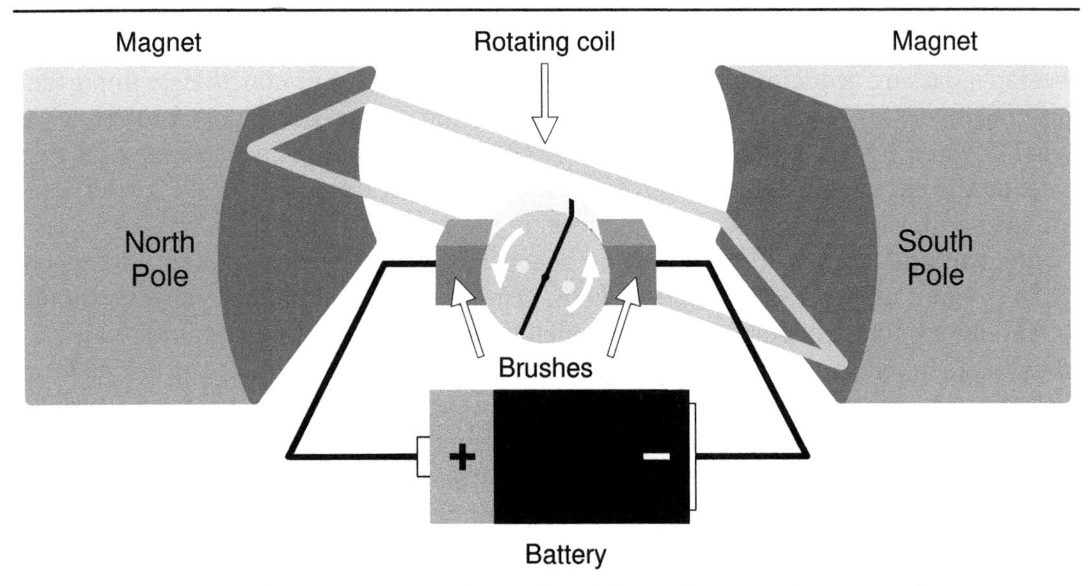

Figure 2.10 An electric motor combines electricity and magnetism to produce movement. A generator uses the same mechanism in reverse to produce electricity from magnetism and movement.

Thus as the dynamic microphone mentioned earlier takes movement and produces a signal, so a moving coil loudspeaker takes an electric current and produces movement. And just as a motor can be a generator and vice versa, so too loudspeakers can be (and occasionally are) used as microphones to pick up sound, and indeed microphones can be made to generate sound like a tiny loudspeaker (although not very loud and probably to the detriment of the microphone). The pickups on an electric guitar are another example of electromagnetic induction in action, and more examples can be found elsewhere in audio technology. All in all the phenomenon plays a crucial role in the technology of this area.

Electromagnetic interference illustrates that just as electromagnetism can be used to achieve many useful tasks, it can also present serious difficulties in audio electronics. Since electromagnetic fields abound in electronics they can often go places and do things which are not wanted. Electromagnetic interference (EMI) is the cause of many a noisy circuit and many a corrupted signal. Whether it be a steady hum, a sudden click, or an unwanted radio broadcast breaking through, noise and interference can often find their way onto an audio signal through the mechanisms of electromagnetic interference.

An understanding of where such unwanted signals come from and how they make their way into circuits is important if they are to be kept out, leading to circuits which produce clean and clear audio signals. In various places throughout this book techniques for grounding, shielding, and cancelling unwanted noise signals in circuits are encountered.

Learning by Doing 2.2

Build an Electric Motor in Ten Minutes

Materials

- Magnet wire (~45cm) – sometimes called speaker wire
- Permanent magnet
- Two large bare metal paperclips – plastic coated types will not work
- AA battery and clip
- SPST-NO switch (optional)
- Breadboard
- Clip leads
- Sticky pad (or a lump of Blu-tack)

Magnet wire or speaker wire (not to be confused with speaker cable) is fine, single strand copper wire insulated with a thin coat of lacquer or varnish. It is the stuff that the coils in electromagnets and moving coil loudspeakers are made from (hence the name). It can look as if it is uninsulated coper wire, so it is important to make sure you have the right stuff. Adjacent strands of uninsulated wire will of course short out when wound into a coil, and so the insulation really is required here. Magnet wire is used when standard wire with plastic insulation is too bulky and too heavy to do a good job in the application at hand.

Theory

As illustrated in Figure 2.10, a basic electric motor consists of little more than a magnet, a coil of wire, and a power source. For smooth consistent operation with DC power, some form of commutator or similar mechanism is needed to repeatedly reverse the connections to the coil, resulting in the forces always working to turn the motor in the same direction. This operation is achieved by the brushes indicated in the figure.

In our simple motor here, we rely on the inherent imperfections of the device in order to crudely mimic this aspect of the operation of a traditional motor. As our coil spins it bumps and jumps, constantly making and breaking the electrical connections to the battery. This (usually) provides the necessary kicks and nudges to keep the coil spinning. I say usually because this little motor can easily get stuck in a position (especially as it starts up) and require a little nudge to get it going. A flick with a finger can provide the nudge, but it is probably easier to add the optional push button switch indicated, so that by clicking the switch off and on a few times the necessary nudge is applied. Once spinning, the motor tends to provide enough random bounce to keep itself going.

It is also worth noting that the motor presents something which might appear worryingly similar to a short circuit between the terminals of the battery. This is generally to be considered a bad thing. For this reason it is probably not a good idea to let the motor sit awaiting a kick start for any period of time, as in this state a significant current is likely to be drawn. It can be instructive to add an ammeter to the circuit, in order to monitor the amount of current being drawn. It is likely to vary quite a bit. In my case I observed an average current of around about 200mA being drawn during steady running, rising to over 800mA when the coil stalled and got stuck in the magnetic field.

Practice

Leaving an inch and a half free at either end, wind the wire into a tight coil. Your finger, a pen body, or even the AA battery can make a suitable former to wind around. Use the two ends to tie the turns of the coil together.

The protruding ends of the coil need to be stripped of insulation. A lighter flame or a hot soldering iron will often remove most of the lacquer but it is a good idea to finish the job with a file or blade, in order to get a good clean conductive surface exposed at either end of the coil.

Next, bend the paper clips as shown, to form two mounting brackets to support the coil and provide electrical connection points. Insert the ends of the paperclips into the breadboard, and mount the coil in the runners. Use the sticky pad or Blu-tack to secure the magnet directly below the coil. Complete the circuit by connecting the battery and push switch in series, as shown (also add an ammeter if you want to measure the current draw). The specified switch is a normally open (NO) momentary type, so the circuit is not completed until the button is pressed. A few clicks should jog the coil into motion. Once it is moving, hold the button down to keep the motor spinning. If it stalls or doesn't quite get going, release the button and try a few more quick clicks before holding steady again.

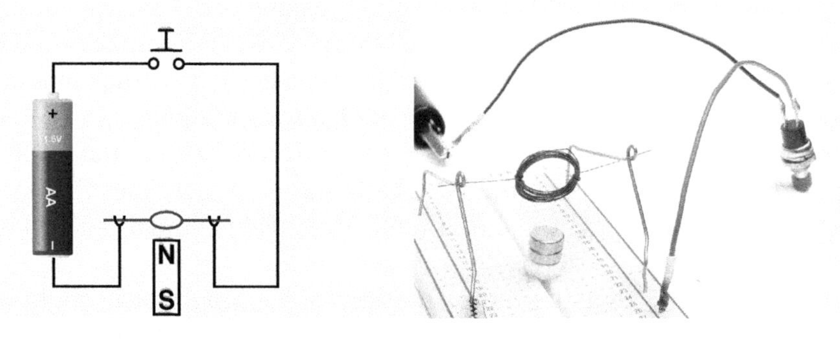

If your motor won't start check that you are getting good connections between the paperclips and the coil by measuring the resistance between the two paperclips (with the battery disconnected). If necessary file the ends of the coil some more, and also file the paperclips where they connect to the coil. Good clean metal surfaces, clear of dirt and metal oxide are a prerequisite for good electrical connections.

Conductors and Insulators

The next chapter in this book is all about semiconductors so it would seem appropriate to finish off this chapter with a section on conductors and insulators, the two extremes to which semiconductors can be thought of as forming a part of the centre ground. Resistance has already been introduced as the quantity that measures how easily electricity flows in a material. Good conductors have very low resistance while good insulators have extremely high resistance. There are materials all along the scale from very low to very high resistance. Table 2.1 lists a few examples of good, intermediate, and poor conductors, with their resistivities listed beside them. Resistivity, measured in ohm-metres ($\Omega \cdot \mathrm{m}$), gives an indication of the characteristic resistive properties for a material. Note that a long piece of wire has a greater resistance between its ends than a shorter piece of the same wire, and so simple resistance in ohms (Ω) is not a suitable quantity to provide a general characterisation for the resistive properties of a particular material. The highlighted Box 2.1 on p. 25 provides an explanation of why ohm-metres ($\Omega \cdot \mathrm{m}$) emerge as the appropriate units for resistivity.

It can be seen from the table why copper is a good choice for making electrical wires. It has an extremely low resistivity meaning that it will allow current to flow without offering any great resistance and so it does not waste energy getting hot and dissipating power. Rubber on the other hand has a very high resistivity and as such forms a suitable basis for the insulation that wires might be wrapped in, preventing unwanted contacts with other conductive objects in the vicinity. Uninsulated conductors are occasionally (but not often) encountered. They are of course not really uninsulated. They are just insulated by air, which is in fact a very good insulator. The only problem with it is that solid, possibly conducting, objects can pass straight through it and touch the conductor in question. However, overhead power lines for instance are often left uninsulated as nothing conducting should normally come in contact with them. The heating elements in a toaster are sometimes also formed from uninsulated conductors, so be careful when sticking a knife in to dig out your toast.

As discussed in the next chapter, a few of those materials in the intermediate section of Table 2.1 turn out to have some extremely interesting properties. This means that they are ideal for making a very useful type of substance called a semiconductor. Silicon is the one most people have heard about, in terms of silicon chips, the things computers are made of. In fact before silicon semiconductors came along, germanium was the

Table 2.1 Some typical resistivities for a range of materials

Material	Resistivity (Ω · m)	Category
Silver	0.000,000,016	Conductor
Copper	0.000,000,017	
Gold	0.000,000,024	
Aluminium	0.000,000,028	
Silicon (doped)[a]	0.003,2	Intermediate
Salt Water	0.20	
Germanium (pure)	0.46	
Fresh Water	200.0	
Silicon (pure)	640.0	
Glass (silica)	10,000,000,000,000.0	Insulator
Rubber	10,000,000,000,000.0	
Wood (dry)	1,000,000,000,000,000.0	
Air	23,000,000,000,000,000.0	

a. The resistivity of doped semiconductor material is highly dependant on the type and concentration of the dopant used. See Chapter 3 for more on semiconductor doping.

element of choice. Germanium based semiconductors are much less common now but are still used in some audio circuits for some of their particular characteristics.

Open Circuits and Short Circuits

If two things or two points in a circuit have a path between them along which electricity can flow then they are said to be connected. If there is no electrical path between them they are disconnected or isolated. A very low resistance connection (close to zero ohms) can be described as a short circuit. Two completely disconnected points can be described as an open circuit. In practical electronics, the terms short circuit and open circuit are usually (though not always) used to describe unwanted or fault conditions. When a fuse is overloaded and burns out it is described as having gone open circuit. Conversely an unwanted electrical connection, perhaps caused by stray solder or a loose wire touching where it shouldn't is called a short circuit.

In building an electronic project two of the most common problems encountered are unwanted short circuits and open circuits.

A **short circuit** (aka short or dead short) can arise if too much solder is used when connecting components. The excess solder can flow onto a nearby connection point creating a solder bridge and resulting in a short circuit between two points in the circuit which should not be connected. Short circuits are also commonly encountered if sufficient care is not taken when connecting stranded wire into a circuit. A single loose strand of copper, finer than a human hair, is all it takes to cause a problem.

2.1 – Units of Resistivity

Resistivity is measured in units of ohm-metres ($\Omega \cdot \mathrm{m}$). It can be instructive to see how these units come about. Clearly resistivity is closely related to resistance, however the latter is a quantity which varies from one instance to another. Measure the resistance of two pieces of copper wire and two different answers result. Clearly resistance itself does not provide a metric describing the inherent properties of the copper making up the wires. The measurement is affected by the length of the piece of wire, and by its gauge or cross sectional area. Increasing the former increases the measured resistance, while increasing the latter decreases the measured value.

These two observations point to the approach to a universal characterisation of the resistance properties of the underlying material. Double the length of a piece of wire and its resistance will double. Each section can be considered as a resistor and, as described in the resistor series rule (see Eq. 12.1, p. 198), resistors in series add.

Similarly, doubling the wire's cross sectional area effectively places two equal sections in parallel, which from the parallel rule (Eq. 12.2, p. 199), can be seen to halve the total resistance.

Thus resistivity (usually indicated by a lowercase Greek letter rho, ρ) can be expressed in terms of the resistance R measured between the opposite faces of a block of the material in question. If the area of the faces between which the measurement is to be taken is A, and the length of the block between these two faces is l (see figure below), then the resistivity can be defined as the resistance measured (R), multiplied by the area of the faces (A), and divided by the length between them (l).

$$\rho = \mathrm{R} \times \frac{\mathrm{A}}{\mathrm{l}}$$

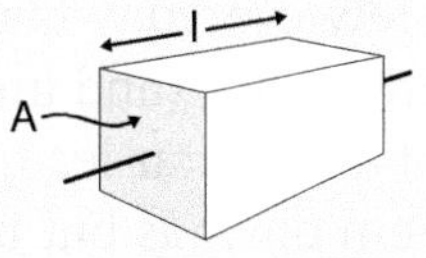

This quantity remains constant as A and l change because the resistance R always adjusts in proportion thus maintaining the result. This yields a constant property of the underlying material – the resistivity. As such the units of resistivity can be determined simply by looking at the units of the quantities used to determine it. R is measured in ohms (Ω), the area A in metres squared (m^2), and the length l in metres (m). Thus with A divided by l yielding metres (m), the final units come out as ohm-metres ($\Omega \cdot \mathrm{m}$) as expected.

An **open circuit** can be introduced if solder joints are not well made. Insufficient heat or moving the joint while it cools can prevent the solder from bonding properly resulting in what is called a dry solder joint. This is a joint which may look good but which provides a very poor electrical connection. It can result in a permanently or intermittently open circuit and a noisy or non-operational device.

The terms are occasionally also used to describe wanted or intended situations, like the state of a switch or a connection point in a circuit. For example a system might be reset by shorting together two normally disconnected points in the circuit. It is common practice when tidying up unused portions of some types of circuit to short unused inputs to ground and to leave unused outputs open circuit. These practicalities are examined in more detail in the sections on circuit building in Part II.

Learning by Doing 2.3

Conductors and Insulators

Materials

- Bulb and bulb holder
- Battery and battery clip
- Clip leads
- Length of wire stripped at both ends
- Various coins
- Pencil (with lead exposed at both ends)
- Guitar lead
- Multimeter with shrouded banana plug and socket

Theory

This exercise looks at what does and does not constitute a good electrical connection. It assesses a few everyday materials as to their conductive properties, and examines what connections are (and are not) present in the standard type of connector found on either end of a guitar lead.

It may seem obvious but it is worth stating that in order for a circuit to work, all parts of the loop which constitute the circuit must consist of conductive elements of one sort or another. A gap at any point in the path breaks the loop and prevents the circuit from working. When tracking down a problem, it is not altogether uncommon to find for instance that a clip lead has been attached to the insulation around a wire, rather than to the length of conductor protruding from its end.

An even more common occurrence is a failure to insert a clip or plug sufficiently into a standard shrouded socket. This once again causes a failure to achieve an electrical connection as required, or perhaps makes an intermittent connection which can be even more difficult to diagnose. Many of the commonest problems encountered by novice circuit builders fall into this category of poor or missing connections, and can easily be avoided with a solid understanding of what can go wrong, and little care and attention to detail in approaching the circuit building process.

Practice

First build the simple battery and bulb circuit illustrated below. Use two clip leads in one of the arms of the circuit so that you can break the circuit here and attach the free ends to various places and things. Initially, with these two clip leads connected, the bulb should be lit. As you proceed, an illuminated bulb indicates a good electrical connection, while a bulb showing no glow indicates a failure to complete the circuit with a sufficiently low resistance. Now to address the items listed in the materials section above, to see how and where they conduct. In each case start by separating the clip leads in our test circuit to give two free ends as described above.

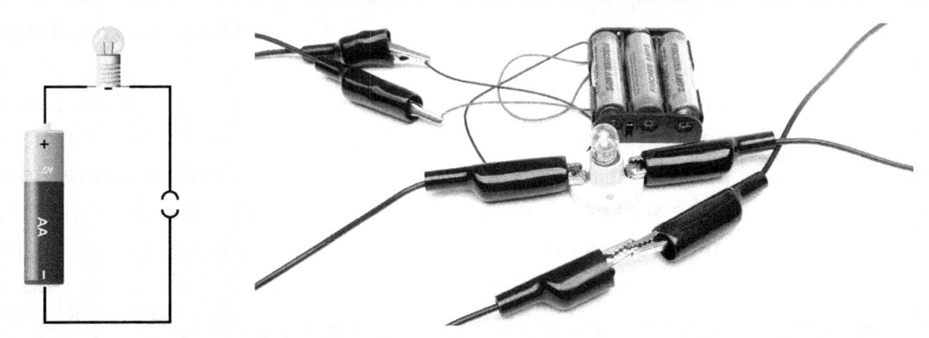

Length of wire stripped at both ends – connect the two free ends to the exposed wire at either end of the length of wire. The bulb lights. Move one clip lead from the exposed wire onto the insulation just beside it. The bulb does not light. Place the two clip leads onto the insulated middle of the length of wire, very close together but not touching. The bulb does not light. Place the two clip leads on top of one another on the insulated middle of the length of wire. The bulb lights. There should be no surprises here.

Various coins – connect the two free ends to the opposite edges of various coins. Does the bulb light? Generally the answer here will be yes. Most coinage is indeed conductive.

Pencil (with lead exposed at both ends) – clipping one or both leads onto the wood results in a negative, but if the clips are touched against the pencil lead then a positive is likely. The 'lead' in pencils is usually actually graphite, which is a good conductor (though not a metal). On the other hand, if you have reached for an artists pencil the 'lead' may be a form of china clay or possibly charcoal, neither of which is conductive. Actual lead, though not as good a conductor as many metals, still works well. A heavy graphite line drawn on paper is also conductive, though in this case it may be difficult to lay down enough material to get a good enough conductor to make the bulb glow visibly – perhaps test it with a multimeter instead.

Anything and everything – try what comes to hand: paper, plastic, glass, fabric... All likely to be non-conductive. What about a drinks can? Yes on the tab, but probably no on the surface of the can (metal food containers are covered in a lacquer seal which is going to be nonconductive), but then go into the edge where the tab opened the can exposing the raw metal and the answer is once again yes.

Guitar lead – the jacks found on either end of a standard electric guitar lead are called TS jacks (T = tip, S = sleeve). Unsurprisingly the pointy bit of metal at the very end of the jack is the tip, and the barrel running down the length of the jack is the sleeve. The two are separated by an insulator (probably visible as a strip of black plastic). The tip at one end is connected to the tip at the other end, and likewise with the two sleeves. Connect your clip leads to the tip and sleeve on either jack and the bulb does not light. Connections from tip to tip, or from sleeve to sleeve do light the bulb. This is one way of testing a guitar lead. A common problem with such cables is that, with use, the solder connections to the tip or sleeve can break, leading to crackly sound, dropouts, or just no signal at all.

Multimeter with shrouded banana plug and socket – take a standard multimeter. Set it to the continuity tester mode. Now it beeps when the leads are touched together. Remove one plug from its socket and, with the two leads touching slowly insert the plug back into the socket. How far in does it need to be inserted before the meter beeps? On many meters it is possible to have the leads holding firmly in place due to the extensive plastic shrouding often found around such connectors, and yet the electrical connection has not yet been made. When using a unit with which you are unfamiliar, always pay close attention to such simple questions as 'How far in does the lead go before it is fully inserted?'

References

K. Brindley. *Starting Electronics*. Newnes, 4th edition, 2011.

A. Hackmann. *Electronics: Concepts, Labs, and Projects*. Hal Leonard, 2014.

P. Horowitz and W. Hill. *The Art of Electronics*. Cambridge University Press, 3rd edition, 2015.

R. Jaeger and T. Blalock. *Microelectronic Circuit Design*. McGraw-Hill, 4th edition, 2011.

A. Sedra and K. Smith. *Microelectronic Circuits*. Oxford University Press, 7th edition, 2014.

3 | Semiconductors

The previous chapter concluded with a section about conductors and insulators. What is discussed there are materials at the two extreme ends of a continuum. In between them lie lots of substances that conduct electricity a bit but not really very well, things like water, graphite, people, and of interest here silicon and germanium. These latter two intermediate materials have some very specific and useful properties the details of which are unimportant but suffice it to say that they are called semiconductors and they are at the heart of all modern electronics. The one of most interest is silicon but germanium (which is only used a little today but which was much more commonly used in early semiconductors) are also mentioned briefly.

Pure Silicon

Silicon atoms form a nice regular lattice structure. As was stated in Table 2.1 this material will conduct a bit of electricity but really not very well at all. It offers quite a lot of resistance. Nonetheless if a pair of terminals are attached across a lump of pure silicon and an EMF is applied (remember, that is a voltage) some small amount of current would indeed flow. Reverse the EMF and the current would flow in the other direction. Attach the terminals on different sides of this little block of silicon and current flows just the same. So far nothing of any great interest or utility is happening. The silicon is acting just like a fairly poor metal, conducting electricity but not very well.

The main reason silicon does not conduct very well is that unlike metals it does not have any electrons which are free to move around easily. All its electrons tend to be bound up fairly tightly in the bonds that hold the silicon atoms together in its neat and tidy lattice structure, as illustrated in Figure 3.1. It takes a bit of energy to encourage electrons to abandon their positions and go roaming giving an electric current. So the first thing which is needed is to make it easier for electrons to move around a bit within the silicon's crystal lattice. Two possible ways of doing this might be to introduce new less tightly bound electrons happy to move about or to provide new free locations that any nearby electron might feel inclined to move into and occupy.

To arrive at useful semiconductor based devices it is actually necessary to do both of these things to produce two new types of semiconductor material called n-type and p-type. The 'n' in n-type stands for negative and this type of semiconductor has mobile

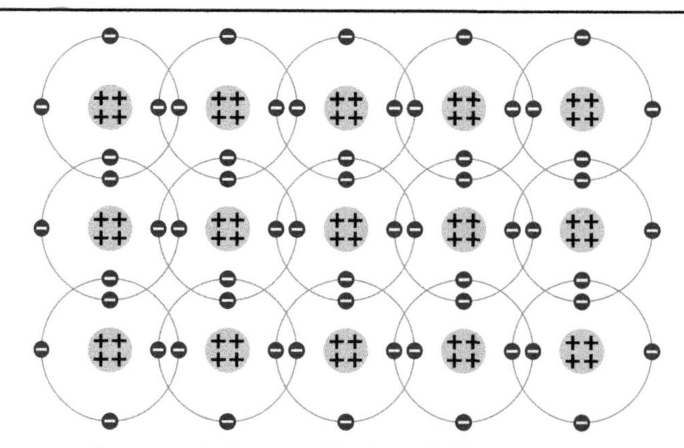

Figure 3.1 A crystal lattice of silicon atoms.

electrons ready and able to move about. Likewise the 'p' in p-type stands for positive and this type of semiconductor has spare room which attracts wandering electrons. These electron friendly locations are usually referred to as holes. Each of these two new materials is much more conductive on its own than the original silicon was. Recall from Table 2.1 that pure silicon has a resistivity of around $640\Omega \cdot \mathrm{m}$. These new materials while still not nearly as conductive as copper will typically have resistivities tens of thousands of times lower than that of pure silicon. The exact resistivity depends on the concentration of these additional electrons and holes.

Doped Silicon

In order to achieve these new more conductive semiconductor materials take a nice well structured lump of silicon and swap some of the silicon atoms in the lattice for other elements. Do not worry too much about what they are, a few different elements can be used. The important thing is that these new elements have a structure similar to silicon but with different electron layouts providing either extra electrons or spare sites for electrons to fill (as previously mentioned these empty locations are called holes). This process of adding non-silicon atoms into the silicon lattice is called doping the silicon, and the other elements are called impurities or dopants. Dopant atoms are quite similar to silicon atoms so they fit into the lattice structure pretty well, see Figure 3.2.

By adding some of these carefully chosen impurities some very useful behaviours can be achieved. What the impurities are doing is either adding a few extra electrons to the mix, or removing a few electrons from the mix (i.e. adding holes). When electrons are added the result is n-type semiconductor material, and when electrons are taken away the result is p-type semiconductor material. It is important to note that the materials stay neutral from a charge point of view because the impurities added have different numbers of protons to balance the different numbers of electrons present, Figure 3.3. Each of these two new materials on their own form fairly good electrical

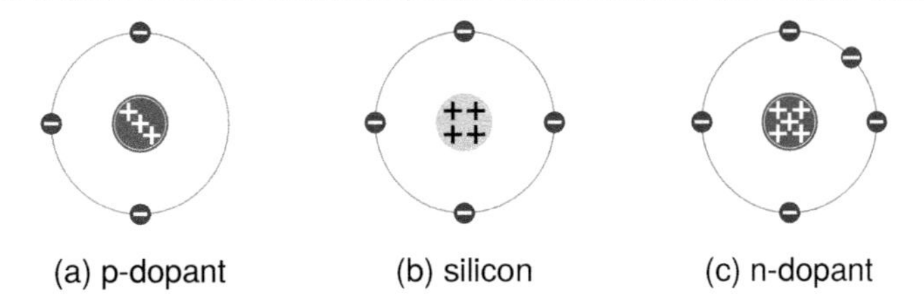

Figure 3.2 Simplified representation of silicon and dopant atoms. The p-type dopant atom is shown with one less available electron and correspondingly one less proton. In the n-type dopant an extra electron and proton are shown.

conductors as in both cases electrons are provided a mechanism for moving through the material easily, but still nothing very new and useful has yet been arrived at.

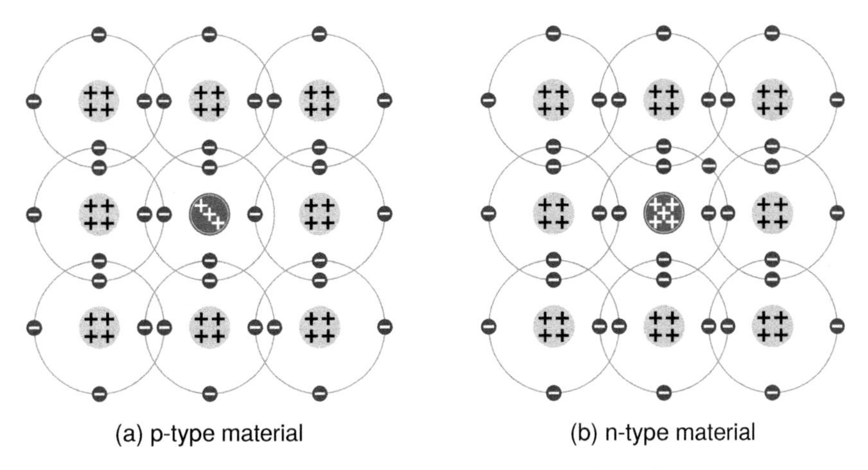

Figure 3.3 Lattices of p-type and n-type doped silicon.

The P-N Junction

Where it starts to get really useful is if a piece of p-type semiconductor and a piece of n-type semiconductor are placed side by side. This forms what is called a p-n junction at the interface between the two materials as in Figure 3.4. All along this junction there is a scattering of mobile electrons on the n-type side and a scattering of holes on the p-type side so unsurprisingly some of the mobile electrons migrate across the junction to fill nearby holes. The problem when they do this is that they leave behind their matching protons in the dopant atoms firmly fixed into the lattice structure. So as a result an electric charge builds up. The n-type side of the junction becomes positively charged (protons which have lost their matching electrons) and the p-type side of the junction becomes negatively charged (wandering electrons move into holes with no matching protons).

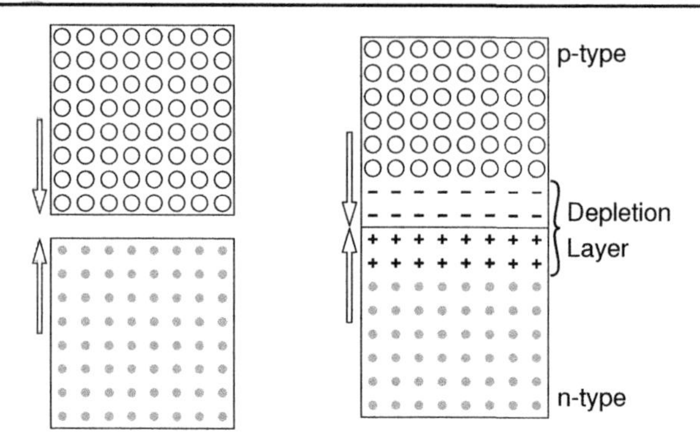

Figure 3.4 Creating a p-n junction.

The resulting electric field slows and eventually stops the migration of electrons across the junction. Remember that opposite charges attract and like charges repel so the increasing negative charge on the p-type side of the junction discourages the migration of further (negatively charged) electrons, while on the n-type side the increasing positive charge tends to encourage the remaining electrons to stay where they are. An equilibrium state is achieved where the effect of the electric field counters the tendency of electrons to move across the junction to fill the waiting holes. The region close to the p-n junction where electrons have abandoned their original positions and filled nearby holes across the junction is referred to as the depletion layer or depletion zone, so called because the available charge carriers (mobile electrons and vacant holes) in the region have been depleted. In a typical p-n junction made from silicon based semiconductors the corresponding build-up of positive and negative charge results in a potential difference of round about 0.6V appearing across the junction.

Now all that has been described so far has happened with the semiconductor materials just sitting on the bench. No electronic circuits have been built and no external electric voltages have been applied. It was observed previously that both n-type and p-type materials were fairly good conductors in their own rights so the question now becomes how good a conductor is this new p-n junction structure going to be? And finally the point has been reached where something useful happens, because the answer to that question is, it depends on the direction in which the external voltage is applied. Consider taking a battery and attaching wires from its terminals to either side of the p-n junction. The results will depend on which side of the junction is connected to the positive side of the battery and which to the negative. Consider each case in turn.

Reverse Biased P-N Junction

Imagine a p-n junction with a battery attached as shown in Figure 3.5 with its positive terminal attached to the n-type material and its negative terminal attached to the p-type material. In this configuration the battery's voltage is applied in the same direction as the internally generated voltage already sustaining the nonconducting depletion layer.

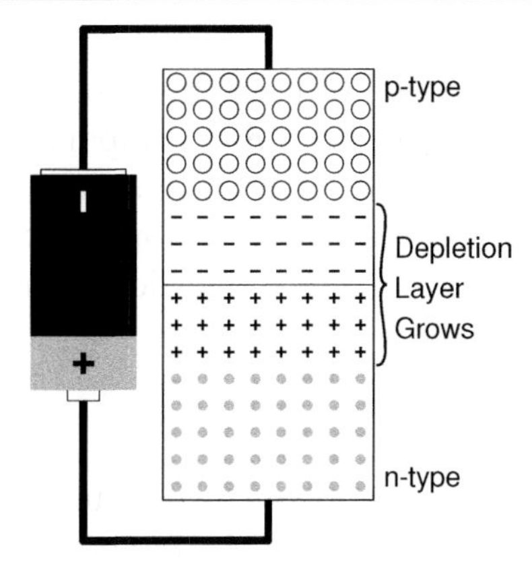

Figure 3.5 Reverse biased p-n junction.

This new external voltage therefore causes the depletion layer to widen. Thus in this situation no current can flow in the circuit because electrons have no way of flowing across the depletion layer. This is called reverse biasing the p-n junction.

Forward Biased P-N Junction

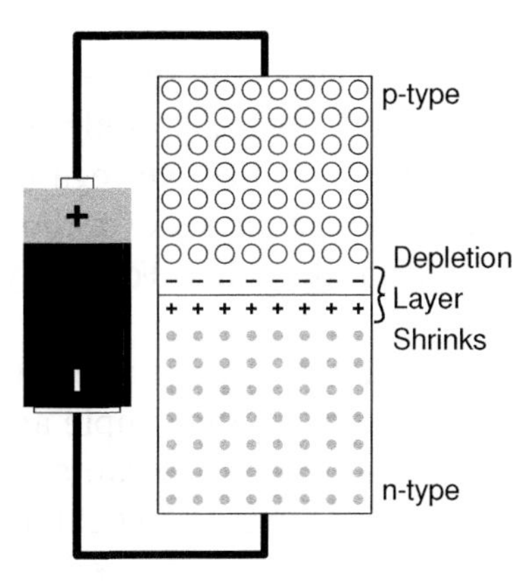

Figure 3.6 Forward biased p-n junction.

Now take the same circuit and reverse the connections to the battery so that the battery's voltage is now applied in the other direction across the p-n junction. The depletion layer now starts narrowing as the applied voltage from the battery overcomes the induced voltage generated by the original migration of charged particles across the junction, Figure 3.6. Once the voltage being supplied by the battery exceeds the internally induced voltage (at round about 0.6V for a typical silicon based p-n junction) current can flow through each piece of semiconductor material just as if the other were not there. This is called forward biasing the p-n junction.

This behaviour leads to the ability to control whether current flows or not depending on the direction of the applied voltage. It makes the p-n junction a very useful structure and is the basis of many and varied electronic components. The operation of some of the most important of these are examined briefly in the next section. This is just intended to give an idea of how things work without looking at real world circuits or applications. Later the much more practical question of how they are used and what can be done with them is addressed.

Diodes and Transistors

The two most important classes of components which utilise the semiconductor materials just introduced are diodes and transistors. Later chapters talk about how to employ these components in various useful ways, but before looking at the purely practical questions of application and performance it will be useful to develop a general understanding of how these important components actually function in order to achieve the various kinds of circuit operation for which they are employed. While a deep understanding of what is going on inside these components is entirely unnecessary a more basic appreciation of what is happening will assist greatly in designing and understanding the circuits to be encountered throughout this book.

Diodes

The simplest semiconductor based device is the diode, and this has already been described in as much detail as is needed. A basic diode is exactly a p-n junction and nothing more and all that has just been said above regarding p-n junctions applies to these very simple devices. The simplest analysis of a diode says that a diode allows current to flow in one direction but not the other. For many applications this is all that is needed.

It was mentioned that the forward bias voltage needs to exceed about 0.6V before a typical p-n junction will start to conduct so the simple analysis is a bit too simple for some cases. This turn on voltage is often used to advantage in audio circuits. For example a very rudimentary fuzz type electric guitar effect can be achieved using little more than one or two diodes. The fuzz distortion is achieved as a result of the audio signal only passing through the diode when the signal's level exceeds the turn on voltage of the diode. This is seen in action later.

There are in fact many different types of diode, some are just slight variations on the basic component described here while others are vastly different in the details of their structure and performance. Most of these are of absolutely no interest here but there are a couple which are likely to be encountered. The most common of these are the light emitting diode (LED), the Zener diode, and the Schottky diode.

LEDs just generate light as a byproduct of the operation of the p-n junction. Apart from this point and the fact that their turn on voltage is a bit higher they are no different from ordinary diodes.

Zener diodes can at first seem a little strange in the way they operate. The standard diode considered so far does not conduct any significant amount of current when reverse biased. However if the reverse bias voltage is increased enough then (as with most things) eventually the diode breaks down and starts conducting strongly. For a normal diode this should happen at a voltage well above anything which will be seen in the circuit because when it happens to a standard diode the component is likely to be destroyed. Zener diodes on the other hand are specifically designed to be operated at this breakdown point and they are in fact used to provide stable voltage references within a circuit because the voltage at which they breakdown can be chosen quite accurately when they are manufactured. So for instance a 6.3V Zener diode is designed to provide a very stable and reliable six point three volts across its terminals when reverse biased. Simple voltage references of this kind are very useful in many circuits.

The Schottky diode is very similar in function to a standard diode except that it exhibits a lower switch on voltage and faster switching, but it also has a higher reverse bias leakage current.

All these aspects of diode behaviour are returned to later when these components are actually utilised in various circuits. Then the significance of the different variations in their behaviours can be appreciated more fully.

Transistors

Transistors are a much more complex subject area than are diodes. There are many different types all with their own special characteristics and areas of application but at the heart of each one still lies some variation on the basic p-n junction examined above. Fortunately once again only a fairly small subset of the large variety available are of interest here. The array can still be a little overwhelming to begin with however. They are introduced here in an orderly fashion so that it is a little easier to keep organised and clear. Figure 3.7 shows how the major types can be broken down.

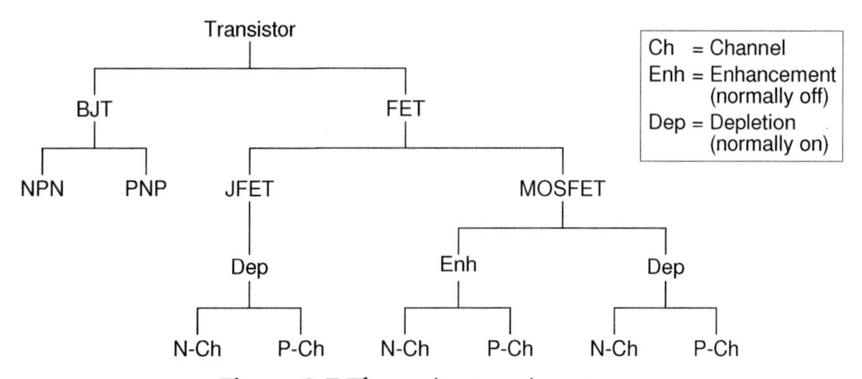

Figure 3.7 The major transistor types.

To start with remember that there are four different types and each type has a positive and a negative variant. The first split is between bipolar junction transistors (BJTs) and field effect transistors (FETs). There are the positive and negative BJT variants called NPN BJTs and PNP BJTs, and that is it for that side of the tree. The FET side gets further broken down into JFETs and MOSFETs, and then MOSFETS get broken down again into normally off and normally on types. Each of the three types of FET comes in n-channel and p-channel flavours. In fact the vast majority of all the transistors likely to be encountered working in audio electronics fall into just two of these eight categories: NPN BJTs and n-channel JFETs.

Whatever the type or variant in question, the basic idea of a transistor remains the same. A transistor has three terminals (as compared to a diode's two) and one way or another what happens at one of those terminals controls the behaviour between the other two. Transistors can operate as switches, buffers, and amplifiers. These are their three basic roles in all the circuits in which they appear throughout this book.

The BJT (bipolar junction transistor) looks a little similar in structure to the diodes already considered. All that is needed is to add another layer of semiconductor material to the stack as in Figure 3.8. Adding another piece of n-type semiconductor gives an NPN BJT and adding another piece of p-type semiconductor gives a PNP BJT.

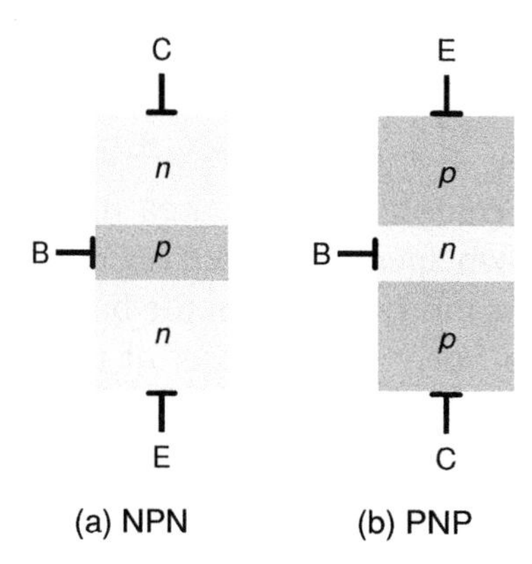

Figure 3.8 The structure of NPN and PNP bipolar junction transistors.

The three terminals of a BJT are called the base (B), collector (C), and emitter (E). The base is the one that does the controlling while the other two are controlled. Since a BJT is essentially two p-n junctions facing in opposite directions it might seem that a BJT should never conduct between the collector and the emitter as one of the two junctions will always be reverse biased, but of course it is a little more complex than this. By altering the voltage on the base terminal which is connected to the central semiconductor layer, the conductivity of the path between the collector and the emitter can be controlled. The details of how this works are not important once the principle is grasped.

The JFET or junction field effect transistor has a somewhat different structure from the BJT (Figure 3.9). Its three terminals are also given different names to those given to

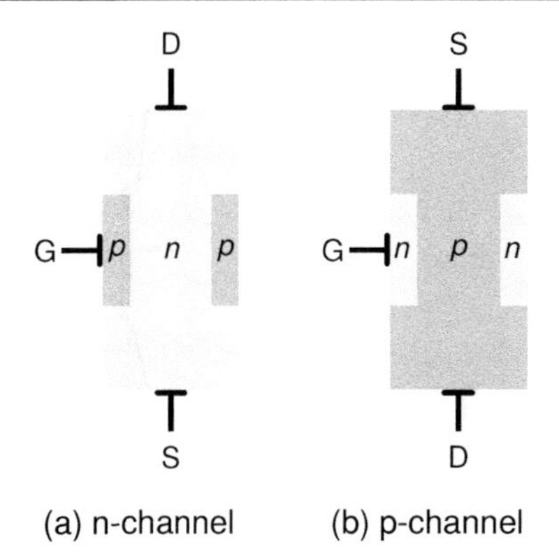

Figure 3.9 The structure of n-channel and p-channel junction field effect transistors.

the BJT's terminals. In this case they are called the gate (G), drain (D), and source (S). These names remain the same for all FETs. The structure of the JFET can be thought of as a single channel of one type of semiconductor material which runs all the way from the drain at one end to the source at the other and the control terminal (the gate) is attached to a region of the other type of semiconductor material which surrounds the channel. The idea in this case is that when a voltage is applied to the gate it can induce the same kind of depletion region as seen in a reverse biased diode. When this depletion layer extends all the way across the channel, conduction is prevented and the transistor is said to be cut off. Reversing the voltage on the gate causes the depletion region to retreat allowing the channel to conduct once again.

So the name explains fairly well how a JFET is made and how it works. The 'J' for junction indicates that there is a p-n junction involved and the field effect bit refers to the electric field which spreads out from the gate junction opening and closing the channel. So the terminals are well named too: the gate controls the flow, and the other two terminals, the drain and source are where the main current flows into and out of the device respectively.

The MOSFET is even more explicit in its name as to the structure of the device. The MOS stands for metal oxide semiconductor which are the three layers involved in making one of these types of transistors as seen in Figure 3.10 and the FET half of the name stands for field effect transistor as before. In this case the oxide layer insulates the gate from the channel so that, while the electric field can still extend out from the gate, opening and closing the channel, no actual current can flow into or out of the gate terminal. In other words the gate terminal has an extremely high input impedance, which can be a useful attribute in certain circuit design situations.

Although the technical details are very different, in terms of operation the MOSFET can be considered as working in a similar fashion to the JFET. Put simply a voltage applied to the gate is used to control the flow of current through the channel between

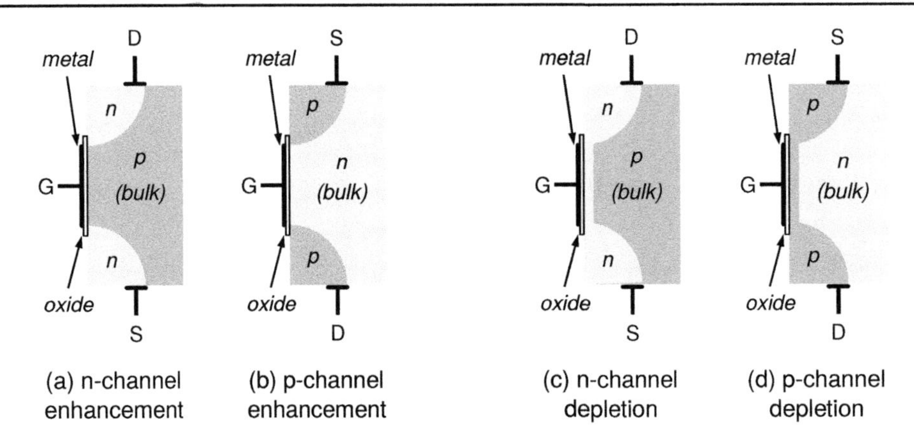

Figure 3.10 The structure of n-channel and p-channel enhancement mode (normally off) and depletion mode (normally on) metal oxide semiconductor field effect transistors (MOSFETs).

the drain and source terminals. As the voltage on the gate is changed in one direction the electric field extends further into the body of the device (labelled bulk in the figures) causing the channel to retreat back towards the regions around the drain and source terminals. And as the voltage on the gate is changed in the other direction the electric field reverses and the conductive channel is encouraged to extend back out linking the drain and source and allowing more current to flow.

The only real difference between the enhancement mode (normally off) MOSFET and the depletion mode (normally on) MOSFET is as the names in brackets imply the first type has no conduction between drain and source when the gate voltage is held at zero volts whereas in the second type the drain source channel does conduct at zero gate voltage. So in the case of the depletion mode MOSFET the voltage on the gate has to swing further in the off direction in order to turn the transistor fully off. Which direction is off depends on whether it is an n-channel or a p-channel device.

Figure 3.10 illustrates these behaviours. No channel is present by default in the enhancement mode type which therefore is normally off. In the normally on depletion mode type on the other hand a channel is shown as being present by default. This is a good way of visualising and remembering the behaviours of these different types of MOSFET.

Germanium Semiconductors

Most semiconductors these days are based on the element silicon. However there was a time when many were manufactured using germanium as the base material rather than silicon. For most purposes germanium devices operate in fundamentally the same way but germanium makes rather poorer semiconductor components than does silicon. A bit more current leaks through the p-n junction under reverse bias conditions. Switching between on and off states tends to be slower and less well defined. In fact germanium generally exhibits poorer performance all round. Germanium diodes do however start to turn on at a lower voltage (typically round about 0.3V) and so they often find favour where very small signals are being handled in homemade radio receivers and such like. They are also used in many audio circuits where their somewhat slower and more gentle approach to switching can lead to subjectively more pleasing results when working with audio signals.

Germanium diodes and transistors are becoming increasingly scarce and difficult to source and correspondingly more expensive. There is however still a significant demand for the components in the niche audio electronics area. Along with vacuum tubes, germanium components hold a special place in the world of audio electronics as components which have been superseded and replaced elsewhere but still have a role to play due to their particular characteristics and how they affect the audio signals which pass through them. Some of this is undoubtedly unfounded or exaggerated but there are places where their special character does make a real (if subjective) difference to the performance of an audio circuit.

References

E. Evans, editor. *Field Effect Transistors*. Mullard, 1972.

A. Sedra and K. Smith. *Microelectronic Circuits*. Oxford University Press, 7th edition, 2014.

4 | Mains Electricity and Electrical Safety

Although this book is concerned mainly with circuits running from low voltage power sources (most typically a single nine volt battery), mains power is never very far away. Indeed the place of the battery can just as easily be taken by a mains power adaptor designed to supply the same DC voltage to the circuit but drawing its energy from the mains supply. When circuits with higher power requirements are encountered, most obviously power amplifiers outputting more than a couple of watts, battery powering quickly becomes troublesome and a mains power supply becomes the only viable option. Additionally mains connected bench power supplies are commonly encountered

Table 4.1 Four of the most widespread and commonly encountered mains power plug and socket connector types indicating the maximum rated current for each, along with where they are most commonly encountered and the line voltage and AC frequency of the mains supply in the jurisdictions indicated

North America 120V 60Hz 15A max		**UK and Ireland** 230V 50Hz 13A max	
Continental Europe 230V 50Hz 16A max		**Australia and New Zealand** 230V 50Hz 10A max	

in electronics labs as they are far more convenient and versatile than using batteries while developing, testing, and experimenting with circuits in the lab.

There are many different mains power supply and connection standards in use across the globe. Table 4.1 illustrates four of the most common and notes the major geographical regions where they may be encountered. Dozens of standard connector types are in use worldwide but for the most part the four shown here are those most likely to be encountered.

The fundamentals discussed in Chapter 2 continue to hold true so that for instance it requires a circuit (a closed loop) in order for current to flow. In the case of mains power the two connections required to form this loop with the mains supply are the live (aka phase) and neutral connections available at the wall socket that any mains powered equipment is plugged into. The third connection present in the standard power outlet is the earth connection and is provided as a safety feature. When everything is working as intended the earth connection does nothing. It only comes into play when a fault occurs somewhere in the system. The role of the earth connection is expanded upon later in this chapter.

MAINS POWER GENERATION AND DISTRIBUTION

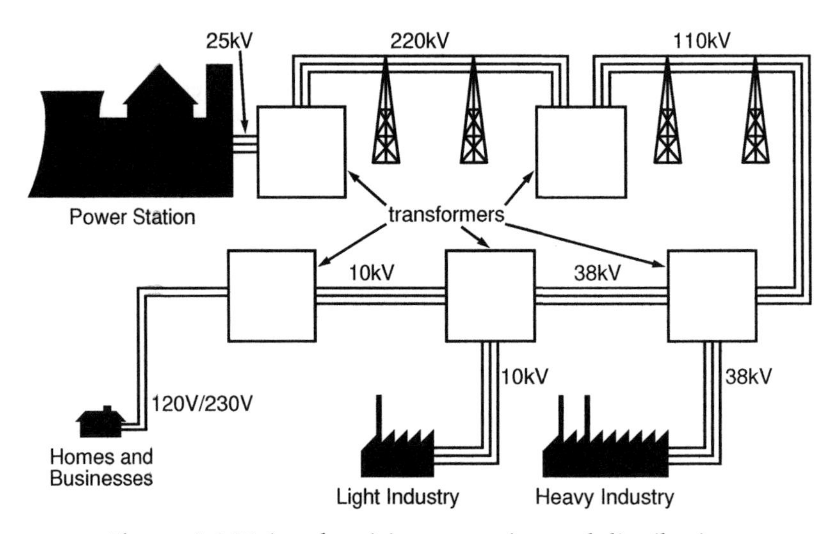

Figure 4.1 Mains electricity generation and distribution.

Batteries utilise a chemical reaction in order to generate their electricity. As has been mentioned already in Chapter 2 one way or another most mains power generation is achieved through the mechanism of electromagnetic induction. The only mainstream exception today is solar power which uses an entirely different process to generate electricity. All the other common methods of mains electricity generation including coal, oil, gas, nuclear, wind, wave, tidal, and hydroelectric just employ different methods to achieve the same goal. That is to turn a turbine connected to a generator. All these generators are essentially identical in the way they work employing the relative motion of a conductor and a magnet in order to generate electricity by way of electromagnetic induction.

Figure 4.1 illustrates the main components of a typical mains power generation, transmission, and delivery system. The voltages quoted in the figure for each section of the system are typical and they do vary quite a lot from one place to another. The only one that is of real interest is the voltage at the very end of the chain. The figures of 120V and 230V quoted here are the two nominal domestic supply voltages encountered across most of the world but different supplies can be encountered in different places.

While DC mains power is possible and indeed exists in isolated locations, all domestic power worldwide is delivered as AC. There are advantages and disadvantages to both AC and DC based systems, and some DC power transmission infrastructure does exist. However currently it is not encountered outside of a small number of highly specialised and highly localised facilities. The standard electricity supply across the globe which shows up at the live connection of a power socket is a sine wave (Figure 4.2) AC signal which cycles at a rate of either 50Hz or 60Hz depending on geographical location. That is to say the voltage swings from positive to negative and all the way back again either fifty or sixty times every second.

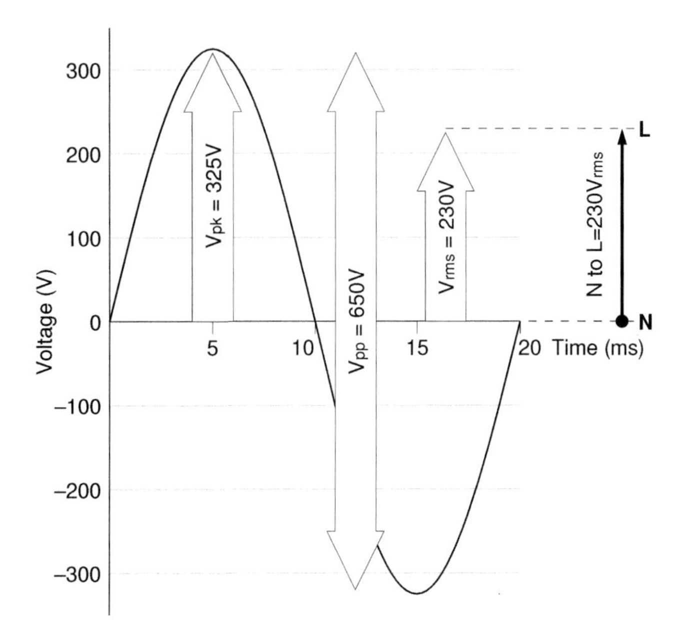

Figure 4.2 Single phase mains power signal indicating the most important characteristics of the mains supply used around much of the globe (outside North America where $120V_{rms}$ is standard).

For the most part all of the above detail can be put to one side. However it is important when plugging a new piece of equipment into the mains to check that its power supply unit (PSU) is suitable for the local supply voltage and frequency especially if the equipment has been sourced abroad. These details will usually be printed somewhere prominent on the unit close to where the mains power enters the chassis. Many modern power supplies are designed to handle anything that is likely to be thrown at them. If a PSU says something like '110–240V, 50/60Hz' it should work with the mains supply anywhere. Alternatively some units have a switch to select between 120 and 230 volts. It is well worth making sure the local mains power supply voltage and frequency is known and confirming compatibility before powering up any new piece of mains powered equipment. Otherwise a loud bang and an expensive repair job may ensue.

Building Wiring

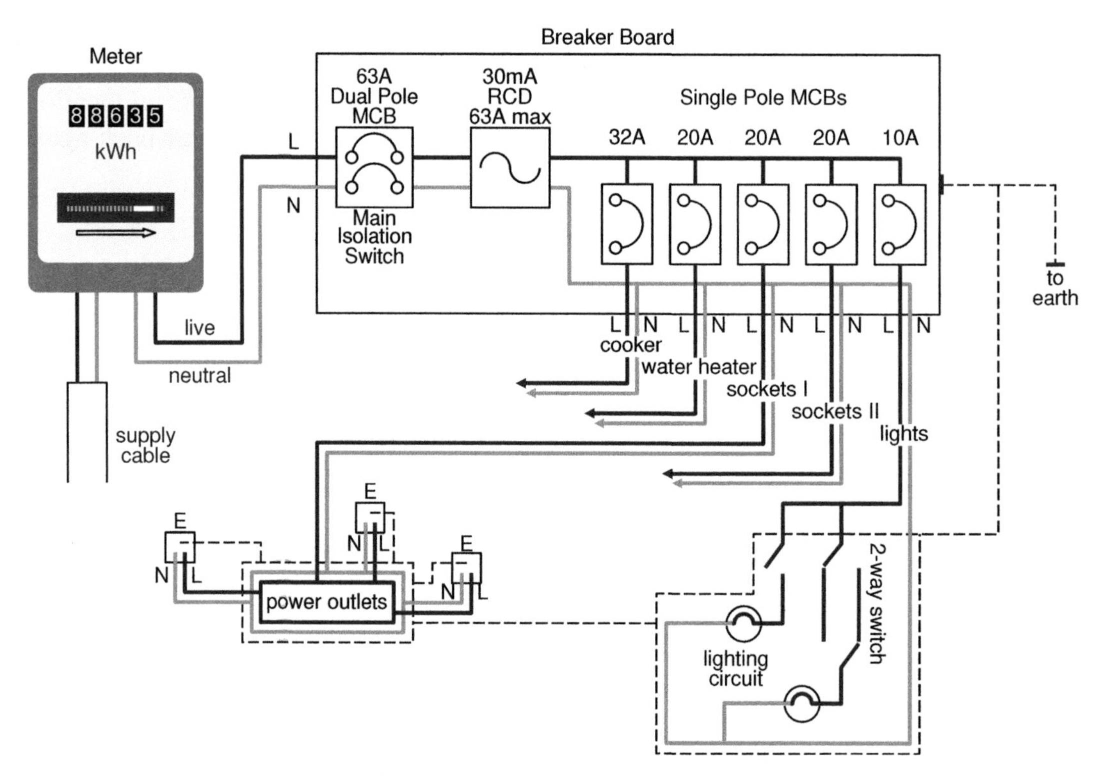

Figure 4.3 An example of the kind of mains wiring found inside a typical residential building or business premises.

Once the mains power lines have reached the building, the internal installation will be configured something like Figure 4.3. Two wires carry the mains supply into the building from the power grid, the live and the neutral. The earth connection present at the power sockets is a local connection into the ground. Its job is to provide a low resistance path for electricity to follow in the case of a fault in the equipment or wiring resulting in the live signal finding its way onto the chassis of the equipment. This ensures that a fuse will blow or a circuit breaker will trip quickly thus minimising the exposure of dangerous voltages.

The incoming mains power supply cable is routed through the electricity meter and then into the main breaker board or fuse box. Here it passes through the main isolation switch typically followed by a residual current device (RCD) and from there spurs are taken off to service the various different zones in the building. Each spur has a corresponding miniature circuit breaker (MCB) on the breaker board, each one rated to an appropriate maximum load current. So for instance lighting circuits tend to require less current than sockets, and high current equipment like cookers and water heaters will usually have their own dedicated MCBs rated at a higher load again. The local safety earth connections are linked into each spur at this point. In older buildings the lighting spurs of the installation often do not include any earthing, but modern wiring practice includes an earth connection in these wiring zones also.

The Fuse Box

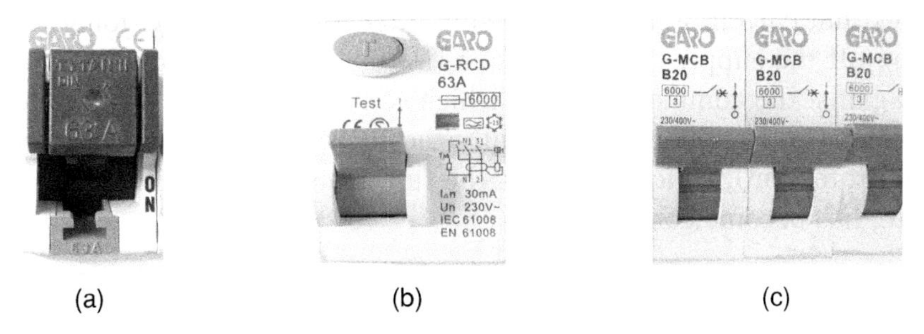

Figure 4.4 Details from the main circuit breaker board in an building showing (a) the main isolation switch, (b) an RCD, and (c) a number of MCBs.

Circuit breakers (see Figure 4.4) have replaced fuses in modern building wiring installations although fuses are still used in individual appliance plugs (and often also inside the equipment itself). The modern MCBs serve exactly the same function as the old fuses did. If more than the maximum rated current flows, the power is cut off. The difference is that with a fuse a wire burns through interrupting the flow of current and so once the problem has been rectified a new fuse needs to be installed whereas with an MCB the switch trips and can simply be reset by pushing it back up once the problem has been dealt with. Like fuses, MCBs protect against overloads where it may be that there is no actual fault but simply the fact that too many things have been plugged into the same circuit drawing too much current and thus presenting the danger of overheating the wiring and the possibility of fire.

In any given mains circuit supplying power to a device there will be a few different points at which a fuse or circuit breaker might be added. Tracing back from the device itself, four or more fuses/circuit breakers may be encountered, typically increasing in rated breaking current at each step upstream towards the start of the line at the building's fuse box. Often a device will be internally fused (sometimes in multiple places). The rating here will be carefully selected by the designer in order to protect the device while allowing sufficient current to flow for its normal operation. The mains plug connecting the device to the building's ring mains sockets may include a standard fuse (usually three, five, or thirteen amps). The particular ring of sockets used will be fused with an MCB at the breaker board, perhaps rated at 20A. A ring may consist of a significant number of sockets around a room or number of rooms. Powering a number of devices which work without issue on their own can easily overload and trip this breaker, cutting power to the whole ring. Similarly, while each ring can accommodate up to 20A, a number of separate rings can together quickly overload the main isolation switch (typically rated at 63A), see Figure 4.3.

RCDs (also known as ELCBs – earth leakage circuit breakers) represent the second line of defence in the safety functions performed at the fuse box. What an RCD does is to monitor the amount of current flowing in the live (aka phase) and neutral wires. So long as the two currents match the RCD is happy but if even a small imbalance appears this indicates that current is going somewhere that it shouldn't be going and the RCD trips, cutting off the supply. Between them, MCBs and RCDs provide a good level of protection against the common dangers which can be encountered when dealing with the mains electricity supply. More recently the RCBO (residual current breaker with overcurrent protection), which combines the functionality of the MCB and RCD into a single device, has become common.

MCBs protect against overloads

RCDs protect against faults

Wiring Safety

Power Outlets

Many different styles of mains socket can be encountered around the world (some of the most common styles are illustrated in Table 4.1) but the principles remain the same everywhere. Poor and dangerous building wiring can be encountered anywhere and it is well worth testing all the power outlets and plug boards before using them in order to ensure that equipment and people are kept safe from harm.

Testing can be performed using a standard digital multimeter (DMM). The DMM is one of the most useful and versatile tools when working with anything electronics related. Further details on their function and use are given in Part II. A simple and convenient alternative for testing mains wiring however is to use a dedicated socket tester like the one illustrated in Figure 4.5. These are small self contained devices which plug directly into a socket and give a visual (and often also an audible) indication if they encounter any of a large number of possible faults in the wiring of the socket.

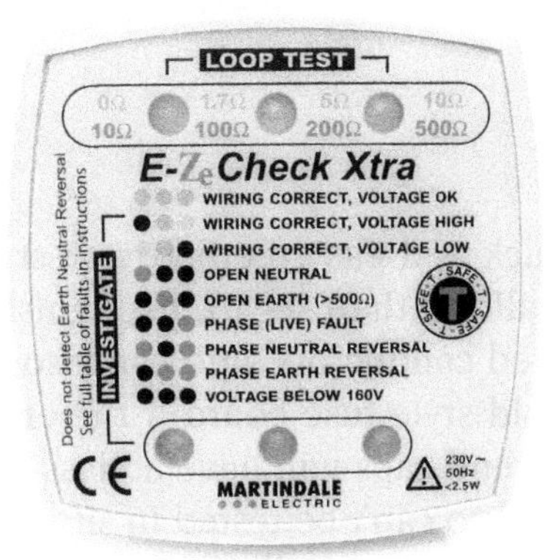

Figure 4.5 The front display of a typical socket tester. The back sports the three pins of a standard mains plug, and it can be inserted directly into a socket for quick and easy testing of the socket wiring for numerous common wiring faults.

Probably the most common wiring issues encountered (and it is surprising just how often they can be encountered) are reversal of the live (aka phase) and neutral wires and a poor or missing earth connection but various other problems are also possible and it is always wise to test for such issues and address any problems found before using a particular mains supply or extension lead. One issue such testers can not detect is the reversal of the neutral and earth connections. If an RCD is fitted then this fault can be assumed not to be present as the RCD would immediately trip whenever anything was plugged into the affected socket. Without an RCD only a full inspection can rule out this particular fault but it is unlikely to be encountered and so it is reasonable to rely on the results of such a standard socket tester to confirm safe wiring in an installation.

Figure 4.6 A phase tester is used to detect the presence of mains voltage.

The phase tester or line tester (Figure 4.6) is another useful tool for performing basic tests on mains wiring. The blade of the screwdriver is inserted into the live terminal of a socket or touched against any other potentially live point. The metal cap on the end of the handle is contacted with a finger and if mains voltage is present at the point being probed the little bulb inside the handle will glow.

LEARNING BY DOING 4.1

USE A PHASE TESTER (AKA LINE TESTER)

Materials

- Phase tester
- Mains power outlet

Theory

A phase tester provides a quick and easy method to test for the presence of a live mains power connection. It allows the user to distinguish between live conductors and neutral/earth/unconnected conductors, and can also sometimes be used for the testing of mains fuses in an old style fuse board if the end caps are exposed (when the board is mounted low the end caps will have a glass window protecting them to prevent accidental contact, and so can't be tested in situ).

A phase tester takes the form of a small screwdriver, see Figure 4.6. Its handle houses a neon bulb and is topped with a metal cap. To use the tester, the blade of the screwdriver is contacted to the conductive surface to be tested and the user places their finger on the metal cap on the end of the handle. If the bulb glows it indicates that the surface is live.

Practice

WARNING – Mains electricity can kill, always exercise caution.

First examine the tester to ensure it is suitable for the supply to be tested. It should have printed on it something like – 'AC 110–250V'. Also check that there is no visible damage; solid plastic insulation should extend down the shaft of the driver to within about 1cm of the end of the blade.

The contact points to be tested may be protected by a cover which will need to be opened. Once the contact points have been exposed hold the end of the phase tester between middle finger and thumb with index finger touching the metal end cap. Insert the blade to touch the contact point and observe the light in the device's handle. Touch the neutral and no indication should appear. Touch the live and the bulb should glow. Similarly in a three phase supply the five conductors present can be tested. The three live phases can be differentiated from the neutral and earth wires.

Wiring a Plug

These days new equipment is shipped with a plug pre-fitted, and indeed these plugs are often of the moulded variety and as such can not be opened and therefore neither require nor allow for any kind of servicing. It is still however a moderately common task to have to fit a new plug either because the old one is damaged or of the wrong type. Furthermore it is a very good idea to periodically check the wiring in all non-moulded plugs first to ensure that they have been wired correctly and thereafter to make sure that none of the screws have become loose over time. This is a much more common occurrence than might be imagined and can lead to anything from crackles and pops in the audio to intermittent or persistent power dropouts in equipment all the way up to irrevocable equipment damage and, in extreme cases, fire.

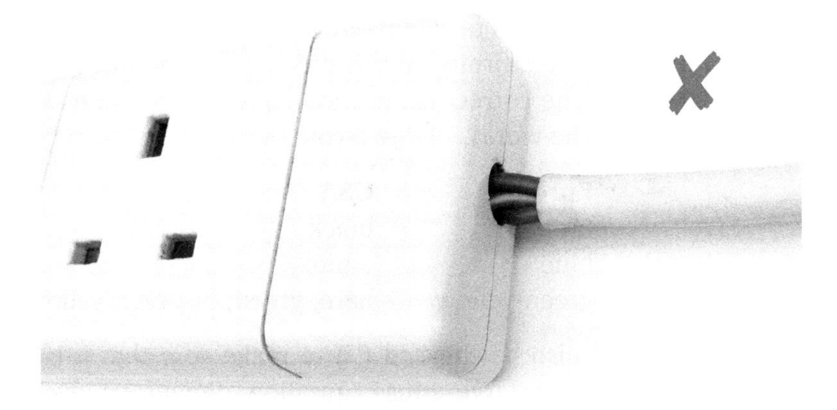

Figure 4.7 Detail of a trailing plug board showing a badly wired mains cable with the outer insulation stripped back too far.

A mains cable wired as in Figure 4.7 where the outer insulation has been stripped back too far and the individual insulated wires are visible should be rewired or removed from service. This issue is equally common on trailing plug boards as shown here and on standard mains plugs. It results in the cord grip being unable to adequately secure the cable and as a result pulling on the cable will over time lead to loosening of the wire terminals with all the consequent dangers outlined above.

The points to check for in a well wired lead are:

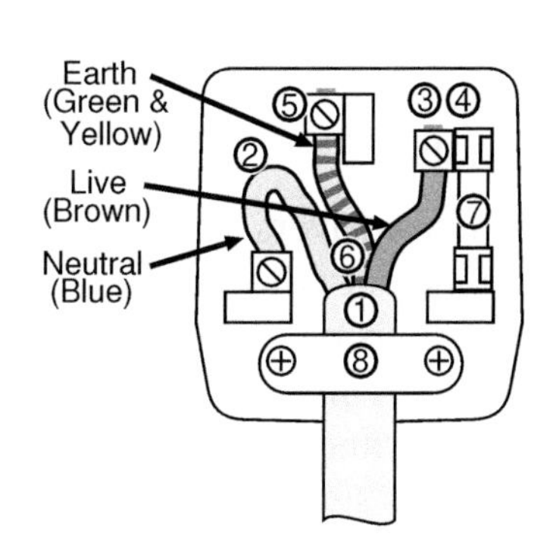

1. Correct length of outer insulation stripped
2. Correct lengths for live, neutral, and earth
3. Correct length of inner insulation stripped
4. No loose strands of copper at terminals
5. All terminal screws tightened sufficiently
6. Correct colour code for live, neutral, and earth
7. Correct size fuse fitted for application
8. Cord grip tightened to prevent slippage

While different styles of plug can differ significantly in the detail of their internal layout the points noted here are generally applicable to most variants. The outer insulation should extend far enough inside the casing (1) that the cord grip can clamp down firmly on it. The live, neutral, and earth wires must be individually trimmed to the correct length so that they reach easily to their respective terminals (2) without any excess wire causing difficulty when closing the casing. Just enough inner insulation should be removed (3) to insert fully into the terminal, and no copper strands should be allowed to escape from the terminal (4) in order to avoid potentially dangerous short circuits.

With the bare copper inserted each terminal screw should be tightened fully (5) and tested to make sure the wire is held firmly. In this process the colour coding should be double checked (6) to ensure the correct connections have been made. A number of colour codes are used around the world. The two commonest are shown below.

	Europe	USA
Live	brown	black
Neutral	blue	white
Earth	green-yellow	bare, green, or green-yellow

The fuse if present should also be checked (7) to make sure that it is appropriate for the device being powered. Too large a fuse can allow dangerous levels of current to flow in the event of a fault causing excess heating and possible equipment damage and fire. As mentioned in Chapter 2, common fuse sizes in mains plugs are three, five, and thirteen amps. The common practice of using thirteen amp fuses as a default replacement should be avoided. If a smaller fuse can be used it should be used. This minimises the chances of low current wiring being exposed to excess current and as a result overheating and possibly burning out dangerously. And finally the cord grip should be tightened (8) and checked, such that pulling on the cable does not pull directly on the terminal connections, running the risk of pulling one or more of the wires loose.

LEARNING BY DOING 4.2

WIRING PLUGS AND SOCKETS

Materials

- Screwdrivers
- Blade or wire stripper
- Various types of wire and cable

Some or all of:

- Plugs and sockets (the non-moulded kind)
- Mains switches (inline and panel mounted)
- Screw terminal speakon connectors
- Terminal blocks

Theory

Even though mains plugs tend to be of the sealed moulded type these days there are still plenty of screw terminals to be found in electrical installations of various kinds. The illustration below shows a few places where they can be encountered. The basic procedure described above for wiring a plug can be used as a check list for the wiring and servicing of any screw terminal electrical contacts. Some of the points will not be relevant in every case (neither a cord grip nor a fuse might be present for instance), but the principle remains the same.

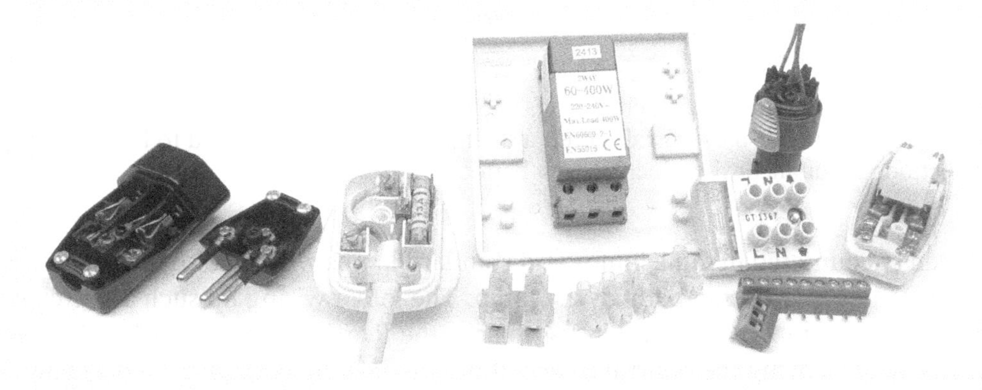

Practice

Start simple with single wires into terminal blocks. Here it is just a case of stripping an appropriate length of insulation, loosening the terminal, inserting the bare wire, and retightening the screw. Always aim for as little exposed wire protruding from the terminal as possible.

Cable with multiple conductors (e.g. mains flex or speakon speaker cable) requires a bit more care. If you don't get the lengths right don't stretch or squash the connections to fit. If necessary trim the cable flush and start again. Where a cord grip is involved, pay particular attention to the length of outer cable insulation stripped so that the grip can do its job.

Fitting a Switch

The simple rule for fitting a mains switch is that it must always isolate the live conductor from the equipment being switched. It is important to realise that a switch on the neutral side will still activate and deactivate the device and so will appear to be operating correctly and safely installed. The problem with a switch on the neutral wire is that while no current will flow when it is in the off position, the live wire will still be connected into the device and so touching the wiring inside the device can result in an electric shock.

Some (especially inline) switches break the connection on both the live and the neutral wires. It can be a more convenient way of inserting a switch into a power cable

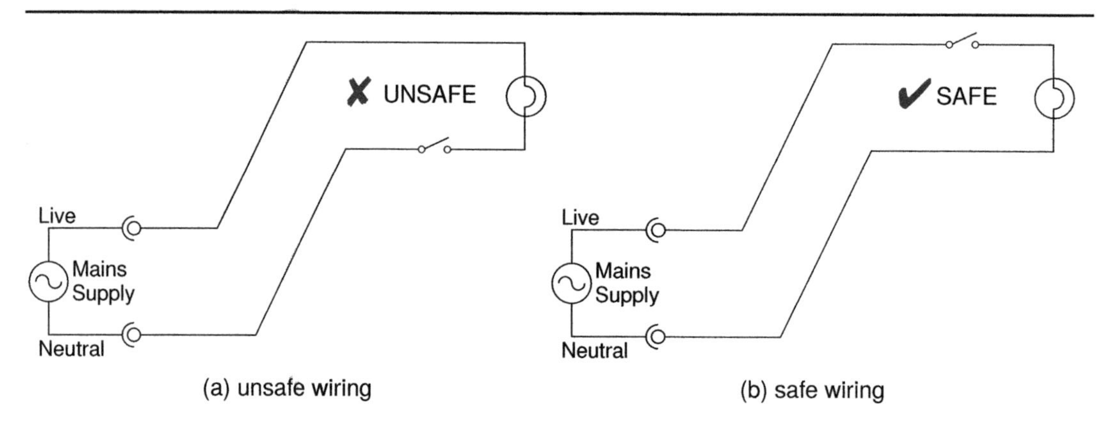

Figure 4.8 Unsafe and safe switch wiring. A mains switch should always be placed in the live conductor, not the neutral. Often a dual pole switch will be used to break both connections simultaneously, which is even better.

and is often encountered in table lamps and standard lamps with an on/off switch fitted in the lead rather than on the body of the lamp. The job of wiring up such a switch can be more easily accomplished than having to split the live (aka phase) and neutral and cut the live while leaving the neutral intact. It also avoids the possibility of accidentally fitting the switch to the wrong wire resulting in the dangers outlined above.

Appliance Protection Classes

Not all electrical equipment includes a connection to the safety earth. There are in fact a number of different appliance protection classes (IET, 2018, p. 26) each of which requires its own set of safety features to be implemented and which may or may not apply to any particular type of equipment. Electrical equipment should carry a legend indicating the protection class which has been implemented in its design and manufacture. This will usually include the display of one of the three symbols shown in Figure 4.9, indicating classes I, II, and III respectively.

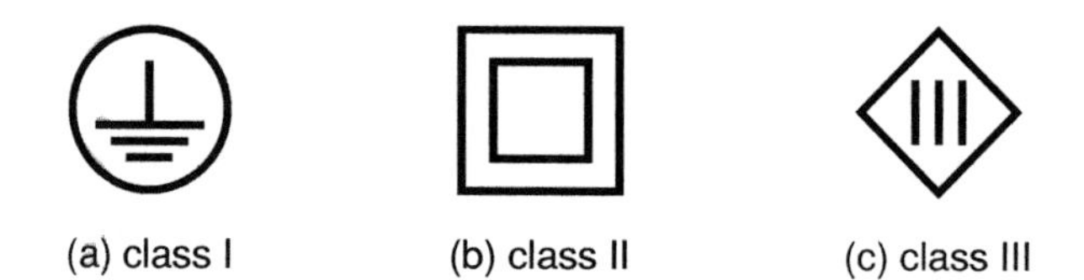

Figure 4.9 Appliance protection class standard symbols.

Class I (earthed) protection is probably the most common for mains powered audio gear. This equipment must have its chassis connected to earth by a separate earth conductor (Figure 4.10). The earth connection is achieved by using a three conductor mains cable, typically ending with a three pronged AC plug which connects into a corresponding AC socket.

The basic requirement is that no single failure can result in dangerous voltage becoming exposed so that it might cause an electric shock and that if a fault occurs the supply will be removed automatically and quickly. A fault in the appliance which causes a live conductor to contact the casing will allow current to flow in the earth conductor which will trip either an RCD or an MCB in the main circuit breaker board, or a fuse in the plug or within the device itself.

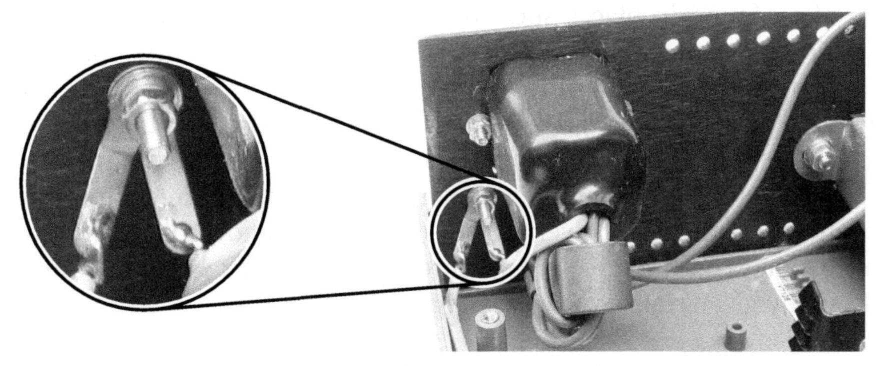

Figure 4.10 A view inside a piece of mains powered equipment showing the back of the mains input socket. The earth connection is clearly seen bolted directly to the metal chassis of the device.

Class II or double insulated electrical appliances are ones which have been designed in such a way that they do not require a safety connection to earth. The basic requirement is that no single failure can result in dangerous voltage becoming exposed so that it might cause an electric shock and that this is achieved without relying on an earthed metal casing. This is usually achieved at least in part by having two layers of insulating material surrounding live parts or by using reinforced insulation. A double insulated appliance should be labelled Class II or double insulated or bear the double insulation symbol (Figure 4.9b).

Insulated AC to DC power supplies (such as mobile phone chargers) are typically designated as Class II, meaning that the DC output wires are isolated from the AC input. The designation 'Class II' should not be confused with the designation 'Class 2'. This latter designation refers to the maximum voltage or current which a cable is designed to be operated with. The 'Class 2' designation can often be seen at the outputs of a power amplifier, where it is particularly intended to warn against using a guitar jack lead to connect from the amplifier and a loudspeaker. A loudspeaker cable may use the same quarter inch TS jack plugs but it will be made using heavier gauge wire in the cable designed to carry a much higher power signal than that generated by an electric guitar.

Class III appliances are designed to be supplied from a separated extra-low voltage (SELV) power source. The voltages present in an SELV supply are low enough (not exceeding $50V_{rms}$ AC or 120V DC, see IET (2018, p. 38)) that under normal conditions a person can safely contact them without risk of an electric shock. The extra safety features built into Class I and Class II appliances are therefore not required.

Three Phase Power

In addition to industrial settings, where three phase power is commonly encountered, it is also often found in places such as in clubs and venues, where a large amount of current is needed to operate all the equipment present. A three phase installation is a way of spreading out the load. Single phase power involves one live and one neutral wire. As the voltage on the live wire cycles up and down, the amount of current flowing alternates in a similar fashion, swinging from a maximum in one direction down through zero and then up to a maximum in the other direction before reversing and repeating. Three phase power provides three independent live conductors, but the sine wave voltages on the three are synchronised so that each is at a different point in the cycle at any given time as shown in Figure 4.11.

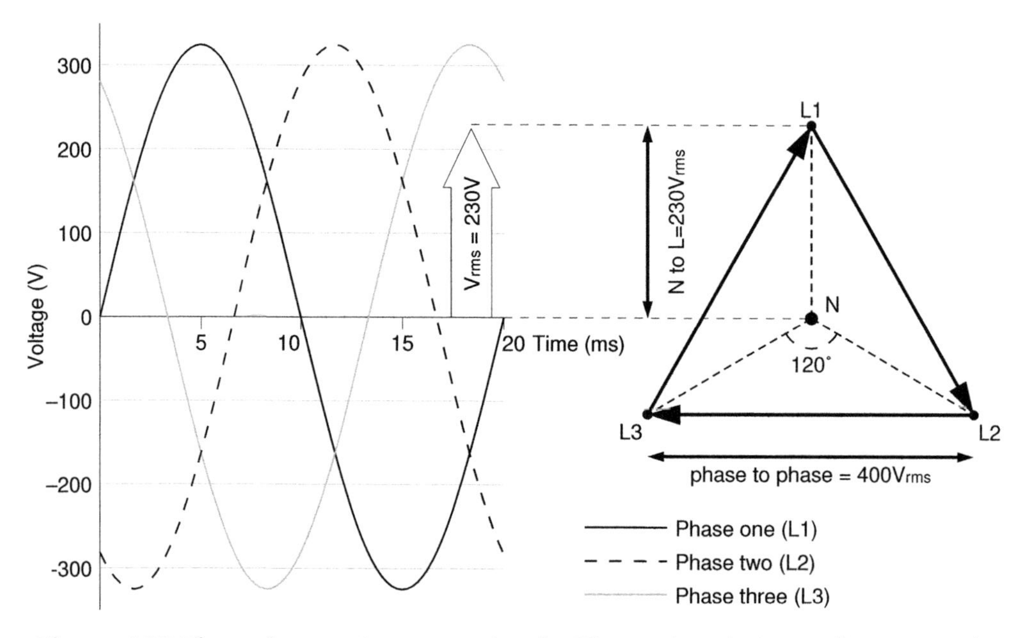

Figure 4.11 Three phase mains power signals. Three wires designated L1, L2, and L3 each carry one of the three sine wave power signals shown.

By doing this the load is spread out so that the peak current flowing is reduced and the overall level of current flow is smoothed out and kept more constant over time. This has advantages for the gauge of the wires needed and the amount of heat generated by the flowing current among other things. The idea is to balance the load between the three phases so as to keep the peak current flowing as low as possible. This is simply done by powering different pieces of equipment from different phases. Each individual phase looks just like a normal single phase power supply as far as the equipment is concerned, but the overall advantage accrues, making the supply of power to high demand facilities more easily manageable.

There are different ways of using three phase power. Instead of connecting appliances between a single one of the three available phases and a common neutral, in some places connections are made from one phase to another and no neutral wire is provided at all. As is illustrated in Figure 4.11, the voltage seen between two phases 120° out of phase with each other is greater than the voltage between any one phase and a common neutral. Wiring between each live phase and a common neutral is called a wye (as in the letter 'Y') configuration. Wiring from phase to phase is called a delta configuration, see Figure 4.12. Such things as three wire wye power, two phase power, and other variations on the same basic themes may also be encountered. In all these cases the goal is the same, to supply large amounts of current in the most efficient and effective manner possible by balancing the load across the different phases thus minimising the demands on any one current carrying wire while also smoothing out the overall amount of current flowing at any instant.

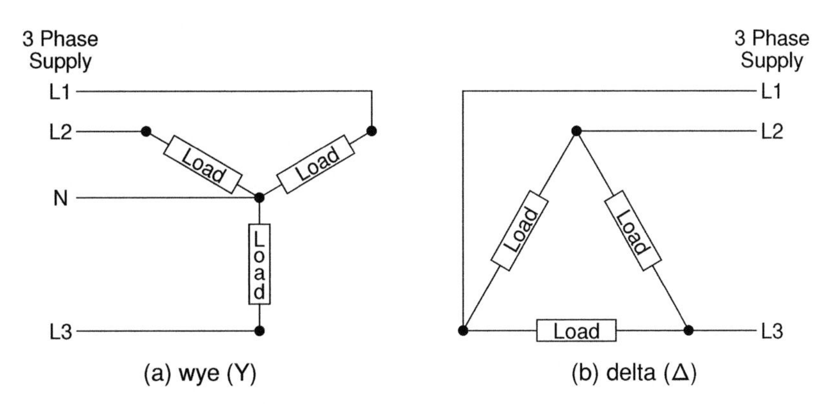

Figure 4.12 Three phase wye and delta configurations.

References

IET. *Requirements for Electrical Installations BS 7671:2018 (IET Wiring Regulations)*. Institution of Engineering and Technology, 18th edition, 2018.

T. Linsley. *Electronic Servicing and Repairs*. Newnes, 3rd edition, 2000.

4.1 – Phase to Phase RMS Voltage Level

In Figure 4.11 the phase to phase voltage has been indicated as being $400V_{rms}$. As explained in Chapter 5, the RMS level of a signal is an important metric, which allows the power delivered by the signal to be calculated.

It is instructive to note that the triangle on the right of Figure 4.11, with the neutral at the centre and the three live phases at the vertices, provides a consistent representation of the relationships between the voltages at the four conductors in a three phase supply. The neutral to live voltage is shown as $230V_{rms}$, this being the standard mains supply in much of the world. (The treatment here is equally valid in other jurisdictions, simply substituting the local mains supply voltage in order to arrive at the corresponding result.)

The geometry of the figure shown thus allows for the phase to phase RMS voltage to be calculated. This phase to phase value comes into play when a three phase supply is being used in the delta configuration illustrated in Figure 4.12. There are a number of ways to approach this calculation. Methods come to mind based on similar triangles, trigonometric functions, Pythagoras' theorem, and the sine formula, but the sine formula, which relates the sides and angles of any triangle, is probably the most straightforward and is the one employed here.

$$\frac{a}{\sin A} = \frac{b}{\sin B} = \frac{c}{\sin C} \quad \text{– sine formula}$$

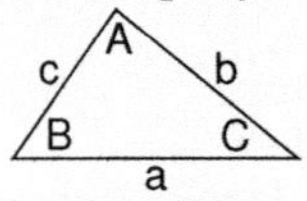

Now consider the sub-triangle with vertices N, L2, and L3 in Figure 4.11. As all of the angles and two of the sides are known, it is a simple matter to calculate the length of the unknown side, which represents the phase to phase voltage which is to be determined.

$$\frac{a}{\sin(120)} = \frac{230}{\sin(30)} \quad \text{– sine formula}$$

$$\Rightarrow \quad a = \frac{230}{\sin(30)} \times \sin(120) \quad \text{– rearrange}^{a}$$

$$\approx 398.4V_{rms} \quad \text{– calculate}^{b}$$

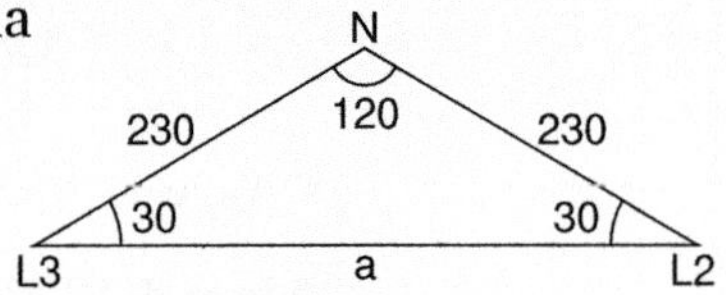

Given the typical variability of mains supply voltages and the tolerances which apply, this can reasonably be rounded to $400V_{rms}$ as a convenient and accurate value for the phase to phase level observed in a typical three phase mains power supply based on a $230V_{rms}$ phase to neutral voltage.

---

a. The $\Rightarrow$ symbol means 'implies', and indicates a logical step in a calculation. The alternative symbol $\therefore$ meaning 'therefore' is often used in a similar way. Strictly speaking the two have different meanings and implications but in less formal mathematical presentation they can be considered essentially equivalent and both will often be encountered.

b. The $\approx$ symbol is used to indicate 'is approximately equal to'.

5 | Signal Characteristics

When building and testing electronic circuits it can be very useful to gauge the performance of a particular design by sending test signals through the circuit and performing measurements on the resulting output waveform. The simplest measurements take the form of a single number, perhaps showing the average voltage level or the frequency of the signal under examination. More elaborate measurements may be presented in graphical form.

Time Domain and Frequency Domain Representations

When it comes to visualising the information in a signal, two primary formats exist as illustrated in Figure 5.1. A time domain representation of a signal (Figure 5.1a) shows how the signal's amplitude changes over time. This would directly correspond to the variations observed in the voltage level which constitutes an audio signal propagating inside a circuit.

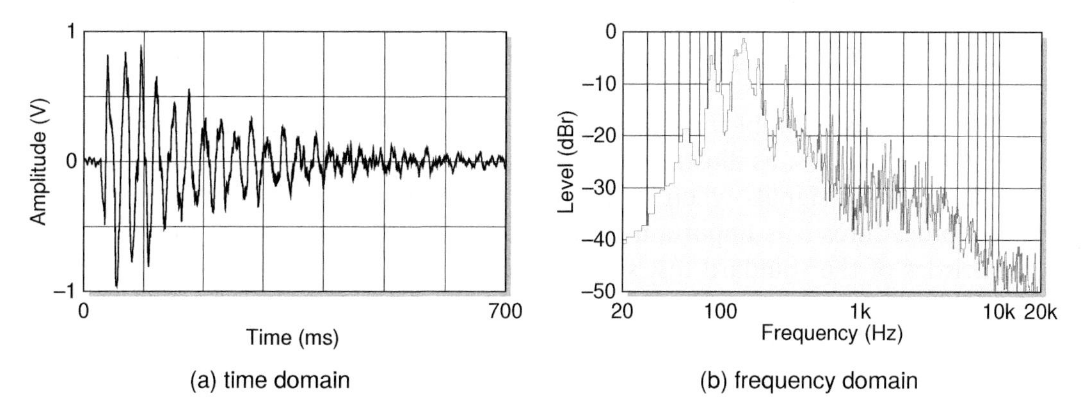

Figure 5.1 An audio signal can be viewed either in the time domain (a), or in the frequency domain (b). These two types of visualisation both represent the same signal, just highlighting different aspects.

The second form (Figure 5.1b), designated a frequency domain representation or frequency spectrum, shows a snapshot of the signal at one point in time. A short snippet of the signal is analysed and broken up into all the individual frequency components which are present at that moment. The amount of energy at each frequency is represented by the height of the line at that position along the X axis of the graph. This representation shows more explicit detail than the time domain view, but only represents one single instant in time.

A real time analysis or RTA view of a signal shows each of these levels dancing up and down as the signal changes over time. This kind of view is commonly found in audio analysis software packages as well as in many digital oscilloscopes. The visualisation will often be simplified by breaking it down into (most commonly either ten or thirty-one) bands, each one representing a range of frequencies as shown in Figure 5.2.

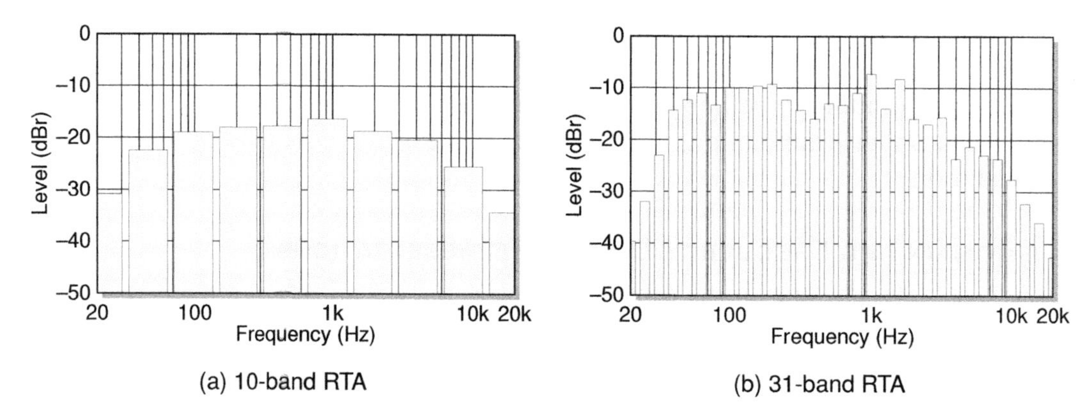

Figure 5.2 A real time analysis (RTA) view might break the audio band up into any number of frequency sub-bands. Ten bands and thirty-one bands are the two most common types of RTA display. These are also commonly called one octave and one third octave displays respectively because each band covers that approximate bandwidth.

Typical audio is a complex signal made up of many constantly varying components at different frequencies, as illustrated in Figure 5.1. As such it is often more instructive to use simpler test signals when performing tests. In order to get the most out of such test procedures it is important to have a good understanding of the nature and characteristics of the standard test signals encountered, and the terms and analytical tools used to describe their various aspects and behaviours.

Signals can be modified in a number of standard ways. The most important of these (phase shift, polarity inversion, rectification, clipping, and crossover distortion) are described next. Following this the commonest test signal types are examined, starting with the most fundamental signal component, the sine wave. After that the related wave shapes of square, triangle, ramp, and sawtooth are briefly examined.

Having looked at signals and their representations in both the time and frequency domains, and considered the commonest types of signal distortion encountered, the final section in this chapter introduces the decibel, and examines the relationship between linear and logarithmic representations of the same data. Decibels are encountered everywhere in audio technology and in electronics in general. They represent a powerful method for measuring and comparing the levels or amplitudes of different signals.

Phase and Polarity

Phase refers to the relative position in time of two signals or two components within a signal. It is useful to remember that while polarity talks about top to bottom changes (amplitude), phase talks about left to right changes (time). For a sine wave a phase shift of 180° looks identical to a polarity flip, but the same is not the case for an arbitrary waveform. These concepts are illustrated in Figure 5.3.

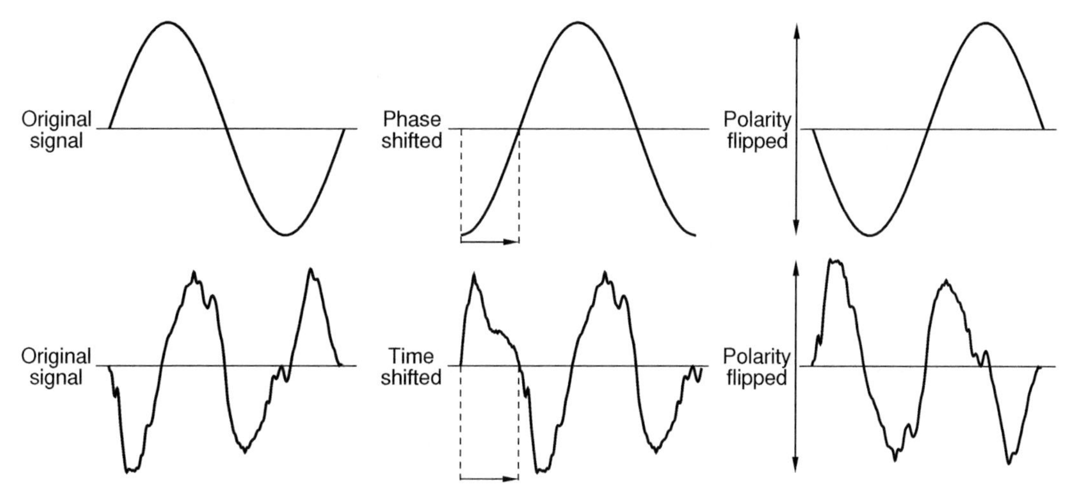

Figure 5.3 Phase and polarity illustrated in the case of a sine wave and an audio signal. If the time shifted audio signal were mixed back in with the original audio signal, comb filtering would result. Each frequency component would experience a different degree of phase shift and so a different level of reinforcement or cancellation, depending on each component's frequency (or more directly its associated wavelength).

While polarity is a binary function (flipped or not flipped) referring to the relationship between the top half and the bottom half of the signal, phase is a continuous function (any number of possible states) referring to the relative position of two signals or signal components in time.

A static phase relationship specifically applies to sine waves of the same frequency and it is not on the whole meaningful to talk about the relative phase of two more complex signals. Two arbitrary signals will constantly move in and out of phase with one another. When both are up at the same time or both are down at the same time they might be considered in phase, and when one is up as the other is down they can be said to be, to a greater or lesser extent, out of phase. This is not however a very useful analysis on the whole. Two sine waves of the same frequency will have a constant phase relationship, being either completely in phase or some specific degree out of phase. Phase is measured in degrees with 360° in a full cycle. If two sine waves of the same frequency are moved relative to each other they will continually move in and out of phase. Phase relationships become very important when multiple signals (be they acoustical or electrical) are being combined or otherwise manipulated. Comb filtering and other unwanted artifacts can result. Some electronic circuits are also capable of altering the phase relationships of various components of a signal. Sometimes this is an unwanted side effect (as in filters and EQs) and sometimes it is used to advantage (as in effects like phasers and flangers).

Unlike phase, polarity is a binary concept. That is to say there are two possible states of polarity, which are often referred to as positive and negative. That portion of a signal which lies above the midline is said to have positive polarity while the portion below the midline is said to have negative polarity. One very common function available on mixing desks and some other audio equipment is a polarity invert or polarity flip function. This process takes the incoming signal and flips it over so that the positive half becomes negative and the negative half becomes positive. This very simple and very useful function is often (but erroneously) referred to as 'phase invert' instead of the correct name of 'polarity invert'.

Rectification

Rectification refers to the removal or reversal of the negative half of any signal, such that no part of the resulting signal extends below the midline. The two basic forms of rectification, called half wave rectification and full wave rectification are illustrated in Figure 5.4. Half wave rectification is when the portion of the signal below the line is set to zero. Full wave rectification is when the portion of the signal below the line is folded back up above the line.

Rectification is a common process in electronics. Indeed another name for a diode is a rectifier because this is pretty much exactly what a diode does. One diode on its own easily achieves half wave rectification while full wave rectification is simply implemented using four diodes in a configuration called a bridge rectifier. The details and applications of these circuits are fully addressed in Chapter 16 on diodes. Rectification is often utilised in distortion and octave up effects due to the nature of the harmonic distortion this kind of processing adds to a signal. It is also a key stage in the operation of a standard linear power supply, converting an AC signal into a DC one. This application is examined in Chapter 16.

Clipping

Clipping is closely related to rectification and is often also implemented using diode based circuitry. Basic symmetrical hard clipping as shown in the second waveform in Figure 5.5 is among the most common (and certainly one of the simplest) approaches used to implement audio distortion as seen in myriad classic guitar effects pedals.

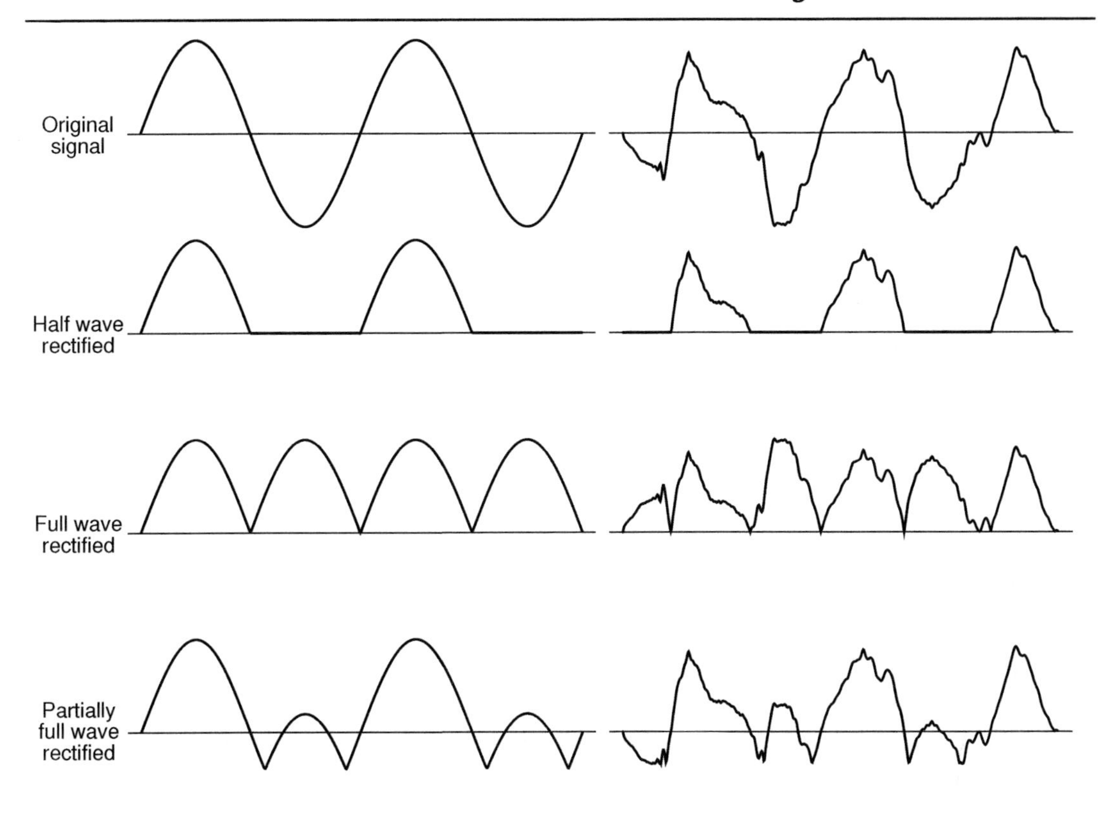

Figure 5.4 Examples of half wave and full wave rectification as applied to a sine wave and an arbitrary signal.

The idea is simply that some portion of the signal peaks are reduced or removed completely. Half wave rectification could be seen as an extreme form of asymmetrical clipping but in general clipping will be applied somewhat more gently. Even a small amount of clipping (only removing the very tips of the largest signal peaks and troughs) can result in a highly distorted sound. As illustrated in the third and fourth waveforms in Figure 5.5 two common variations on the theme exist. Asymmetrical clipping is where one side of the signal waveform is clipped more heavily than the other, which introduces a different balance of odd and even harmonics, and thus a different character to the sound. The idea in soft clipping is that rather than simply lopping off the tops and bottoms of the waveform they are instead rolled off more gently, producing less high frequency harmonics in particular and thus resulting in a somewhat less harsh form of distortion.

Clearly there is much scope for experimentation here. Different amounts of asymmetry and varying degrees of hard/soft balance in the clipping introduced will result in a broad palette of possible sounds. Again, these possibilities are more fully explored in the chapter on diodes, as well as in some of the project work presented in the final part of the book.

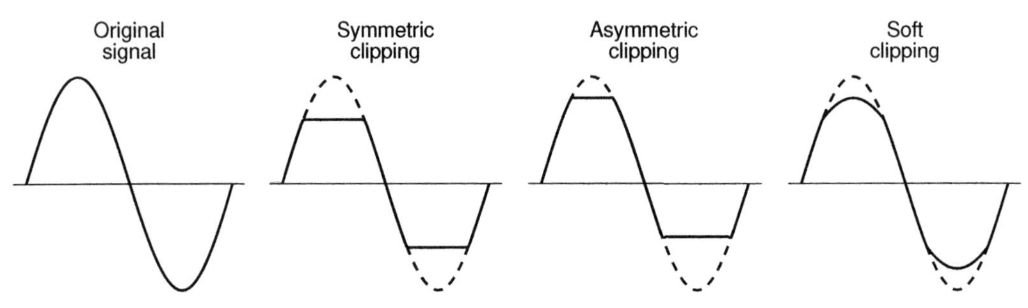

Figure 5.5 Various types of signal clipping. The dotted lines show the shape of the unclipped signal in each case for comparison.

Crossover Distortion

Closely related to the distortion introduced into a signal due to clipping is the concept of crossover distortion, as illustrated in Figure 5.6. This is when, rather than modifying or removing the positive and negative peaks of a signal as in clipping, it is instead altered at the point where it crosses the midline. This is referred to as crossover distortion.

Crossover distortion can take many varying forms, depending on the mechanism generating it. The two extremes shown in Figure 5.6 illustrate either end of the spectrum, but various other kinds of steps and deformations are possible. This kind of signal modification is less commonly utilised in distortion effects (although by no means unheard of). On the other hand, it is a problem commonly encountered in amplification circuits which use some form of symmetrical circuit architecture to perform amplification of the positive and negative portions of a signal. If the circuit is not correctly designed and appropriately configured or calibrated, then crossover distortion can be introduced at the point where the signal amplification task is handed over from one

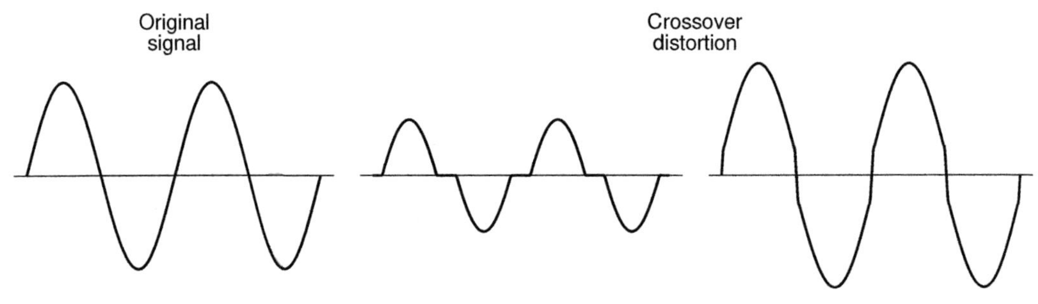

Figure 5.6 Crossover distortion can manifest in a number of different ways.

side of the circuit to the other. This most often takes a form close to the first of the two variants shown.

The Sine Wave

The sine wave is the simplest of all possible signals. As illustrated in Figure 5.7b all the energy in such a signal is concentrated at a single frequency. This makes measurement and comparison of sine wave signals very easy to perform. As such it is an effective test signal to use when building and experimenting with audio circuits.

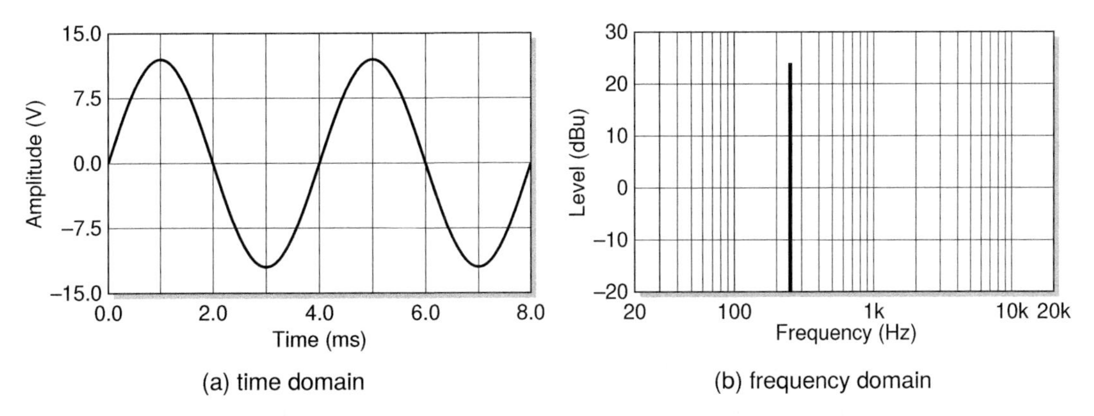

Figure 5.7 In the case of a sine wave all of the signal's energy is concentrated at a single frequency. Hence the familiar time domain wave shape in (a) corresponds to a single line in the frequency domain (b).

In Figure 5.8 some of the key characteristics of a sine wave are shown. In the horizontal or X direction, which can represent either time (period) or distance (wavelength), the main metric is the length of one cycle. For an electrical signal travelling in a circuit this will be measured as a time (perhaps one millisecond per cycle). This is called the waveform's period and is usually represented by the lowercase Greek letter tau (τ).

Distance measures may arise more frequently when considering an acoustic sound wave travelling through the air, in which case the quantity measured would be the waveform's wavelength, typically represented by the lowercase Greek letter lambda (λ). In the case of an acoustic rather than an electrical signal it is often of more interest to know the wavelength, although a measure of the period can be useful in acoustics also.

Frequency and wavelength are related to each other by the speed of propagation of the signal (whether acoustic or electrical). For acoustic signals the speed in question is the speed of sound (about 340m/s). Electrical signals are a form of electromagnetic energy. Electromagnetic energy in a vacuum propagates at the speed of light, about

$3 \times 10^8 m/s$ (after all light is electromagnetic radiation). In a wire, electricity propagates at a considerable fraction of this speed; still very fast indeed.

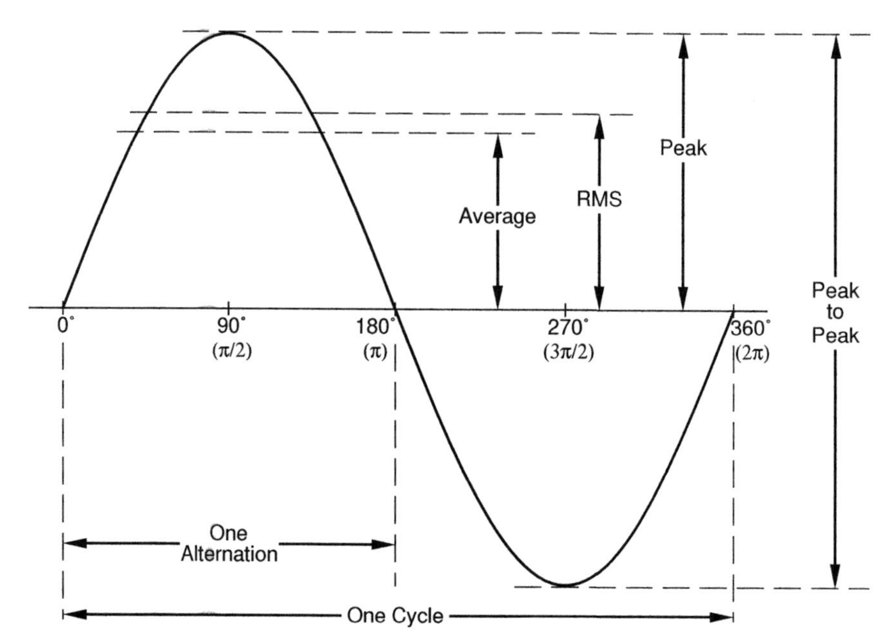

Figure 5.8 Measurements on a sine wave. The labels in brackets along the X axis ($\pi/2$ etc.) show the equivalent angles measured in radians rather than degrees.[a]

The actual speed can vary considerably depending on various characteristics of the wire or cable in question. In both the acoustic and electrical cases the relationship between speed (c), wavelength (λ), and frequency (f) is defined by Eq. 5.1. So audio electrical signals, with frequencies in the range 20Hz–20kHz, have very long wavelengths. As such it can be seen that the wavelengths of electrical signals in general only become comparable (and therefore significant) for very long cables (e.g. telephone lines or cable TV) or very high frequencies (such as in digital systems).

$$c = \lambda \times f \tag{5.1}$$

In the context of electronics however the period is the measurement of interest and thus that remains the focus here. The period has units of seconds per cycle, and so for instance the period of the sine wave shown in Figure 5.7 can readily be decerned from the time domain graph as being four milliseconds per cycle. The 'per cycle' portion of the unit is commonly omitted when quoting the period of a waveform, and so the

a. The difference between degrees and radians can be thought of as being similar to the difference between inches and centimetres, just two ways of measuring the same thing. Radians are the natural units to use in some mathematical manipulations, and this illustrates why expressions involving π appear in a number of equations which follow.

period in this case might be reported as four milliseconds or simply 4ms. The implicit 'per cycle' should not be forgotten completely as it can be useful when recalling the relationship between period and frequency as discussed below.

A figure more familiar to most than the period of a signal would be its frequency. It is however often the period which is directly available through measurement and so it can be useful to remember that there is a very straightforward relationship between these two quantities: period is one over frequency, and frequency is one over period.

$$f = \frac{1}{\tau} \quad \text{and} \quad \tau = \frac{1}{f} \tag{5.2}$$

An examination of the units of these two quantities quickly confirms this relationship. Frequency has units of cycles per second (cps), also known as hertz (Hz), while as observed earlier period has the opposite units of seconds per cycle. As such the simple reciprocal relationship between these two quantities should come as no surprise. In electronics it is often easier to measure the period of a signal (for instance with an oscilloscope), but it is often more useful to know its frequency, and so this relationship is very important and often used.

In the vertical, Y direction on Figure 5.8 four amplitude metrics are marked: peak, peak-to-peak, average, and RMS. The peak and peak-to-peak values should be fairly self explanatory. They simply represent the maximum excursion of the signal up from the centreline, and from top to bottom of the waveform respectively. For a sine wave the peak-to-peak measurement is exactly twice the peak value. This is not necessarily the case for more complex signals, which may often be asymmetrical above and below the centreline.

Strictly speaking, the average value of a sine wave is zero, since the waveform is symmetrical around the centre or zero line. What has been marked as 'average' on Figure 5.8 might more accurately be described as the half-wave average or rectified average, where the portion of the signal below the zero line is reflected up to lie above it (see later in this chapter for a discussion of signal rectification). Here it could most simply be characterised as the average signal level as measured between 0° and 180° (or indeed 0° and 90°, which would yield the same result due to the symmetry of the waveform). This rectified average level is a simple thing to measure for an electrical signal, requiring only a straightforward circuit in order to perform the task. Unfortunately, its ease of measurement notwithstanding, it is a far less useful metric than the final one marked on the graph, the RMS or root mean square.

The RMS is the value required in all calculations and equations which allow electrical circuits to be analysed and their behaviour predicted. The reasons why this is the required measurement relate to the amount of electrical power in the signal, but is not further addressed here. The name RMS itself is entirely self explanatory, it is calculated by taking the root of the mean of the square. In other words first square the signal, then calculate the mean (or average) of this, and finally take the square root of this

5.1 – Signal Wavelengths

From Eq. 5.1 the relationship between the speed, frequency, and wavelength of a signal is known. As such, given the speed of propagation, it is a simple matter to calculate the wavelength for any given frequency by simply rearranging this equation:

$$c = \lambda \times f$$

$$\Rightarrow \quad \lambda = \frac{c}{f}$$

Audio is generally taken to fall within the frequency range 20Hz to 20kHz so considering these two limits illustrates the extremes of the range. First consider the case of acoustic sound propagating through air. Sound travels in air at approximately 340m/s.

Wavelengths in air:

At 20Hz $$\lambda = \frac{340}{20} = 17\text{m}$$

At 20kHz $$\lambda = \frac{340}{20{,}000} = 17\text{mm}$$

Electrical signals propagate in a wire at a significant fraction of the speed of light in a vacuum, typically between about 60% and 95%. The speed of light in a vacuum is roughly 3×10^8m/s (three hundred million metres per second).

Wavelengths in wire (assume speed = 2/3 speed of light in a vacuum):

At 20Hz $$\lambda = \frac{2 \times 10^8}{20} = 10^7\text{m} = 10{,}000\text{km}$$

At 20kHz $$\lambda = \frac{2 \times 10^8}{20{,}000} = 10^4\text{m} = 10\text{km}$$

Thus it can be seen that even an electrical signal at the highest audio frequencies will have a wavelength of the order of ten kilometres. This has implications for questions of interference due to reflections of electrical signals. Signals at audio frequencies will not typically exhibit reflection related issues until cable lengths become several kilometres, whereas higher frequency signals (including digital audio signals) can encounter difficulties at much shorter cable lengths if care is not taken in the design of the system.

calculated mean. It is a simple enough process to describe but not nearly so simple to actually implement when making an electrical measurement.

There are however simple mathematical relationships which connect the peak, average, and RMS levels of a sine wave. Note that these relationships apply only to sine waves and do not extend to arbitrary signals. Eq. 5.3 states that the average is about six tenths of the peak level, while Eq. 5.4 shows the RMS as about seven tenths of the peak. These first two can easily be combined to arrive at Eq. 5.5, which gives the relationship between the sine wave's average and RMS levels.

$$V_{avg} = V_{pk} \times \frac{2}{\pi} \approx V_{pk} \times 0.636 \tag{5.3}$$

$$V_{rms} = V_{pk} \times \frac{1}{\sqrt{2}} \approx V_{pk} \times 0.707 \tag{5.4}$$

$$V_{rms} = V_{avg} \times \frac{\pi}{2\sqrt{2}} \approx V_{avg} \times 1.11 \tag{5.5}$$

This final multiplication factor of approximately 1.11 is often used to convert a measured average voltage value into an estimate of the RMS level of a signal. As previously stated this conversion will be accurate if the signal being investigated is a sine wave, but a significant error can result for waveforms of other shapes. This point is expanded upon in the next section. It is worth noting here that inexpensive voltmeters usually measure the average signal level (which is an easy thing to do), and then scale the result before reporting, in order to approximate an RMS reading. More advanced voltmeters will often carry the moniker 'True RMS' to indicate that they implement a more complex design in order to directly measure the RMS signal level rather than estimating it based on a more straightforward average measurement.

Complex Signals

All signals no matter how rich and complex can be decomposed into and analysed as a collection of sine waves of different frequencies, amplitudes, and relative phase relationships as illustrated in the frequency domain plot in Figure 5.1b. By mixing together sine waves at frequencies all along the X axis, with amplitudes given by the height of the plot at each location, the original signal could be reconstituted. Decomposing a complex signal into its constituent sine wave components is a powerful tool for analysis and manipulation much utilised in digital audio systems.

Sine waves provide a simplicity which facilitates easy of analysis but this same simplicity also imposes limitations on their usefulness in testing and characterising the behaviour of audio electronic circuits. The real world audio signals which these circuits will encounter once deployed have a far more rich and complex character. In order to best understand how such circuits will respond and react test signals which more closely resemble true audio might be of some benefit. However the inherent complexity of a regular audio signal renders detailed analysis using such signals difficult or impossible.

For these reasons various test signals of intermediate complexity are often employed when performing test and measurement procedures, more complex than sine waves but simple enough to allow for meaningful observations and measurements to be made. The commonest of these are the square wave, the triangle wave, and the ramp and sawtooth waves.

Square, Triangle, Ramp, and Sawtooth Waves

Each of these waveforms is composed of simple mathematical combinations of a series of sine waves. The specific spectral makeup of these wave shapes is described widely throughout the literature, and can be of particular interest in the design and use of subtractive synthesisers and related circuits. These details are not of immediate concern here and so are left for now. Figure 5.9 gives as much of an overview of the makeup of these signal shapes as is required in the context of this book.

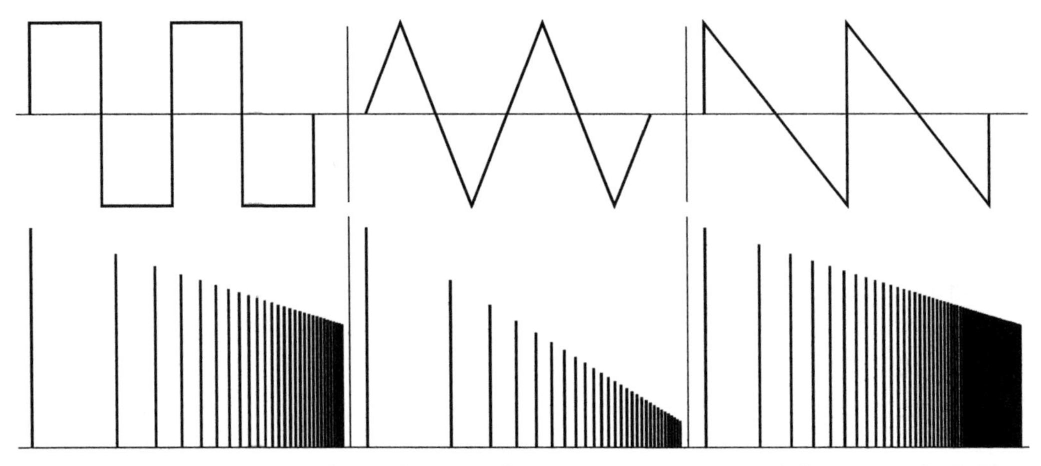

Figure 5.9 Square, triangle, and sawtooth waves in the time and frequency domains. Square and triangle waves are composed of the same frequency components, just in different proportions. The sawtooth involves twice as many frequencies, but in similar proportions to the square wave. The associated ramp wave shape is just a polarity flipped sawtooth, and as such has the same spectral makeup.

It is worth noting the corresponding relationships between peak, average, and RMS levels for these wave shapes, and comparing them to the relationships previously given for the sine wave. In the case of a square wave the relationships are extremely simple and fairly easy to deduce by looking at the shape of the square wave graph. All three values are in fact equal.

Square wave – $V_{avg} = V_{rms} = V_{pk}$

In the case of the triangle, ramp, and sawtooth waves things are a little more complicated, but the answers come out the same for all three wave shapes. This is not all that surprising as the three shapes are very closely related, as can be seen both from their time domain and their frequency domain representations, as illustrated in Figure 5.9.

Triangle/ramp/sawtooth –

$$V_{avg} = V_{pk} \times \frac{1}{2}$$

$$V_{rms} = V_{pk} \times \frac{1}{\sqrt{3}}$$

The mathematics behind these relationships is quite straightforward, although it does require some basic calculus. There is no need to get into these details. For those who may be interested, the treatment is presented in the highlighted Box 5.2.

5.2 – Peak, Average, and RMS Levels for Standard Wave Shapes

The expressions quoted in the main text for the calculation of average and RMS voltage levels for the various wave shapes are given there without justification or proof. The derivations are not in fact too complex, and are provided here by way of illustration. All the shapes involved are simple, symmetrical, and can all be approached in the same fashion.

For the average (recall this is in fact the rectified or half wave average) all that is needed is to calculate the area under the curve divided by the time period for the half wave considered. In mathematical terms this involves calculating the definite integral between 0 and T, and dividing it by T, for an appropriate T.

The second calculation is just slightly more involved. RMS stands for root mean square, and is calculated exactly as the name suggests, by taking the root of the mean of the square. Thus in the case of the required RMS values the function must be squared prior to integration. The integral is then divided by T as before, and finally the square root of the result is calculated to arrive at the RMS value.

The sine wave average case

$$v = V_{pk} \sin \frac{2\pi t}{\tau} \qquad \text{– sine wave function of period } \tau$$

$$\Rightarrow V_{avg} = \frac{V_{pk} \int_0^{\tau/2} \sin \frac{2\pi t}{\tau}\, dt}{\pi} \qquad \text{– average calculation}$$

$$= \frac{V_{pk} \times \left[- \cos \frac{2\pi t}{\tau}\right]_0^{\tau/2}}{\pi} \qquad \text{– integral of } \sin x \text{ is } - \cos x$$

$$= \frac{V_{pk} \times \left(\left[- \cos \pi\right] - \left[- \cos 0\right]\right)}{\pi} \qquad \text{– evaluate}^{a}$$

$$= V_{pk} \times \frac{2}{\pi} \qquad \text{– QED}$$

The sine wave RMS case

$$v = V_{pk} \sin \frac{2\pi t}{\tau} \qquad \text{– sine wave function of period } \tau$$

$$\Rightarrow V_{rms} = \sqrt{\frac{V_{pk}^2 \int_0^{\tau/2} \sin^2 \frac{2\pi t}{\tau}\, dt}{\pi}} \qquad \text{– rms calculation}$$

$$= \sqrt{\frac{V_{pk}^2 \times \left[\frac{1}{2}\left(\frac{2\pi t}{\tau} - \frac{1}{2}\sin 2\frac{2\pi t}{\tau}\right)\right]_0^{\tau/2}}{\pi}} \qquad \text{– integral of } \sin^2 x \text{ is } \tfrac{1}{2}\left(x - \tfrac{1}{2}\sin 2x\right)$$

$$= V_{pk} \times \frac{1}{\sqrt{2}} \qquad \text{– QED}$$

The square wave case

The square wave case is trivial as the level is constant over the half cycle.

Average

$$v = V_{pk} \qquad \text{– square wave function (over +ve half cycle)}$$

$$\Rightarrow V_{avg} = \frac{V_{pk} \int_0^{\tau/2} 1\, dt}{\tau/2} \qquad \text{– average calculation}$$

$$= \frac{V_{pk} \times \tau/2}{\tau/2} \qquad \text{– integral of 1 is x}$$

$$= V_{pk} \qquad \text{– QED}$$

RMS

$$v = V_{pk} \qquad \text{– square wave function (over +ve half cycle)}$$

$$\Rightarrow V_{rms} = \sqrt{\frac{V_{pk}^2 \int_0^{\tau/2} 1^2\, dt}{\tau/2}} \qquad \text{– RMS calculation}$$

$$= \sqrt{\frac{V_{pk}^2 \times \tau/2}{\tau/2}} \qquad \text{– integral of 1 is x}$$

$$= V_{pk} \qquad \text{– QED}$$

The triangle, ramp, and sawtooth wave cases

The calculations for triangle, ramp, and sawtooth waves can be considered in an identical fashion due to the similar symmetry of these three wave shapes.

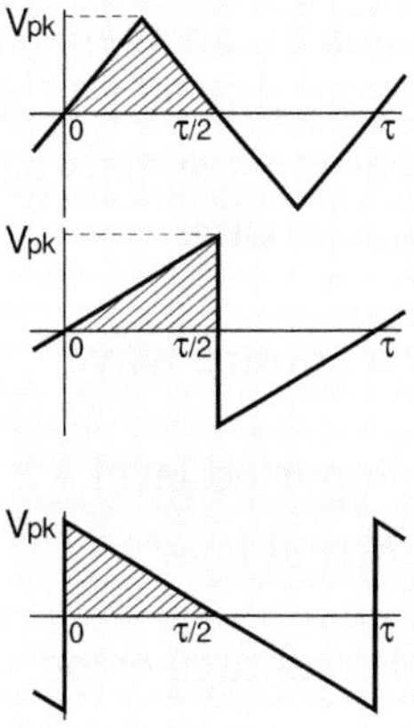

Average

$$v = \frac{V_{pk} \times t}{\tau/2} \quad \text{– the ramp function}$$

$$\Rightarrow V_{avg} = \frac{\frac{V_{pk}}{\tau/2} \int_0^{\tau/2} t\,dt}{\tau/2} \quad \text{– average calculation}$$

$$= \frac{\frac{V_{pk}}{\tau/2} \times \left[\frac{t^2}{2}\right]_0^{\tau/2}}{\tau/2} \quad \text{– integral of x is } x^2/2$$

$$= V_{pk} \times \frac{1}{2} \quad \text{– QED}$$

RMS

$$v = \frac{V_{pk} \times t}{\tau/2} \quad \text{– the ramp function}$$

$$\Rightarrow V_{rms} = \sqrt{\frac{\left(\frac{V_{pk}}{\tau/2}\right)^2 \int_0^{\tau/2} t^2\,dt}{\tau/2}} \quad \text{– RMS calculation}$$

$$= \sqrt{\frac{\left(\frac{V_{pk}}{\tau/2}\right)^2 \times \left[\frac{t^3}{3}\right]_0^{\tau/2}}{\tau/2}} \quad \text{– integral of } x^2 \text{ is } x^3/3$$

$$= V_{pk} \times \frac{1}{\sqrt{3}} \quad \text{– QED}$$

a. Evaluating a definite integral running from a to b: $[F(x)]_a^b = F(b) - F(a)$.

As mentioned in the previous section, while the RMS level is what is usually wanted, the rectified average is easier to measure. Many multimeters therefore measure this and use it to estimate the RMS value. The estimation, which makes an assumption that the signal being measured exhibits sine wave characteristics, can result in a significant error in the reported value. The highlighted Box 5.3 shows how to calculate the errors which will ensue when measuring the standard wave shapes.

5.3 – Average-Responding Error Levels

Based on the relationships derived for the square, triangle, ramp, and sawtooth waves it is easy to calculate by how much an average-responding voltmeter will misrepresent the actual signal level when presented with signals of these shapes.

Recall that while the more elaborate true RMS meter directly measures the value of interest, an average-responding meter will measure the average level and then multiply this by 1.11 as its best guess of what the RMS value is. The assumption is that most signals being measured are sine waves or exhibit broadly sine wave characteristics.

For a square wave

Reported level – $V_{rep} = V_{avg} \times 1.11$

Actual level – $V_{rms} = V_{avg}$

Percentage error – $Err = 100 \times \left(\frac{V_{avg} \times 1.11}{V_{avg}} - 1\right) \approx 11\%$ high

For a triangle/ramp/sawtooth wave

Reported level – $V_{rep} = V_{avg} \times 1.11$

Actual level – $V_{rms} = V_{avg} \times \frac{2}{\sqrt{3}}$

Percentage error – $Err = 100 \times \left(\frac{V_{avg} \times 1.11}{V_{avg} \times \frac{2}{\sqrt{3}}} - 1\right) \approx 4\%$ low

Decibels

Decibels are a large and complicated topic, however an in-depth treatment can be sidestepped here. Instead some useful rules and guidelines are provided without getting into the more technical and mathematical background needed to fully explore the subject. For a more comprehensive examination of the use of decibels in audio technology and electronics see for instance Brixen (2011). In the current section just the details needed to understand and usefully employ decibels in the context of this book are presented.

When the size of an electronic signal is measured directly the result will most often be expressed in volts. Audio signals might typically range in size from tens of microvolts (at the noise floor), up to perhaps some hundreds of volts (for the output of a high powered amplifier). These extremes would correspond to a factor of about ten million between the smallest and largest signals which might be encountered. Using these voltage values directly when measuring, comparing, and visualising audio signals presents two particular issues: the range of numbers employed becomes unwieldy and the values encountered do not relate well to how the ear perceives the loudness of sounds. Using decibels however the same range can be covered by just 140dB.

Thus decibels provide an alternative way of expressing levels and level changes in audio signals that addresses these two difficulties effectively. The primary goal is to provide a system where the unit of measurement relates naturally to the perceived changes in loudness or level of the signal. To achieve this a mathematical function called the log or logarithm is employed (a graph of the log function is given in Figure 5.10). A deep understanding of the properties and behaviour of this function is not required. For the most part it is sufficient to know which button to press on a calculator in order to apply the operation. Some familiarity with the broad shape of the function is however helpful in gaining an appreciation of how decibels behave.

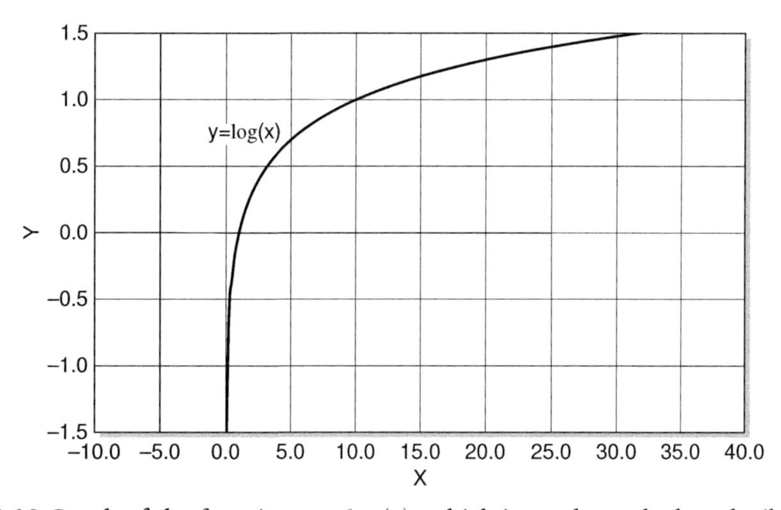

Figure 5.10 Graph of the function $y = \log(x)$, which is used to calculate decibel values.

The most pressing difficulty with using voltage values directly when measuring a signal is that the same voltage change in a signal level (say a one volt change) does not correspond to the same change in the perceived loudness for different signals. Changing a quiet signal by one volt will have a dramatic effect, while the same one volt change in a loud signal could go virtually unnoticed. On the other hand, a change of say three decibels in each of the two aforementioned signals will result in very similar changes in the perceived levels of the signals.

A small change in a small signal and a big change in a big signal are required in order to result in the same perceived change in the loudness of the two signals. Therefore a measurement scheme which reflects this nonlinear relationship will make for a more useful and user friendly way of talking about levels and level changes. Returning to the graph in Figure 5.10 helps us to understand how the decibel achieves this goal. The voltage level of the signal can be thought of as being plotted horizontally along the X axis while the decibel values range up and down on the Y axis. Now consider the two cases mentioned previously.

A small signal sits on that part of the graph close to the origin where the line moves up and down very quickly with just small movements left and right. Thus in this region of the graph a small voltage change (along the X axis) will correspond to a relatively large decibel change on the Y axis.

Conversely, a large signal sits farther to the right, on the part of the graph where it takes a large left-right movement to achieve even a small up-down change. Thus for a larger starting signal, a larger voltage change is required in order to achieve the same given decibel change in the signal level. This is exactly the behaviour which is required in order to give values which correspond well to the way in which the human ear registers changes in loudness.

The general equation used for calculating decibel values from raw voltages is given in Eq. 5.6. As can be seen from the equation, two voltages (here called V_1 and V_2) are involved in any decibel calculation – decibels represent ratios or differences.

$$\mathsf{dB} = 20\log\left(\frac{V_1}{V_2}\right) \tag{5.6}$$

All the examples given in the rest of this section should be entered into a calculator as they are encountered, in order both to confirm the answers shown and to gain familiarity with entering such calculations correctly. There are a few different ways in which modern scientific calculators work. It is well worth confirming and solidifying familiarity with a number of different devices.

As an example consider two voltages, $V_1 = 20$ volts and $V_2 = 10$ volts. Inserting these two values into Eq. 5.6 gives:

$$\mathsf{dB} = 20\log\left(\frac{V_1}{V_2}\right) = 20\log\left(\frac{20}{10}\right) = 20\log(2) \approx 6\mathsf{dB}$$

and so it can be seen that the difference between twenty volts and ten volts is six decibels or, put another way, twenty volts is six decibels larger than ten volts. It is also very useful to remember that if V_1 and V_2 are swapped the calculated decibel value simply switches sign; in this case the six decibels becomes minus six decibels. This makes sense since if twenty volts is six decibels larger than ten volts then ten volts is going to be six decibels smaller than twenty volts, i.e. -6dB.

$$\mathsf{dB} = 20\log\left(\frac{10}{20}\right) = 20\log(0.5) \approx -6\mathsf{dB}$$

Eq. 5.6 allows the decibel relationship between any two voltage levels to be calculated. It does not however in this form provide any method for specifying actual signal levels (entering one volt and two volts into the equation also results in an answer of 6dB, i.e. two volts is six decibels larger than one volt, just as twenty volts is six decibels larger than ten volts). This is useful for comparing levels or specifying changes in level

5.4 – The Form of the Decibel Equation

Decibels evolved as a method for comparing power-like quantities. The equation for a power ratio measured in bels is simply:

$$\text{bel} = \log\left(\frac{P_1}{P_2}\right)$$

However one bel turns out to represent quite a large ratio, while a ratio one tenth as large equates to a very convenient unit step size. The prefix multiplier deci- (see Appendix A) corresponds to one tenth, and so the decibel emerges naturally as the standard unit in this context.

$$\text{dB} = 10\log\left(\frac{P_1}{P_2}\right)$$

This is the form of the decibel equation used for power and power-like quantities. To understand how Eq. 5.6 is derived from this, it is only necessary to observe the relationship between power (in watts) and voltage (in volts), as specified by Watt's law in Eq. 9.2. At first glance it might seem that volts are directly proportional to watts, since $P = I \times V$. However the I in this equation is not independent. By Ohm's law (Eq. 9.1), if V changes then I changes, the resistance itself (R) being a constant property of the system. So it is actually the final variant of Watt's law which provides the required piece of information, $P = V^2/R$. Power can thus be seen to be proportional to voltage squared, $P \propto V^2$. The decibel equation can thus be rewritten in terms of voltage rather than power.

$$\text{dB} = 10\log\left(\frac{V_1^2}{V_2^2}\right) = 10\log\left(\frac{V_1}{V_2}\right)^2$$

A power inside a log function can be brought outside as a factor ($\log x^n = n \log x$) and so this can be rewritten into the final form given in Eq. 5.6.

$$\text{dB} = 20\log\left(\frac{V_1}{V_2}\right)$$

– 'Signal A is 12dB louder than signal B.' or 'Turn track two down by 8dB.' – but it does not provide a method for indicating what the levels are.

In order to be able to specify an actual voltage level in decibels ('Set the main volume to X.'), it is necessary to first define a reference level or zero point. This will be the voltage which corresponds to zero decibels on the scale. The use of a standard reference level is indicated by adding a suffix to the decibel unit. There are a number of common reference levels which might be encountered. Two in particular are very commonly found in audio work. Where an uppercase V is added to the units, as in dBV, the reference level is taken to be one volt, i.e. $0\text{dBV} = 1.0\text{V}$. If a lowercase u is encountered (dBu), the zero point is defined to be $0\text{dBu} = 0.775\text{V}$. It does not really matter where these reference levels originally came from but it is useful to remember them.

In order to convert Eq. 5.6 for calculations of dBu or dBV all that is required is to replace the V_2 by the appropriate reference level/zero level.

$$\text{dBu} = 20\log\left(\frac{\text{V}}{0.775}\right) \tag{5.7}$$

$$\text{dBV} = 20\log(\text{V}) \tag{5.8}$$

This provides the tools needed to quote signal levels in decibels, in addition to signal level changes, as already explained. It is important to remember that pure decibels (dB) refer to level changes while modified decibels (dBu, dBV, etc.) correspond to actual signal levels. Thus a signal's level might be changed by +6dB or −12dB, whereas the level might be equal to +4dBu or −10dBV. These two particular levels (+4dBu and −10dBV) are encountered often. They are two signal levels commonly used when designing audio equipment. In order to calculate the voltage levels which they correspond to, Eq. 5.6 needs to be inverted so that instead of entering two voltages to get a decibel value, one voltage and a decibel value are entered in order to yield a voltage. This inverse function is shown in Eq. 5.9.

$$\text{V}_1 = \text{V}_2 \times 10^{\left(\frac{\text{dB}}{20}\right)} \tag{5.9}$$

Returning to the first example and reversing it, this equation provides an answer to the question, what voltage is six decibels smaller than twenty volts?

$$\text{V}_1 = \text{V}_2 \times 10^{\left(\frac{\text{dB}}{20}\right)} = 20 \times 10^{\left(\frac{-6}{20}\right)} = 20 \times 10^{-0.3} \approx 10.0\text{V}$$

This calculation confirms the expected operation of the inverse function, returning the original value for V_2 from the earlier example. Similarly it is now possible to calculate the signal voltage levels which correspond to the commonly used −10dBV and +4dBu levels mentioned above. Recall that V_2 is the reference voltage level for the particular decibel variant used. So for dBV, $V_2 = 1.0\text{V}$ while for dBu, $V_2 = 0.775\text{V}$.

$$\text{V}_{(-10\text{dBV})} = 1.0 \times 10^{\left(\frac{-10}{20}\right)} = 10^{-0.5} \approx 0.316\text{V}$$

$$\text{V}_{(+4\text{dBu})} = 0.775 \times 10^{\left(\frac{4}{20}\right)} = 0.775 \times 10^{0.2} \approx 1.23\text{V}$$

Notice that (as illustrated in Figure 5.10) the logarithm of one is zero ($\log(1) = 0$) so if $V_1 = V_2$ then $V_1/V_2 = 1$, and dB equals zero. This corresponds to the statement that any level is exactly 0dB different from itself.

In addition to dBu and dBV it is worth being aware of one other modified decibel type commonly encountered. dBr stands for decibels relative, and is often seen used in measurement results and most especially on graphs. It means that the decibels in question have been measured relative to some arbitrary zero level which should then usually be specified.

Two common ways in which these units are used are to show an output level relative to an input level, or to show measurement levels at various frequencies relative to the measurement at 1kHz. In the first case what is being said is that whatever input level is sent into a circuit, the output will be this much different. The second example is used to show the frequency response of a circuit; with a response of 0dB at 1kHz, it shows how many dB up or down the response is at any other frequency. The graphs in Figure 5.1 and Figure 5.2 include axes labelled in dBr indicating that the signal levels are measured relative to an unspecified maximum level. Notice that the values are all negative in these examples, indicating that all the recorded levels are smaller than the maximum possible level by the numbers of decibels shown.

Example Calculation

Returning to the graphs of Figure 5.7, it proves an instructive exercise to confirm mathematically that the two plots do indeed represent the same sine wave in both their frequencies and their amplitudes.

Eq. 5.2 states that frequency is one over period. The period (time for one cycle) can quickly be determined from the graph in Figure 5.7a as being $4\text{mS} = 4/1000\text{Sec}$. Therefore in accordance with Eq. 5.2 the frequency of the time domain sine wave shown in Figure 5.7a must be $1000/4 = 250\text{Hz}$. Correspondingly the frequency peak shown in Figure 5.7b sits neatly between the 200Hz and 300Hz grid lines indicating a consistent frequency representation between the two graphs.

Turning to the amplitude of the sine wave in question the decibel equations just introduced allow for comparison of this property across the two graphs. In Figure 5.7a the sine wave's peak level reaches approximately 12V. It is important to remember that decibel levels correspond to volts RMS and so first this peak voltage must be converted to an RMS voltage. As the signal under consideration is a sine wave, Eq. 5.4 can be used to perform the conversion.

$$V_{rms} = V_{pk} \times \frac{1}{\sqrt{2}} = \frac{12}{\sqrt{2}} \approx 8.485V_{rms}$$

It is then a simple matter to convert this RMS value into a level in dBu using Eq.5.7.

$$\text{dBu} = 20\log\left(\frac{V}{0.775}\right) = 20\log\left(\frac{8.485}{0.775}\right) \approx 21\text{dBu}$$

Again returning to the plot in Figure 5.7b the frequency peak can be seen to rise just above the 20dBu grid line confirming that the represented amplitudes are also consistent between the two plots.

Linear and Logarithmic Scales

It can be instructive to consider the four possible variations which can arise in graphs with respect to the use of linear and logarithmic scales for the X and Y axes. The four variations depend on whether each axis is plotted with a linear or logarithmic scale. Figure 5.11 shows the same three functions plotted using each of these four variations. (In this case the three functions are actually those for the impedance versus frequency of three fundamental components, which are examined in detail in later chapters. These graphs are encountered again in context at that point.)

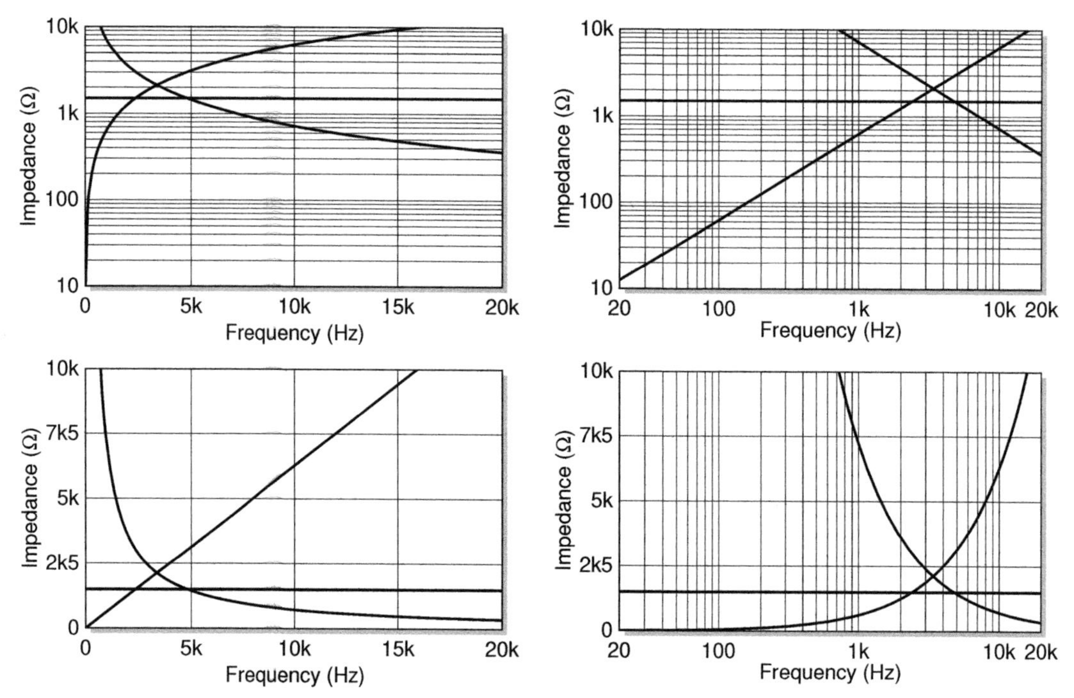

Figure 5.11 Comparison of linear and logarithmic scales.

At least three of these combinations are commonly encountered in the literature on capacitors and inductors and each variant has its place. In general terms the approach using logarithmic scales on both axes is probably most often likely to be the best option in this context. All things else being equal, straight lines are to be preferred as they are easier to interpret and easier to extrapolate. Additionally human perception generally operates in logarithmic terms and so log scales make sense from this point of view also. The frequency steps between musical notes proceed logarithmically as do steps of equal loudness (decibels) and of course the E-series preferred values used to define component sizes also proceed in a logarithmic fashion. As such it often makes most sense to use logarithmic scales to plot such quantities.

However a logarithmic scale can never include zero and so if behaviour at the origin is to be displayed explicitly on a graph then linear scales are necessary. For instance the fact that an inductor's impedance (theoretically) reaches zero ohms at a signal frequency of zero hertz becomes obvious from a graph using linear scales for impedance and frequency whereas with logarithmic scales this can not be shown. The important thing is to appreciate that both methods of plotting values are encountered regularly, and to gain a sufficient understanding of them and their relationship in order to make best use of the graphical information being presented.

And finally always recall that a scale labelled in decibels is logarithmic by default. The labelling is linear because the linear to logarithmic transformation has already been applied. The underlying quantity (most often voltage in the case of the present work) is transformed from a linear to a logarithmic scale in the calculation which converts voltages into decibels as for instance is used in the frequency domain plots in Figure 5.1, Figure 5.2, and Figure 5.7 at the beginning of this chapter. Thus these graphs are displaying their data using logarithmic scales on both axes despite initial appearances. The two time domain plots in the same figures are by contrast linear in both axes.

It is unsurprising that logarithmic relationships appear so naturally in so many places. Fundamentally they reflect the fact that small changes in small things and big changes in big things result in similar perceived amounts of subjective change.

References

G. Ballou, editor. *Handbook for Sound Engineers*. Focal Press, 4th edition, 2008.

E. Brixen. *Audio Metering: Measurements, Standards and Practice*. Focal Press, 2nd edition, 2011.

S. Gelfand. *Hearing: An Introduction to Psychological and Physiological Acoustics*. Informa Healthcare, 5th edition, 2010.

M. Mandal and A. Asif. *Continuous and Discrete Time Signals and Systems*. Cambridge University Press, 2007.

B. Metzler. *Audio Measurement Handbook*. Audio Precision, 2nd edition, 2005.

M. Russ. *Sound Synthesis and Sampling*. Focal Press, 2nd edition, 2004.

W. Sethares. *Tuning, Timbre, Spectrum, Scale*. Springer, 2nd edition, 2005.

F. Stremler. *Introduction to Communication Systems*. Addison Wesley, 3rd edition, 1990.

Part II

Practical Electronics

6 | Component Overview

The final section of this book, Part III, provides an in-depth component reference, examining in some detail the electronic devices most frequently encountered in audio electronic circuits. It discusses what they do, and how they are used, and it introduces some common applications and typical circuits in which they are found. It provides background information and guidance as to the use and the uses of these components.

In advance of this, and in order to facilitate the discussion presented in the current section, this present chapter opens Part II with a brief outline of the components presented in more detail in Part III. This provides some context to allow a better grasp of the material presented in the chapters which follow, delving into the practical topics which form the central theme of this book. A number of less common components are also introduced here, which are not expanded upon in the more detailed component chapters of Part III, but which do nonetheless merit a brief mention. There also exist many electronic devices too numerous and too specialised to be mentioned here at all.

Components covered in detail in Part III are given no more than a one or two line description here, while a short paragraph is provided on those components which are not further examined elsewhere. The devices presented here are introduced under the headings of passive components, semiconductor devices, vacuum tubes, transducers, and miscellaneous components. A more comprehensive list of components and general circuit elements is included in Appendix C, which covers such things as plugs, jacks, power sources, and meters, along with various other common schematic diagram symbols and conventions like ground points and wire connections.

Passive Components

A passive electronic component might be defined as one which does not require external power in order to operate, or one which does not use one signal in order to control another. The category is somewhat flexible but is generally taken to include at a minimum: resistors, capacitors, inductors, and transformers. Various other devices fit the criteria but may be more logically placed in another category.

Resistors – drop voltage and limit current.

Variable Resistors – split the voltage across them. There are various types which may generally or specifically be referred to as: potentiometers, pots, faders, linear faders, trimmers, trim pots, and presets.

Capacitors – block low frequencies and pass high frequencies. A division worth noting is between polar and nonpolar types.

Inductors – pass low frequencies and block high frequencies. In many respects inductors can be considered to be the opposite of capacitors. In some applications an inductor is referred to as a choke.

Transformers – transform voltages up and down.

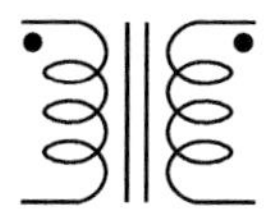

Semiconductor Devices

In addition to the two commonest and most widely used semiconductor devices, the diode and the transistor, this section includes a variety of less commonly encountered discrete semiconductor based devices, as well as the more generic symbol often used to represent standard DIL integrated circuit packages. Many more specific IC symbols are also commonly encountered, some of which can be seen in Appendix C.

Photoresistors (aka light dependant resistors or LDRs) – have a resistance which varies depending on the amount of light striking them.

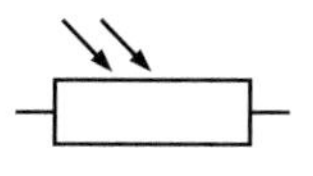

Thermistors – have a resistance which varies depending on the temperature of the device. Not often found in audio electronics, thermistors are commonly used in power electronics to detect overheating in a circuit or device. Thermistors come in two varieties depending on how the resistance changes with changing temperature. In positive temperature coefficient (PTC) devices the resistance rises as the temperature rises, and falls as the temperature falls. The PTC symbol has two arrows pointing in the same direction. In a negative temperature coefficient (NTC) thermistor the resistance falls as the temperature rises, and rises as the temperature falls. The NTC symbol has two arrows pointing in opposite directions (see Appendix C).

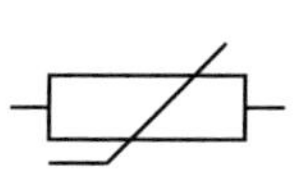

Varistors – are also referred to as MOVs (metal oxide varistors), or VDRs (voltage dependant resistors). These devices exhibit a very high resistance until a specified threshold voltage is applied, at which point the resistance falls sharply. A varistor might sometimes be more generically referred to as a TVS or transient voltage suppressor (c.f. GDT in Appendix C).

Diodes – allow current to flow in one direction but not the other. The diode is the most fundamental semiconductor device. There are lots of variants, for instance Zener diodes and Schottky diodes. Diodes are sometimes also called rectifiers.

Light Emitting Diodes (LEDs) – are diodes which produce light.

Laser Diodes – produce coherent, columnated (i.e. laser) light, rather than the less controlled emissions from a standard LED. Laser diodes have a wide variety of applications including such things as CD players and laser pointers.

Photodiodes – are diodes in which the semiconductor junction is exposed to allow light to fall upon it. When illuminated, photodiodes generate electricity. This component is the basis of much solar cell technology. They can also be used in various light detection tasks from infrared to ultraviolet.

Transistors – amplify, buffer, and switch electrical signals. They come in many different types, shapes, and sizes.

Phototransistors – are transistors which are controlled by light falling on an exposed semiconductor junction, rather than the usual electrical control signal applied to the base terminal. Phototransistors can be used as the detection element in CD and other optical players. Light reflected back from the disc strikes the exposed semiconductor junction and the resulting signal is interpreted in order to extract data from the disc medium.

DIACs, Thyristors, and TRIACs are related semiconductor devices which are often used in combination. One of their commonest applications is in the control circuitry used to implement adjustable lighting dimmers.

DIACs – are a form of semiconductor switch. They can conduct in both directions, but only once a threshold voltage or breakover voltage has been exceeded. DIAC stands for diode for alternating current. DIACs are two terminal devices and often look much like diodes. Sometimes a stripe around the middle of the body differentiates a DIAC from a diode (whose stripe is at one end, indication the cathode), but this is by no means always the case. The terms DIAC and SIDAC (silicon diode for alternation current) are sometimes used interchangeably, although they are not usually considered strictly the same thing. Convention differs between various sources.

Thyristors – or SCRs (silicon controlled rectifiers) can be thought of as diodes which will only start to conduct when they are triggered by a signal on their gate terminal (the third terminal, coming off at an angle). Smaller thyristors often look much like small signal transistors, but thyristors are often physically bigger devices used to control large electrical currents.

TRIACs – are much like the standard thyristor above, except that they operate in both directions (comparing the two symbols illustrates this point). Also like thyristors, TRIACs are three terminal devices, often physically similar to transistors. TRIAC stands for triode for alternating current.

Integrated Circuits – provide a complete circuit in a prefabricated package. Opamps are probably the commonest and most obvious example in audio electronics, but there are thousands of others, both within the audio field and beyond.

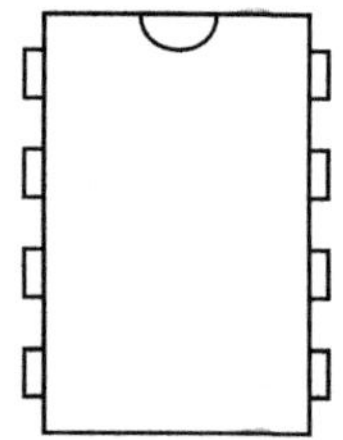

Vacuum Tubes

Vacuum tubes (also called tubes, valves, or thermionic valves) were the forerunners of semiconductors (diodes, transistors, etc.). While generally rendered obsolete in the wider world of electronics, they still find favour in some audio applications because of the particular colour they can add to a sound signal. Only diode and triode tubes are mentioned here. Other more elaborate variations exist, but these are beyond the scope of the material presented in this book.

Diode Tubes – were the precursor to semiconductor diodes and, like these, they similarly allow current to flow in one direction but not the other.

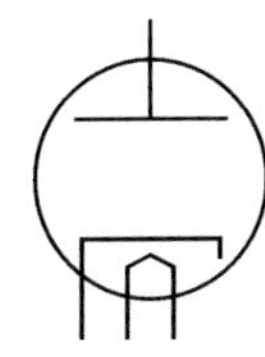

Triode Tubes – were the precursor to transistors, and like transistors they can be used to amplify, buffer, and switch electrical signals.

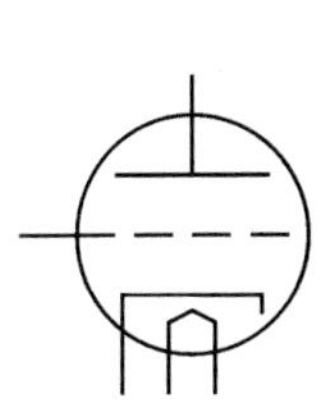

Transducers

Transducers transform signals from one form into another. Two broad subcategories are sensors and drivers. Other categories exist but are of less interest here. Sensors (including pickups and microphones) transform acoustic sound or mechanical vibrations into electrical signals. Drivers (most commonly, but not exclusively loudspeakers) transform electrical signals into mechanical movement (and then usually into sound).

Microphones – transform sound waves into electrical signals. The two commonest types are dynamic and condenser microphones, but there are others.

Coil Pickups (aka electric guitar pickups) – detect the vibration of metal guitar strings nearby. They consist of a coil of wire and a magnet.

Piezo Transducers – transform vibrations into electrical signals, and vice versa. They can be used as pickups or as drivers.

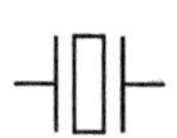

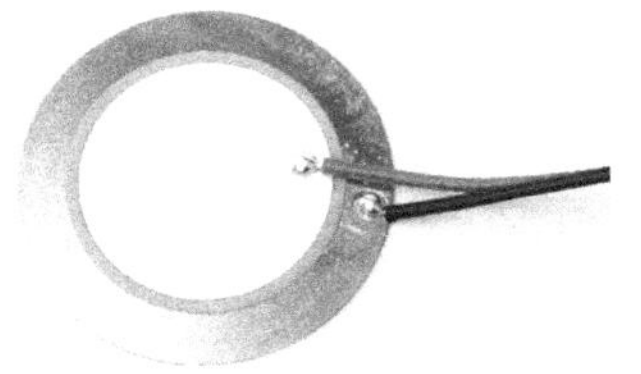

Loudspeakers – transform electrical signals into sound waves. The moving coil loudspeaker is the commonest. It consists of a coil of wire mounted in a magnetic field and attached to a cone.

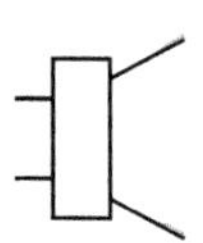

Miscellaneous Components

There are many other electronic components available, however the focus here remains on those which are most likely to be encountered in audio electronics projects and audio technology in general.

Switches – provide the ability to route, connect, and select electrical signals.

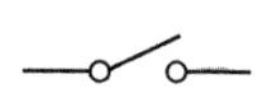

Relays – use an electrical signal to actuate a mechanical switch.

Jack Connectors – are amongst the commonest connector types used for audio. Electric guitar leads and headphone jacks are two examples.

RCA Connectors/Power Connectors – and other generic connector types may be represented by this symbol. Often the choice of connector to use in an application is more a packaging rather than an electrical decision.

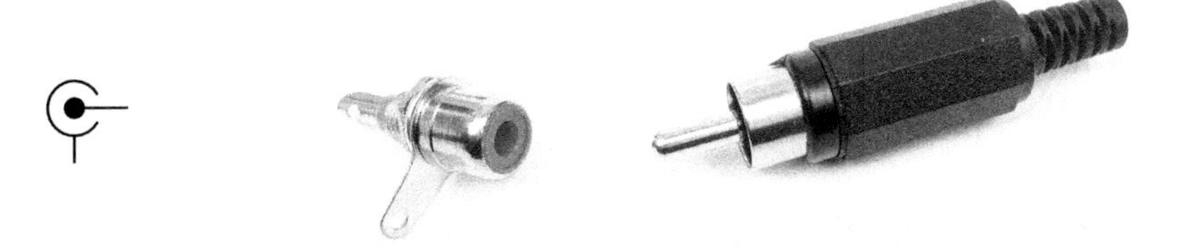

Batteries – can be used to provide power to many of the circuits encountered here.

Meters – can be used to visually monitor and represent an electrical signal.

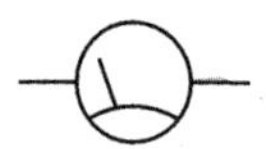

Lamps – are most obviously used for indicators and illuminators, but can also be used as control and stabilisation elements within circuits, in a similar fashion to thermistors.

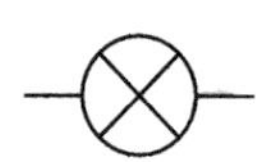

Optocouplers (aka opto-isolators) – provide electrical isolation between two parts of a circuit or system. In audio applications they are often employed as much for the distortion they introduce as for their intended purpose. The illustration shows a home-made optocoupler comprised of an LED and an LDR wrapped up in heat shrink tubing.

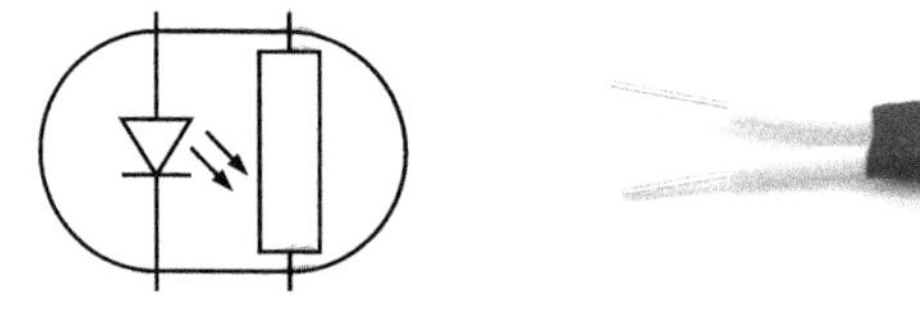

Reverb Pans – transmit their signal through long springs in order to introduce reverberation distortion into an audio signal.

Learning by Doing 6.1

Component Specification Sheets

Materials

All that's needed is a device with an internet connection and some storage.

Theory

When working with a new component it is always a good idea to see what the manufacturer has to say about it. The maker's component specification sheets, or data sheets, often contain far more information that you will ever need, and it is by no means essential to read, learn, and inwardly digest all the information presented there. On the other hand, data sheets can provide a trove of valuable information, and often include 'typical applications' circuit ideas which can prove very useful in designing your projects, as well as in understanding the function of a particular part in other peoples.

Practice

Create a folder on your computer called 'Data Sheets' and start a collection. Make it a habit where possible to find and save a data sheet for each new component you encounter. Part numbers are often the best way to find data sheets. Performing a search on 'diode' or 'LED' is unlikely to lead directly to what you are looking for. However searches on '1N4148' or '2N3904' will most probably result in a pdf data sheet in the first few results returned. Data sheets can also be found on manufacturer's websites, and online stores often provide a link to them on many of their individual product pages. Collections of data sheets are sometimes compiled by manufacturers into data books (e.g. Motchenbacher and Connelly, 1993; RCA, 1950; TI, 1982). Data sheets are particularly useful for transistors and integrated circuits, but it can be worthwhile to accumulate a few for more basic components like resistors, capacitors, and switches too.

References

K. Brindley. *Starting Electronics*. Newnes, 4th edition, 2011.

C. Motchenbacher and J. Connelly. *Low-Noise Electronic System Design*. Wiley, 1993.

RCA. *Receiving Tube Manual*. Radio Corporation of America, 1950.

M. Schultz. *Grob's Basic Electronics*. McGraw-Hill, 11th edition, 2011.

TI. *The TTL Data Book*. Texas Instruments, 5th edition, 1982.

E. Young. *Dictionary of Electronics*. Penguin, 1985.

7 | Circuit Diagrams

Circuit diagrams in the broad sense can cover a wide range of different circuit or system representations. At the most abstract, block diagrams can be used to conceptualise the high level functional blocks in a system, and the connections and signal flow between these blocks. Such representations may be referred to as block diagrams, functional block diagrams, signal flow diagrams, connection diagrams, etc. Moving to a more detailed design, the schematic diagram includes full details of which components are actually required in order to build the circuit in question, and the precise connections between all these components. Circuit schematics or schematic diagrams are often also referred to simply as circuit diagrams, and when this term is encountered in a general context, this kind of diagram is usually what it is being used to refer to. A schematic diagram defines how a circuit is to work, but does not provide a full blueprint to building it. The third and final class of circuit diagram adds physical layout information to the mix in the form of a breadboard, stripboard, or printed circuit board (PCB) layout diagram. These provide a representation which can be directly followed by a builder in order to construct the circuit, and are the usual end product of the initial circuit design process, before building, testing, and modification commences.

Block Diagrams

Block diagrams can be a very useful tool in the early stages of the design process, allowing for the general architecture of a circuit to be analysed and assessed. They are also often employed in order to provide a high level description of an existing system to users, who do not need a detailed understanding of the inner workings of the system but only require a firm grasp of the general operation, the options and controls available, and what connections and setup are required or optionally available. Block diagrams are useful when explaining a circuit to a new user. They can also be used to help guide the construction, servicing, or repair of larger systems which might be composed of multiple physical circuit boards and ancillary connections to other devices. Figure 7.1 illustrates the typical functional blocks and signal flow in a simple synthesiser. It clearly differentiates between different kinds of control signals and audio signals, and indicates the major controls and options available in each section of the device.

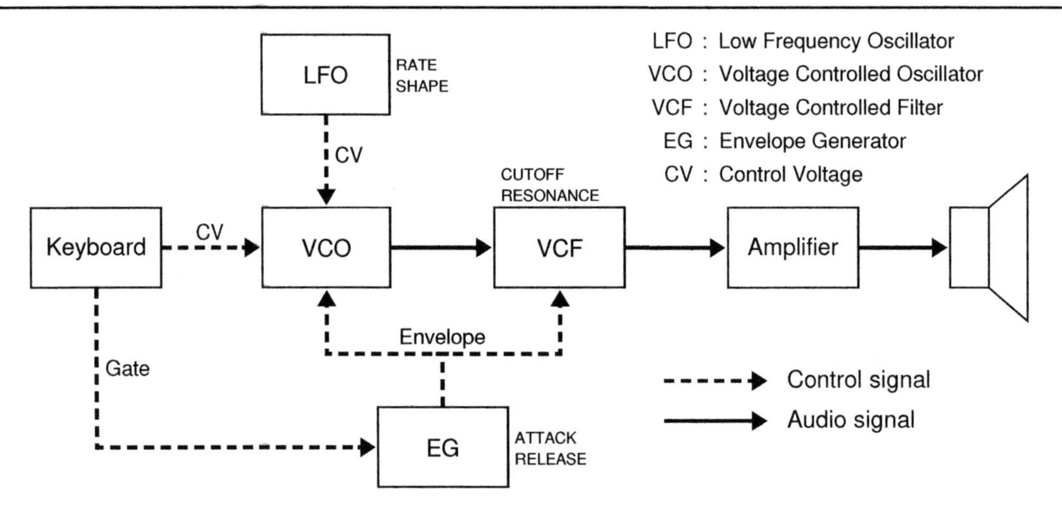

Figure 7.1 A typical block diagram.

Functional block diagrams and signal flow diagrams, although useful, are not usually required for the relatively simple kinds of circuits which are the main focus of this book. Additionally, such diagrams are generally relatively straightforward and self explanatory in structure and interpretation, and as such require little more in the way of explanation or examination here.

Schematic Diagrams

More often than not a project starts out with a schematic diagram. The correct and unambiguous interpretation of such diagrams is essential if a circuit is to be successfully constructed. The component symbols introduced in Chapter 6, along with the broader collection presented in Appendix C, and still more symbols and variations in common use generally, form the core of the schematic diagram representations which must be recognised and interpreted. Many additional conventions are gathered around these basic symbols in order to provide a comprehensive and (relatively) consistent methodology for the schematic representation of electronic circuit designs. Often some amount of creative license is employed by the drafter of a schematic diagram, but once the reader has gained a moderate amount of familiarity with the general conventions, interpreting the intended meaning is usually a straightforward matter. Occasionally a bit of research or experimentation may be needed if a schematic has been drafted without sufficient attention to detail, but on the whole this is rare.

The simple circuit introduced in pictographic form in Chapter 2, and reproduced here in Figure 7.2a, can be drawn in a more formal fashion as shown in Figure 7.2b. Once the two symbols used to represent the battery and the bulb are known, this schematic representation of the circuit should be easily interpreted. All the information needed in order to build the circuit is presented. Notice in particular that some components, such as the bulb here, have connection points or terminals which are interchangeable. It does not matter which way round such a component is connected into a circuit. In other components (such as the battery here) the terminals must be differentiated. The positive and negative terminals of the battery are not generally interchangeable in a circuit. Although in this particular case the battery connections could be flipped without altering the circuit this is not generally the case and the orientation of such components in a circuit must be carefully observed.

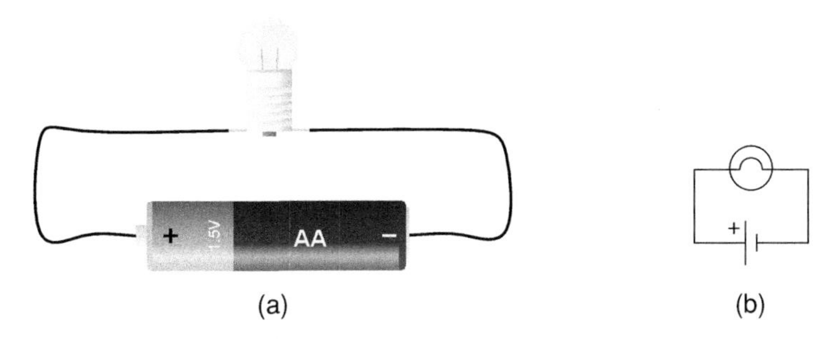

Figure 7.2 Pictographic versus schematic representation of circuit.

Schematic Layout Conventions

As the circuits being represented get bigger and more complex it is important to make sure that the schematic diagrams drafted to represent them are as clear and consistent as possible. A number of conventions exist which can help in this respect. Ignoring such conventions does not alter the accuracy of the schematic diagrams, it just results in schematics which are less easily followed and interpreted. Oftentimes it is not possible to follow the conventions strictly, but by and large the more closely they are observed, the easier the tasks of drafting and of reading the schematic become.

The most broadly applicable conventions guide the overall layout of the diagram. Inputs are on the left, outputs are on the right, and signals travel left to right through the circuit. This is not always strictly possible, but in general applying this convention makes it much easier to follow the signal flow in a complex circuit. The second, related convention stipulates that the power supply voltages generally go from positive to negative moving from top to bottom within the schematic diagram. Very commonly for simple circuits, perhaps powered from a single nine volt battery, a line across the top of the diagram connects to the positive terminal of the battery, with a similar line along the bottom connecting to the negative terminal. Finally, components and the traces connecting them are, in as much as it is possible, laid out horizontally and vertically. These three general points are illustrated in the circuit represented in Figure 7.3.

Power connections can in fact be represented in a number of different ways. The simple circuit in Figure 7.2b explicitly places a battery rather than incorporating the power rails as described above. Several options are illustrated in Figure 7.4. The

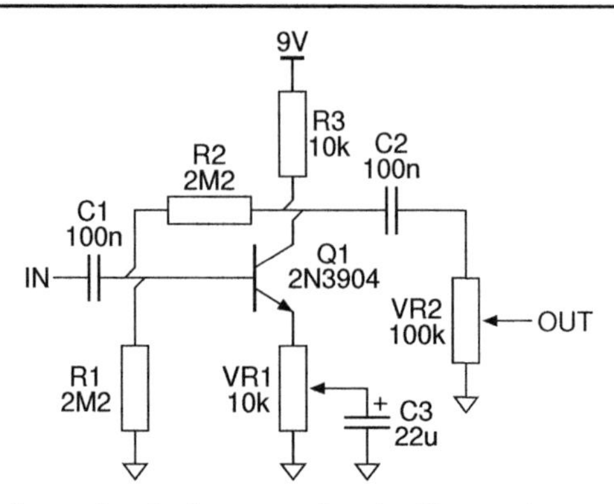

Figure 7.3 Schematic for a simple booster circuit, illustrating conventional positioning and routing of inputs, outputs, and power connections. Inputs enter from the left. Outputs exit to the right. Positive power supply voltages enter from the top and negative voltages connect from the bottom. This circuit is revisited in Chapter 17.

first three contain some implicit suggestion as to what the intended power source is. The first two are a battery and a power jack for connection of an external PSU. The third indicates a general DC source and is suggestive of the kind of adjustable bench power supply found in an electronics lab. The final pair – which utilise labelled power terminals, symbolic ground points, and power rails – make no such suggestions and are entirely generic.

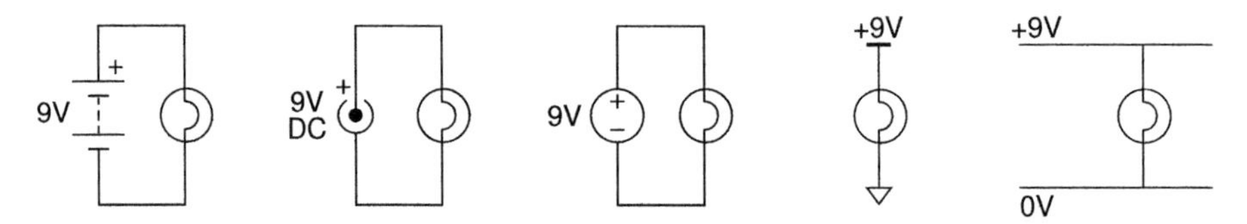

Figure 7.4 Various alternative representations of power connections.

While there may be good reason in a particular case to specify for instance a battery or a power adapter, such choices can generally be assessed and finalised when the circuit is actually being built. The abstract power supply variants carry no direct implications and are by and large interchangeable, their choice being based on questions of convenience and clarity.

Sometimes particular power connections are not explicitly shown in a schematic diagram at all. The drafter of the schematic is in this case assuming that the builder knows that certain power connections are needed in order to complete the circuit. Opamps and other ICs are commonly included without power connections being shown. Figure 7.5 shows equivalent opamp subcircuits. In the first case the power supply connections are explicitly shown, while in the second the connections are assumed. There is no difference between the circuits being represented by these two alternative diagrams.

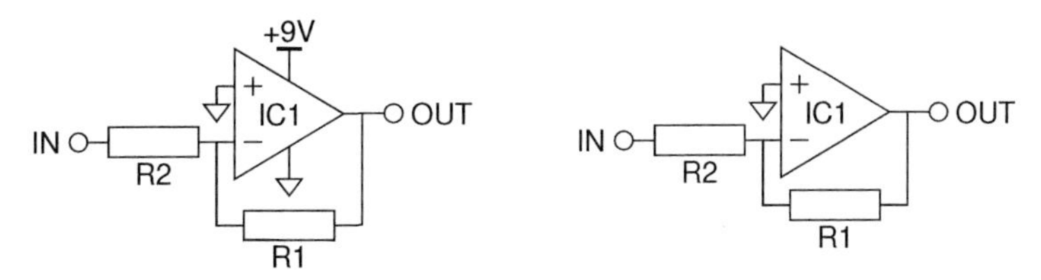

Figure 7.5 Opamp power connections may or may not be shown.

A dual opamp is a commonly found eight pin DIP IC. It contains two totally separate opamp circuits, with just a single pair of +/− power connections serving both. When such a device is used in a circuit, often one of the pair of opamps will be represented with a symbol which includes the power connections, while the second opamp will omit these connections. The same situation pertains in the realm of vacuum tubes, where one of the commonest tube types still in use today is the dual triode – a single physical tube containing two triode valves. Valves require a heater in order to work, but often a single pair of +/− connections will supply the heaters for both valves. As such, a circuit diagram using one of these tubes will often include the heater connections with the symbol for one of the valves, while leaving the second valve without. It is even occasionally the case that one of the two connections is shown on each of the two valve symbols (Figure 7.6).

Figure 7.6 Vacuum tube heater connections.

Input and output connectors can be similarly implicitly or explicitly represented on a schematic diagram (Figure 7.7). At the simplest level, a line labelled, for instance, IN might be all that appears. Alternatively, a more explicit TS jack symbol might be used. In this case the tip (T) portion is connected to the line carrying the signal into (or out of) the circuit. The sleeve (S) portion may be left undecorated, or it may have an explicit connection to signal ground indicated. All these representations correspond to exactly the same final circuit connections. When not shown explicitly, it is assumed that the necessary ground connection is understood.

Generic Components and Component Substitutions

There are often various alternatives which may be substituted for a particular component without affecting the performance of the circuit to any significant degree. Transistors and diodes are the two component types most commonly subject to these kinds of flexible circuit diagram representation. Diodes will often simply be labelled Si for

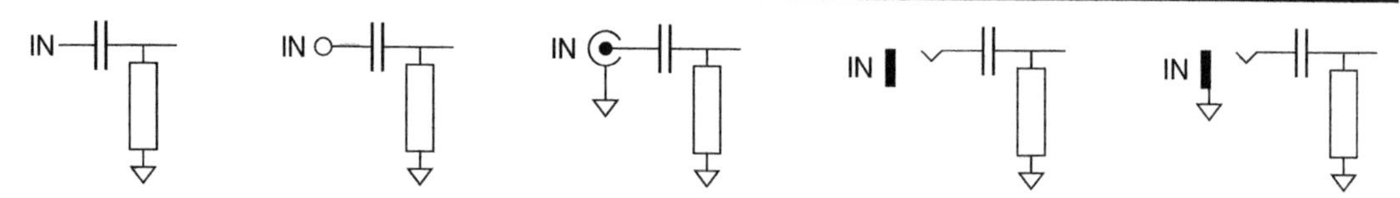

Figure 7.7 Input and output jacks can be represented at varying levels of detail. All of the above might be encountered, representing the same connections on the final circuit.

silicon or (less commonly) Ge for germanium, indicating that most typical diodes made from the specified material may be used. Similarly, transistors can be labelled with something like 'any high gain NPN'. Many years ago *Elektor* magazine instituted a system whereby noncritical diodes and transistors in the circuits they published would be labelled DUS, DUG, TUN, or TUP as appropriate (see Table 7.1) rather than specifying particular part numbers for the components to use (Elektor, 1974).

Table 7.1 DUS, DUG, TUN, TUP acronyms

Acronym	Meaning
DUS	Diode Universal Silicon
DUG	Diode Universal Germanium
TUN	Transistor Universal NPN
TUP	Transistor Universal PNP

The magazine article included details of the minimum allowable specifications for diodes and transistors to qualify as suitable devices to use in such situations. These details are reproduced here in Table 7.2 and Table 7.3. Also included in the original *Elektor* article were lists of commonly available devices which fitted into each of the four specified categories. While this system is not in common usage today it is still encountered occasionally, and in any case it does provide a very useful illustration of how to approach the more general question of identifying suitable substitute components for specified devices which are not readily available to the circuit builder.

Table 7.2 DUS and DUG diode specifications

	Type	V_R max	I_F max	I_R max	P_{tot} max	C_D max
DUS	Si	$\geq$25V	$\geq$100mA	$\leq$1μA	$\geq$250mW	$\leq$5pF
DUG	Ge	$\geq$20V	$\geq$35mA	$\leq$100μA	$\geq$250mW	$\leq$10pF

Notice that the values in all but two of the numerical categories across the two tables are marked $\geq$, with just the values in I_R max and C_D max labelled $\leq$. This is separate from and not to be confused with the min and max designations in the column headings. Thus in the second table I_c max and h_{fe} min are both quoted as $\geq$

Table 7.3 TUN and TUP transistor specifications

	Type	V_{ceo} max	I_c max	h_{fe} min	P_{tot} max	f_T min
TUN	NPN	≥20V	≥100mA	≥100	≥100mW	≥100MHz
TUP	PNP	≥20V	≥100mA	≥100	≥100mW	≥100MHz

values, indicating that both specifications should have values greater than or equal to the quoted levels. The only two specifications quoted here where lower is better are a diode's maximum reverse leakage current (I_R max) and its total capacitance (C_D max).

General Schematic Diagram Labelling, Markup, and Organisation

A few other commonly encountered conventions are also worth mentioning. When a device (such as the dual opamp or dual triode tube mentioned above) consists of two or more sections within the same device, the different sections may appear in different parts of a circuit far removed from each other within the layout. In this case it is standard practice to label the sections with the same initial label and to append an a, b, c etc. in order to indicate the association between the otherwise unconnected parts of the diagram (notice the labelling of the two opamps in Figure 7.16). This practice is also commonly encountered with multi-pole switches (see Chapter 15) and various types of integrated circuits in addition to dual opamps. For instance the individual gates in a quad XOR or a hex NOT IC (see the 4000 series chips described in Chapter 18) are often scattered throughout a circuit, and labelled appropriately in order to indicate their common physical location.

As previously mentioned, power and ground connections are often individually labelled in order to avoid tracing lines all around a complex circuit diagram making it much more difficult to read. Other points can be similarly labelled for the same reason. One common example is a generated mid voltage levels (see the points labelled +4V5 in Figure 7.16), or other reference voltage (often designated V_{ref}). In fact any difficult to route connection may be omitted, simply labelling its two ends instead (see the two points labelled A, also in Figure 7.16). This approach should not be used too frequently or the resultant scattering of labels quickly becomes even more difficult to follow than the snaking lines which they replace would have been.

A large circuit can also be broken into convenient subcircuits, with the points at which they interconnect similarly labelled. In Figure 7.16 a small section has been hived off and placed to the right of the main circuit, separated from it by a dashed line. Small sections of power supply circuitry like this are often drawn in isolation, in order to keep the overall layout as tidy and uncluttered as possible. Larger circuits will by necessity be broken up across several pages, but the magnitude of projects considered in this book will usually fit comfortably on a page. The noise gate example which rounds out this chapter is around the top end of the size of circuits likely to be encountered until a significant amount of experience has already been gained in practical audio circuit building.

A clear and unambiguous differentiation must be maintained between crossing lines which represent four wires connecting together at a point, and crossing lines which simply intersect in the diagram on their ways to their respective destinations. Leaving such a crossing point undecorated is very bad practice, although this can be encountered. Sometimes it is intended to mean that there is a connection at this point and sometimes not. It is usually obvious from the wider context of the circuit, but it introduces an unnecessary and highly undesirable level of uncertainty. Table 7.4 illustrates some common conventions used to indicate these two situations in schematic diagrams. The schematics in this book adopt the first pair as their standard representation of these two situations.

Table 7.4 Crossing wires on schematic diagrams

Connected	Unconnected	Ambiguous

In fact it is good practice wherever possible to avoid bringing four connected wires together at a point. Introducing a small offset wherever possible can help to remove any doubt or confusion. Both sides of resistor R11 in Figure 7.16 feature unambiguous four point connections. (In fact to the right of R11 is actually a six way connection since point A leads to another three components at its far end.) For comparison, the point to the right of R8 in the same diagram illustrates a usage of the first connected wires convention illustrated below. See also Figure 11.7 for several examples of unconnected crossing lines.

Integrated Circuit Representations

ICs can be represented on a schematic diagram in a number of different ways. The most direct representation for a DIP IC consists of an outline of the physical chip with its two rows of legs showing the connections to be made to each. Alternatively, a simple box can be used with connections entering at whatever points are most convenient. In this case each connection must be accompanied by a corresponding pin number in order to specify where on the chip the connection is to be made.

For some types of IC (opamps being the most obvious case, but also including logic gates, and some other specific IC types), special symbols have been adopted in order to represent the functions performed by that particular chip. Some of these are included in the table in Appendix C. Figure 7.8 shows examples of the three alternative styles of IC representation.

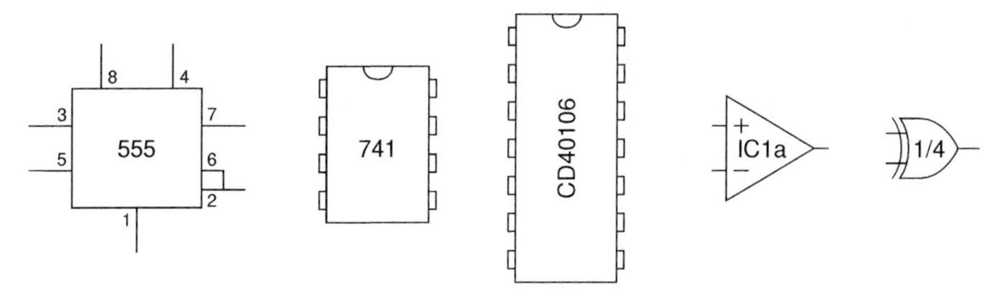

Figure 7.8 Different schematic representations for ICs.

Standard Omissions

A number of instances have already been mentioned where omissions are made, on the assumption that the builder will know to add these connections in the final circuit. Power and ground connections, including the sleeve connections of input and output jacks, have been discussed. A few other standard circuit elements are also commonly omitted or left up to the builder's discretion. A circuit may be battery powered or it may take its power from an AC adapter, or it might be deemed desirable to allow for either option. In this latter case the usual arrangement will mean that any battery present will be disconnected if a jack is inserted into the power socket. This requires the use of a switching socket for the AC adapter, appropriately wired. Such details are rarely included in the schematic for the kind of generic audio circuit targeted here. A TRS jack at the input can often be found in guitar effects pedals, wired up to perform the function of an on/off switch. The details of implementing this, along with other options mentioned here, are covered in the standard building blocks discussed in Chapter 15.

More involved powering options are also available. Reverse polarity protection can be valuable when sensitive circuitry is involved. More elaborate transient protection is also possible, although this kind of advanced protection is more likely to be found on expensive equipment than on the kind of low end projects which are the primary focus here. The added cost and complexity is simply not justified in the case of smaller, less expensive circuits and projects. Such circuity may occasionally be encountered however, and so it is worth being aware of its existence.

It can also sometimes be beneficial to apply some tidying up to a circuit, which again is not usually shown on the circuit schematic. If a multi-section IC is used, such as one of the quad XOR or hex NOT 4000 series ICs mentioned earlier, it is quite likely that not all sections will end up being used. In this case it can be worth while terminating the unused sections so as to minimise the chances of noise or oscillations being generated, or unregulated current draw diminishing battery life. A good rule of thumb to start with is to tie unused inputs to ground, and leave unused outputs floating. This is not a universal approach but it is a good start in the absence of any other guidance. (Manufacturer data sheets sometimes offer advice in this regard.) Often a mid-level V_{ref} voltage rail is a better place to tie the inputs in a circuit with a single rail power supply.

For instance, unused opamps in a dual or quad package are probably best wired as a voltage follower, as shown in Figure 7.9, with IN+ wired to V_{ref} and OUT wired back to IN−. This keeps everything sitting nice and steady at the reference voltage (ideally half way between the power rails), preventing oscillations and minimising current draw.

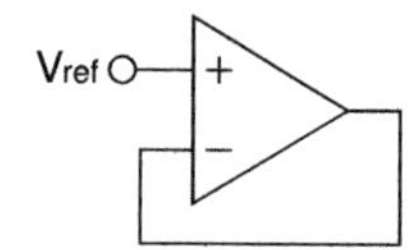

Figure 7.9 Wiring unused opamp sections.

Audio effects circuits will typically require a bypass option so that they can be switched in and out of the signal path. Such bypass wiring may or may not be shown, but in any case it is always an option as to whether and how to implement it. Alternative schemes, along with their advantages and disadvantages are also covered in Chapter 15. One of the variants presented includes a status LED in order to indicate what state the bypass switch is in. Such status LEDs, along with power indicator LEDs are often desirable but seldom included explicitly on a schematic. The choice of whether and how to add them is always available, and such decisions are best not left until the end of the circuit layout design phase which is described next.

Breadboard Layouts

Laying out a circuit, whether on breadboard, stripboard, or printed circuit board (PCB), is as much an art as a science. Practice, experimentation, patience, and attention to detail are the best route to generating a successful layout. The PCB option is a more advanced approach to circuit building and is not covered here, but some guidance towards the drafting and interpretation of breadboard and stripboard layouts is provided. Layouts are produced using symbols which broadly reflect the physical form factors of the various components to be used in the final construction, rather than the abstract symbols encountered previously in schematic diagrams. These layouts, unlike a schematic diagram, are direct physical representations of the actual circuit to be constructed. As such there is less scope for interpretation, and thus less use for abstract drawing conventions. The layout of complex commercial circuit designs is an extremely involved business. Often many factors need to be taken into account; noise, interference, high frequency performance, accuracy and precision, size and shape of the circuit board, heat dissipation and power management, and numerous other considerations can all feed into the design process. The audio circuits being considered here require little of this level of detail in order to arrive at an effective and efficient final circuit layout. Indeed the breadboard and stripboard circuit building options do not lend themselves to such exacting standards in their design requirements. Even the next step up, simple single sided PCB circuit layouts, provide only limited flexibility in this regard. Professional multilayer designs where these considerations come into play are far beyond the scope of what is presented or required here.

Breadboards enable simple plug and play circuit construction. Components can be added and removed without the need for soldering or otherwise fixing the connection points. They provide an excellent framework for experimentation and testing. The downside is that breadboard based circuits are not robust. Components can work loose, and wires can move around and touch, making unwanted, possibly intermittent, connections. As such they can be difficult to debug if sufficient care is not taken in their construction, and they are entirely unsuitable for permanent projects. Breadboards are for prototyping, learning, and experimenting; once a circuit is finalised, a soldered version is the best option for a permanent build. This entire procedure is examined in Chapter 10 – Constructing Circuits.

Breadboards come in various shapes and sizes, but by far the most common format is laid out as illustrated in Figure 7.10. Each numbered row in the centre of the board consists of five connected insertion points, allowing for up to five components to be connected together by inserting legs and wires into the five holes. The gap down the middle of the board is just the right size for a DIP IC (as introduced in the component list in Chapter 6) to span the central reservation. Four longer busses run vertically from top to bottom, two on each side of the board. There will usually be a gap about every five steps along these busses but the connection runs all the way from one end to the other. These busses are most commonly used to connect power into the board, but there is nothing special about them and they can in fact be used for any connections (in Figure 7.17 the first of the four busses is used to facilitate a tricky set of intra-circuit connections, while the other three are used in their more familiar roles, two providing board-wide connections to 0V and the fourth connecting to +9V).

Figure 7.10 A blank breadboard template, ready for populating.

Symbols like those illustrated in Table 7.5 can be used to design a layout for a circuit. Attempting to go straight from a schematic diagram to building a physical breadboard version of the circuit is likely to prove a very frustrating endeavour. Things will quickly become confusing and difficult to follow. Much better to first design a layout by drafting a diagram, as described here. The actual building process is then a simple matter of copying the finalised design onto the actual breadboard. The symbols presented here are not standard representations, but rather just aim to be graphically recognisable as, and with a comparable footprint to, the actual components they are intended to indicate. This table includes only on-board components. Off-board components such as jacks, pots, and switches are represented only by the flylead wires going to their connection terminals. This is the style preferred here. Others may choose to include graphical elements to represent off-board components also.

Table 7.5 Circuit layout component symbols used in this book

Name	Symbol	Description
Resistors	2k2 10k 1M	In common with their schematic diagram symbol, resistors are mostly represented here on breadboard and stripboard layouts by a simple rectangle. When a layout requires the two leads to be placed very close together the rectangle symbol no longer works well, and instead a circle is centred over one of the connection points with a lead extending to the other, nearby point. The idea is to represent the resistor as if it were standing on end with its top lead bent down through 180° in order to make the second connection. In a tight layout this is often what is actually done in order to reduce the footprint of resistors and other axially leaded components.
Photoresistors (aka light dependant resistors, LDRs)	λ LDR λ	The shape of the photoresistor's symbol reflects the shape of the type of LDR most commonly encountered (see Figure 12.16). The Greek lambda is a symbol often used to designate LDRs in circuit diagrams, although this is by no means a standard. LDRs have typical dark and light resistances, but they do not have a single defining value which it might make sense to label them with. Circuit notes will often give an indication of good working values in a given circuit, but since they are highly lighting dependant, some circuit calibration is often needed for best performance.
Capacitors	1n 2n2 4n7 1u 2u2 4u7	Although capacitors come in many types, the most important thing to specify is whether one is polar or nonpolar. In the diagrams presented here, nonpolar types are indicated using the oval symbols, while the round symbol with a dark segment is used for polar types. The dark segment is positioned over one of the legs and indicates the negative connection. Non polar types by contrast do not have a positive and a negative side, and can be connected either way round. In tight placing the leads are hidden under the symbol but their positions are unambiguously located on the breadboard or stripboard grid below them.

Table 7.5 (continued...)

Name	Symbol	Description
Inductors	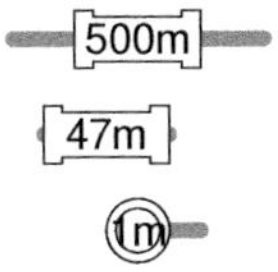 500m 47m 1m	Inductors are often made by wrapping a wire around a bobbin with flanged ends. The outline of the symbol used here reflects this oft encountered form factor. In common with the resistor symbols shown above, once again the third example here is suggestive of a component mounted vertically when the two leads are to be inserted into the board very close to one another.
Transformers	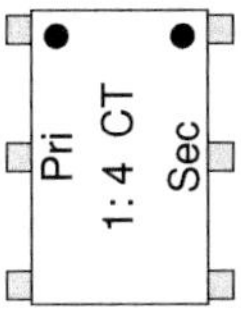 Pri 1: 4 CT Sec	Transformers come in many shapes and sizes. The symbol given here works well for the commonly encountered centre tapped audio transformers most likely to be found in DIY audio circuits. The primary winding is to the left, with the secondary to the right. Both windings are centre tapped (the middle connections on either side), with the dots indicating the connections to the start of each winding. Differentiating start and end is important in maintaining the desired polarity in the signals passing through the transformer. The labelling here indicates the turns ratio (1 : 4) and that the transformer provides centre taps (CT).
Diodes	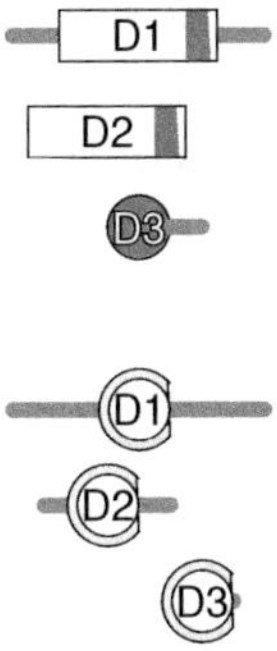 D1 D2 D3 D1 D2 D3	The diode symbol unsurprisingly sports a stripe on one end, matching the common markings on a real diode. In common with the polar capacitor described above, diodes must be inserted into a circuit in the correct orientation. The third symbol, representing an upright component, is similar to the equivalent resistor symbol. It is important to interpret this symbol with a little more care however, as orientation matters. The circle in this case is coloured in dark. This is intended to suggest that the end with the dark stripe is uppermost, and therefore should be connected to the point offset from the centre, with the non-striped end falling directly below the body of the symbol. The second set of three symbols here represent LEDs. Again orientation is important, and the flattened off side where one of the two leads originates indicates the LED's negative lead (equivalent to the stripe in the standard LED).

Table 7.5 (continued...)

Name	Symbol	Description
Transistors	Q1 Q2 Q1 Q2	The two sets of outlines here actually represent the two most commonly encountered three pin semiconductor packages. Most transistors will come in one of these two forms but other semiconductor devices can also be encountered packaged in these standard packages. There are three pins and it is always important to get them the correct way around. The first package type is called TO-92 and the second is designated TO-220. In the case of TO-92 orientation is easy to specify by indicating in which direction the flat face should point. For the TO-220 package it is not so straightforward. The thick line on one side of the symbol is used here to indicate the back of the device. TO-220s are relatively high power dissipation devices, designed to have heat sinks attached. The thick line indicates the face to which a heat sink is attached.
DIP ICs	DIP	DIP IC stands for dual-in-line package integrated circuit. The dual-in-line bit means there are two parallel lines of legs, positioned on either side of the device's body. The semicircle at the top allows the device to be placed the correct way around, which is of course essential. Some DIP ICs have a dot at pin one instead of, or as well as, the semicircle (pin one is at the top left in this picture, and pins are numbered counterclockwise from pin one, down the left side and back up the right).

As a very basic example, consider the circuit shown in Figure 7.11. It consists of two components, a resistor and an LED in series, connected between +9V and 0V. There are many possible layouts which would work for such a simple circuit. Figure 7.12 shown a couple of possibilities. It also illustrates a few of the commonest errors made by inexperienced breadboard builders. Notice also how the 0V and +9V connections are distributed to the busses on both sides of the board. This is not really necessary for simple circuits, but it is a common practice where more elaborate layouts are concerned

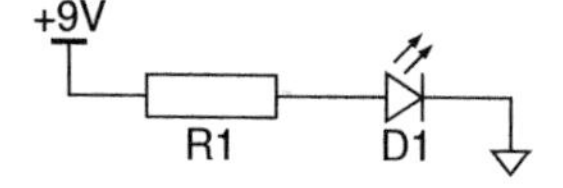

Figure 7.11 a simple circuit.

(see for instance Figure 11.9 (p. 177)). As mentioned earlier, the noise gate example at the end of this chapter does not follow this procedure, as one of the busses is actually used for other connections.

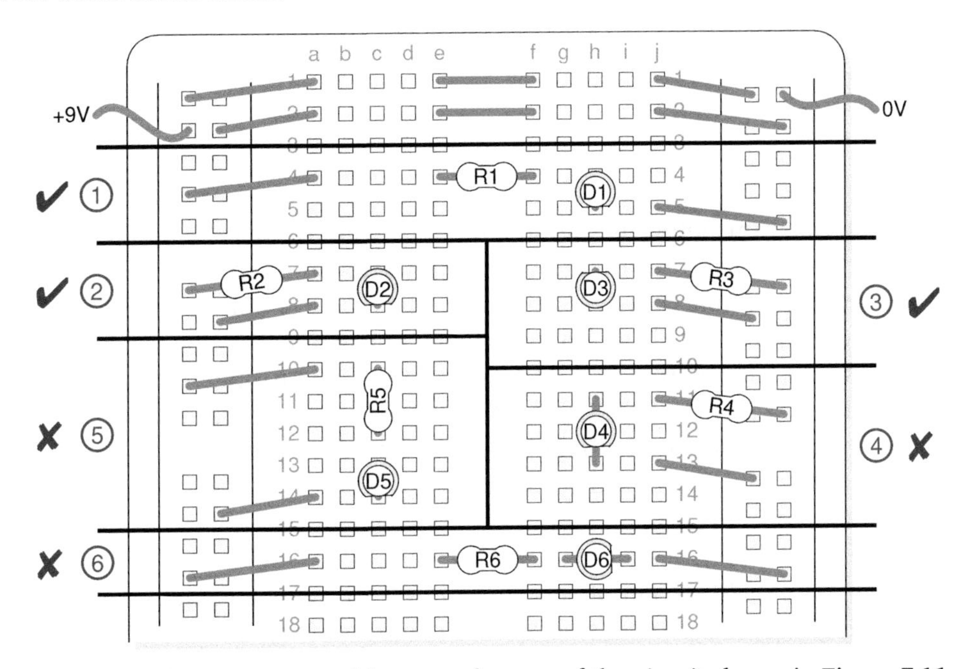

Figure 7.12 Some correct and incorrect layouts of the circuit shown in Figure 7.11.

Figure 7.12 shows six different attempts to layout the circuit, numbered (1) through to (6). Numbers (1) to (3) are all valid layouts, although it should be noted that (3) reverses the order of the resistor and LED. This makes no difference in this very straightforward circuit. It is often the case that two simple components in series with nothing else connected where they join can be reversed without ill effect. Notice that in the noise gate schematic in Figure 7.16, SW1 comes before R13 on the way down to ground. However in both the breadboard and the stripboard layouts which follow the order has been reversed to allow for a simpler layout design.

Layouts (4), (5), and (6) illustrate different layout errors. In (4) the diode is connected the wrong way round. Care must be taken to make sure that directional components such as diodes and polar capacitors are oriented correctly. Numbers (5) and (6) illustrate two related issues which arise from the visually tidy results of placing components in a line. Inserting two components in line vertically as in (5) does not result in a connection between them, they must be inserted on the same row. Similarly, it is almost never going to be correct to insert more than one of a component's legs into the same row of five, as is done in (6). In this case it short-circuits the legs of the diode, bypassing it and effectively removing it from the circuit.

Learning by Doing 7.1

Schematic Diagram to Breadboard Layout

Materials

Some squared paper or one of the layout drafting packages mentioned in the main text is all that is needed.

Theory

The schematic diagram below is the simplest of a family of very popular circuits which can be traced back to a design usually referred to as the Bazz Fuss. It is a simple, effective, and remarkably versatile fuzz effect. The circuit and its variants are examined in a bit more detail in Chapter 17 on transistors. For now no knowledge of the circuit's operation is required. All that is needed is to understand the meaning of the various elements of the schematic diagram and to come up with a viable breadboard layout for the circuit. Building it will be left for now and revisited as an exercise in Chapter 10 – Constructing Circuits.

Practice

The circuit consists of five components in addition to input and output jacks, and power. Of these, the resistor needs no special consideration, nor do the two capacitors, apart from the observation that one of these is polar, and needs to be connected the correct way round. The diode and transistor form the heart of the circuit.

Diodes must be inserted in a particular orientation in order to work, and the symbols used to represent diodes must indicate this orientation clearly and unambiguously. The standard schematic symbol for a diode can be viewed as an arrow with a line on the end of it. The end of the symbol with the line on it is the same end as the end of the component with the stripe on it (and obviously, this stripe appears on the layout symbol also, as shown in Table 7.5).

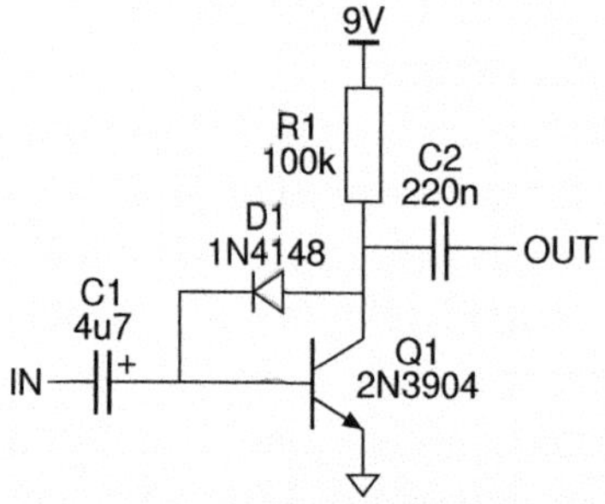

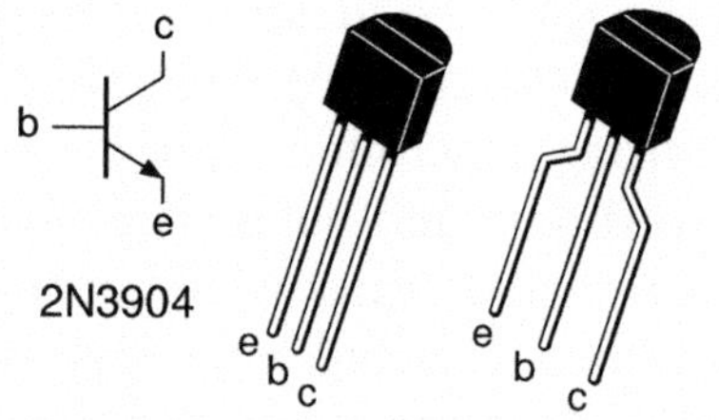

Transistors have three pins. All that is needed for now is to be able to tell which pin corresponds to which connection on the schematic. This information can vary from one transistor type to another, so it is always important to check. The device specified in the schematic here is a 2N3904. The picture on the right shows which pin is which in this case.

This is all the information you need in order to generate a breadboard layout for the Bazz Fuss. See Learning by Doing 10.1 (p. 156) for one solution to this task. The layout presented there represents a possible answer, there are of course many other equally valid layouts. Try working out a couple of alternative layouts before looking. The aim should be to develop a tidy, easy to follow, relatively compact configuration. While keeping things compact may not seem crucial here, it is important to remember that as the circuits you lay out become larger and more complex, space on the breadboard can quickly start to run out.

Layout Drafting Options

It is quite a simple matter to hand draft breadboard (and stripboard) layouts using squared paper but a software option is usually going to be easier. There are many software packages available which will do the job but a good free solution is a program called DIYLC ('Do-It-Yourself Layout Creator'). Freely downloadable versions are available for PC and Mac. Note that this is just a drawing package, it does not run circuit simulations as some more elaborate software does. Chapter 8 mentions a few of the more accessible options in this department.

Stripboard Layouts

Once a project has been successfully breadboarded, and a final version has been settled on, the next step is to develop a stripboard layout for soldering up into a permanent circuit. (Various alternatives are discussed in Chapter 10 but stripboard is the medium used throughout this book.) Stripboard consists of a rigid circuit board material with a matrix of holes drilled in it, and parallel copper traces running across the board connecting together each row of holes. Figure 7.13 shows two images of the same piece of stripboard, viewed from either side.

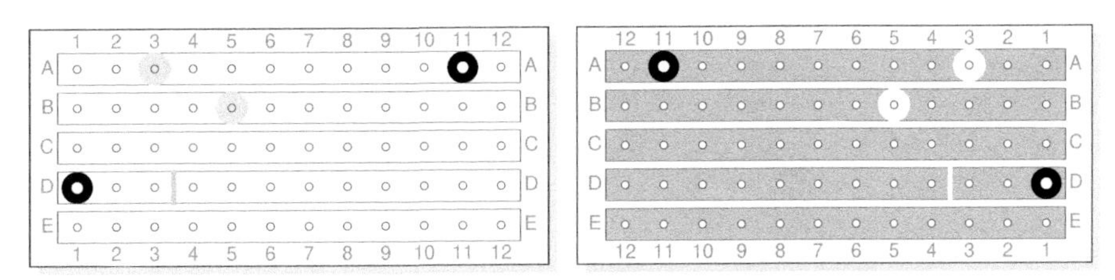

Figure 7.13 A piece of stripboard, showing views of the component and copper clad sides of the board side by side. Note that column numbering is mirrored left to right between the two views.

On the left is the component side, and to its right the copper side. Components are inserted on the component side with their legs sticking through the holes to emerge on the copper side, where they are soldered into place. While most of the layout work is done on the component side, it is useful to maintain a view from the copper side in tandem because breaks in the copper strips are almost always needed in order to complete a design, and these must be correctly located on the copper side. As the board is turned over and back it is very easy to get mixed up as to what is where. If the board is turned over like the page of a book, perspective changes from the left hand view to the right, as illustrated. Notice that the numbers 1–12 along the top and bottom boarders of the board are mirrored in the copper side view on the right, running 12–1.

Three types of feature have been indicated on this figure, and their locations match between the two views. The black ring along the left boarder at 1D is mirrored over to the right in the second view but is still at position 1D, and likewise for each of the other features indicated. Black rings indicate mounting holes drilled through the board. Light rings indicate breaks in the copper strip located over holes, and light bars indicate breaks cut into the copper between holes. The former are easier to work with but they do use up a hole, while the latter don't require to use a hole but can be quite tricky to apply, and are easily overwhelmed by solder during the building process. The on-hole breaks can be made using a drill bit or a dedicated little hand tool called a spot face cutter. It is important to check that the hole is cut deep enough to complete the break. It is easy to leave a tiny sliver of copper connecting across the intended break point. A craft knife or other sharp point can be used to scrape a between-hole break into the copper strip. The layouts presented here generally stick to on-hole breaks for simplicity's sake.

Figure 7.14 shown an alternative arrangement with the copper side located below instead of beside the component side. Notice that this time it is the letters along the

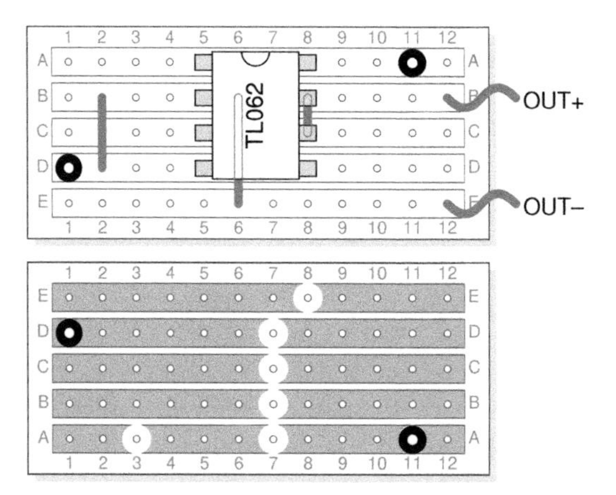

Figure 7.14 Stripboard with the copper side view placed below the component side view. Here the row letters are flipped top to bottom, rather than the column numbering left to right as in Figure 7.13.

rows which are flipped, rather than the column numbers as before, but once again the breaks and holes retain the correct indices between the two views. This time only the drilled holes are actually placed on both views. The component side can quickly get cluttered, and it is better to leave the breaks out in this view. Using the indices it is a simple matter to check that they are in the correct locations, even on a large layout. In this case notice for instance that the column of four breaks from 7A to 7D run under the DIP IC labelled TL072. This is a common thing to see, as the two sets of pins on a DIP must be isolated from each other.

Two other features are illustrated on this board: jumper wires and fly leads. Jumper wires connect two points on the board. They are most often run vertically but can be routed at an angle if need be. As illustrated, they can (with care) be run underneath a DIP. An outline is used here to show such a routing. In addition, a connection between two immediately adjacent points (as in 8B to 8C here) can be achieved simply by applying sufficient solder so as to bridge the gap. As such the connection here can be overlayed on other connections, in this case two pins of the IC. While these two options can be convenient in tight layouts, they have been avoided in the projects presented in this book in order to maintain maximum clarity.

Fly leads are wires coming off the board to other components, usually such things as battery clips, connector jacks, switches, and pots. The convention adopted here is simply to label all such connections, rather than using graphical symbols for the components in question. This leads to tidier layouts as it avoids the necessity to bring together the various leads connecting to any given off-board component. Careful labelling ensures that all such connection points are unambiguously connected to the correct terminals on each of these offboard components. Note that these jump lead and fly lead connections, and all the notes associated with them here, apply equally to breadboard layouts, where they can be used in exactly the same fashion, with the exception that the two routing options (under components and shared connection points) do not apply to breadboards.

LEARNING BY DOING 7.2

SCHEMATIC DIAGRAM TO STRIPBOARD LAYOUT

Materials

Some squared paper or one of the layout drafting packages mentioned in the main text is all that is needed.

Theory

This stripboard layout practical uses exactly the same starting point as the previous breadboard layout exercise. The schematic diagram below is this time to be turned into a stripboard layout.

Practice

The circuit notes from Learning by Doing 7.1 (p. 110) apply equally here. Most especially, pay attention to the orientation of the diode and the polar capacitor, and refer to the transistor pinout below for connections to the transistor.

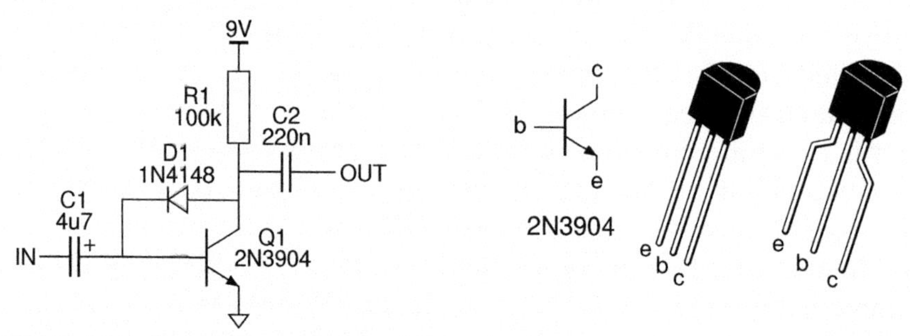

This is all the information you need in order to generate a breadboard layout. See Learning by Doing 10.2 (p. 159) for one solution. As before, the layout presented there represents a possible answer, and there are many other equally valid layouts. Try working out a couple of alternative layouts before looking.

A More Elaborate Example

To round out this chapter an ambitious circuit layout task will be examined. The circuit in question implements a noise gate. Much audio electronic circuitry can be quite noisy, some electric guitar effects pedals are particularly bad offenders. A noise gate is often added to the end of such a noisy signal chain in order to keep the noise level down when the instrument is not being played. When the signal level coming into the noise gate drops below a certain threshold, the gate closes and the noise is prevented from emerging at the output of the signal chain. The idea is that the threshold is set just above the noise floor so that when the guitar is played the threshold is breached and the gate opens. Obviously a gate can only be effective when the threshold can be adjusted to a band between the level of the noise floor and the quietest signal level which is needed to open the gate.

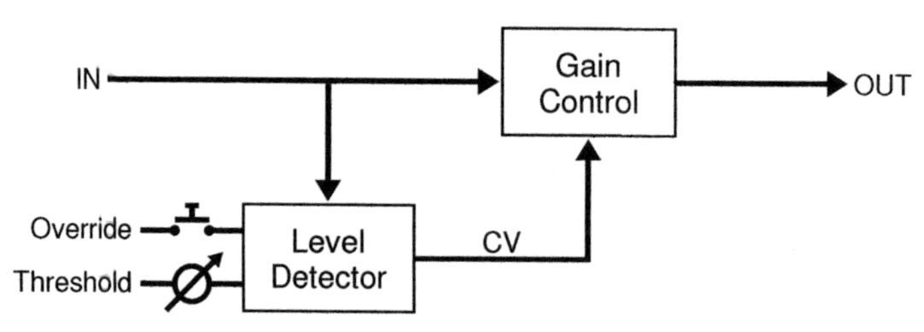

Figure 7.15 Noise gate circuit block diagram.

The circuit here provides two controls. A threshold control is of course needed so that the gate can be configured for the particular levels of signal and noise present. An override switch is also implemented. Usually as the guitar sound dies away below the set threshold, the gate will close. Sometimes a player wants the long decay to be preserved. In this case the override switch is kept pressed as the sound dies away, for as long as the decay is desired. On release the gate will close, preventing further low level signal from passing. A block diagram of the system is given in Figure 7.15, showing signal flow, functional blocks, and user controls, but no implementation details. This is the level of information the user needs in order to get the most out of the device. To build the circuit, a little more concrete detail is required.

Figure 7.16 shows the complete circuit diagram for the noise gate. As described earlier, such things as powering options and bypass circuitry are omitted. In fact these will not be included in the layouts here either, as they would be implemented as off-board connections in the final build. Here only the circuit board layouts are considered (the complete procedure to build a working project is addressed in Chapter 10). This is a particularly tricky circuit to address for a number of reasons. It is quite large for the class of project being considered (although not the largest), but in addition it entails some quite complex connectivity.

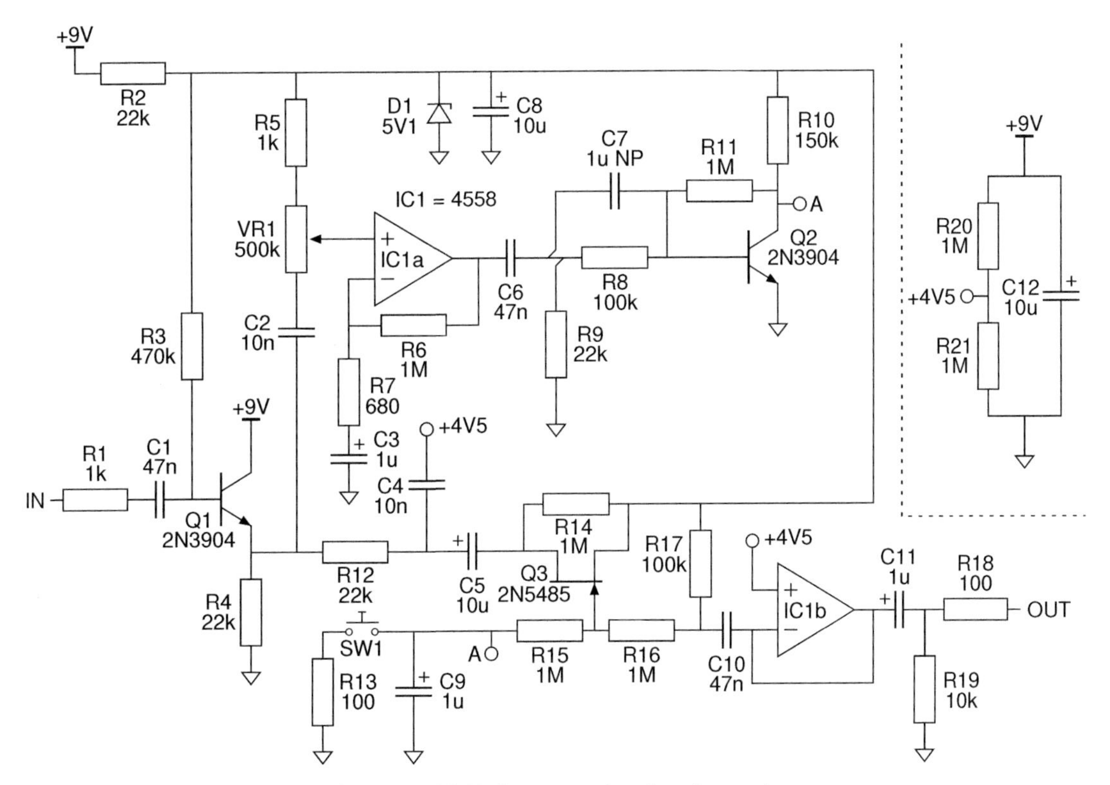

Figure 7.16 Noise gate circuit schematic.

The two opamps, IC1a and IC1b, are implemented through a dual opamp IC. Using two single opamp chips is perfectly valid, and variants of this circuit exist which do exactly that. Since the aim here is to tackle the challenge rather than to avoid it, this simplifying design choice has not been embraced. Another complex routing challenge encountered here starts at R2 in the schematic in Figure 7.16 and snakes around the circuit connecting together a total of nine disparate components. Another six way connection point hangs from the collector of Q2. The ground buss is traditionally the busiest connection point and in this case a total of 15 nodes come together here. Routing multiple multi-way connections neatly can prove a difficult task.

Figure 7.17 illustrates a possible breadboard layout for the noise gate circuit. As noted, this circuit presents a number of challenges when it comes to developing a viable layout. As such it represents a good example to examine. It is important to note that there is no step by step formula which will reliably result in a breadboard layout. A certain amount of trial and error is inevitably involved. There are two useful approaches

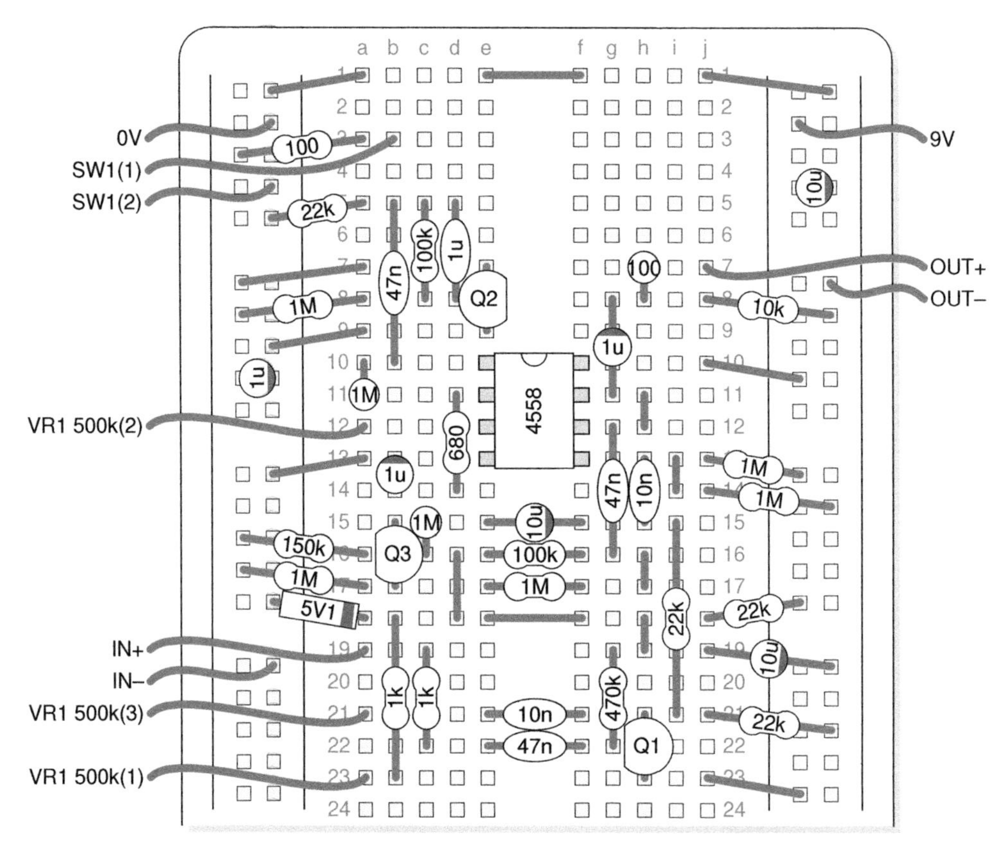

Figure 7.17 Breadboard layout for the noise gate circuit.

to getting a layout started; either start from the input and work through the signal flow, or start with the most interconnected component (in this case IC1) and work out.

When attempting a difficult layout false starts are inevitable and can be useful in clarifying where the difficulties lie. It can be most productive to abandon a layout which has hit a blockage, and start again using what has been learned to improve the chances of a successful conclusion next time around. When laying out this circuit, I started my first attempt working out from IC1 but couldn't arrive at a layout I was happy with. It started to get very messy. This attempt was scrapped, and I tried again starting this time at the input. The second attempt lead to the layout presented here. Having identified the two most complex connection points in the first pass (the nine and six way connection points mentioned above), I highlighted these on my circuit diagram, and kept an eye on how they might layout as I worked.

Given the number and distribution of ground connections (15 in all), it was clear that a vertical buss on each side of the board dedicated to these would be required (the three jumper wires across the top of the board make this connection), but since there were fewer +9V connections (six in total) it was decided to try to keep these to one buss and reserve the fourth and final buss for other purposes. With the power connection to IC1 on the right it made sense to reserve the buss on this side, and leave the second left hand side buss free. This ended up being used for the six way connection point hanging off the collector of Q2. The nine way connection point occupies four rows around the board. Three jumper wires can be seen linking these up: d16–d18, e18–f18, and h18–h19. Identifying these key points in the circuit makes it easier to follow the various connections and confirm the accuracy of the layout.

Having built and tested the breadboard version, and decided that a permanent version is wanted, the next task is to arrive at an equivalent stripboard layout. It is possible to convert a breadboard layout almost directly into a stripboard version (the busses present difficulties but these are not insurmountable). However such a layout will not result in terribly efficient use of space on the stripboard, and is likely to be somewhat inelegant. Better to start from scratch. Figure 7.18 illustrates a possible stripboard layout for this circuit. Stripboard layouts tend to entail components almost exclusively being laid out vertically. Breadboards tend to result in much more of a mix, as connections to the busses and across the central gap lay out horizontally, while the rest tend to run vertically. It is by no means a requirement that fly leads exit the board from around the edge, but aiming for this does make a tidier layout, and also results in a slightly easier layout to build. The final project in Chapter 11 has a stripboard layout (see Figure 11.15, p. 185) where a compact layout has been favoured over following this general rule. Some might say this layout is rather too compact. It certainly does make for a challenge to solder it all up in such a small footprint. At this level, it is simply a question of personal preference.

As already stated, these layouts represent quite a difficult and challenging layout design task. Once a bit of practice and experience has been gained laying out smaller

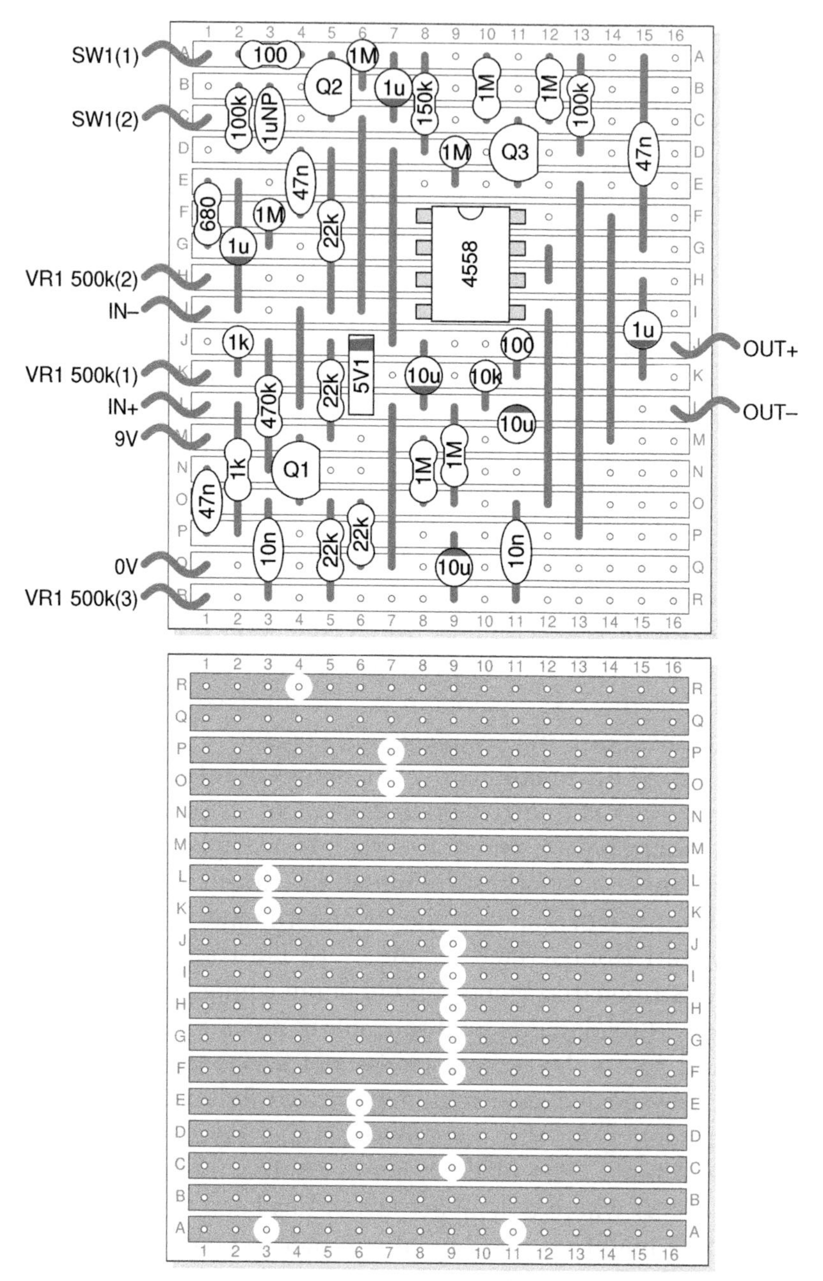

Figure 7.18 Stripboard layout for the noise gate circuit.

and more straightforward circuits, the reader might consider the challenge of returning to this project and attempting to complete a breadboard and a stripboard layout of their own. It might take a few attempts before a satisfactory result is achieved. It is unlikely to look too much like the examples given here. There are many possible layouts, and most certainly layouts which complete the task more elegantly or simply.

References

C. Coombs. *Printed Circuits Handbook*. McGraw-Hill, 6th edition, 2008.

Elektor. Tup-tun-dug-dus. *Elektor*, 1(1):9–11, 1974.

A. Sedra and K. Smith. *Microelectronic Circuits*. Oxford University Press, 7th edition, 2014.

E. Young. *Dictionary of Electronics*. Penguin, 1985.

8 | Electronics Lab Equipment

A few key pieces of equipment allow for a wide range of circuit test and measurement tasks to be accomplished with ease. A good level of familiarity with the equipment available to you can make a huge difference to the success and the enjoyment of audio circuit building. The basic tools of the trade come in a wide variety of shapes, sizes, and price points, but for the most part, while bells and whistles vary widely, the primary functionality provided is consistent and quickly mastered.

Make a new folder on your computer beside the one called 'Data Sheets' from Chapter 6, and label it 'Manuals'. Whenever a new piece of gear crosses your path, find and save the user manual and any other useful documentation you come across. Your equipment will differ from the kit presented in the examples here, but the basics will remain the same. Once you can identify the primary controls and connections, it should not prove difficult to transfer that knowledge to a new piece of equipment of the same type. Four key tools are highlighted, with others mentioned more briefly.

Bench PSUs, or bench power supply units, provide the power for circuits as they are developed and tested. Most of the circuits considered here can be powered from a standard 9V battery or a (non-adjustable) power brick, but the bench PSU provides more flexibility and a number of other advantages.

Function generators produce test signals ideal for injecting into circuits in order to see how they perform under various conditions. By using a carefully controlled input signal rather than an audio track for testing, the performance characteristics of the circuit in question can be much more easily and much more precisely assessed.

Multimeters of various types and form factors allow for basic measurements of voltage, current, resistance, and continuity. They also usually provide for the simple testing of components like diodes and bipolar transistors, and some can incorporate an impressive array of additional functionality.

Oscilloscopes allow the user to visualise and characterise audio and other signals as they pass through circuits under test. They can provide for more detailed analysis than is possible with a basic multimeter.

Audio analysers and advanced test gear such as electronic loads, spectrum analysers, and distortion meters can be very useful in more advanced work, but are beyond the scope of what is needed for the basic tasks considered here.

Software tools covering a wide range of tasks and applications exist. Programs to help in the design of both circuits themselves and circuit board layouts can be a valuable aid. Software interfaces to test and measurement hardware also exist. The softscope is a particularly common application; a software interface providing oscilloscope functionality via an audio sound card or similar hardware frontend.

Workbench tools in general are a given, and will not be considered further here. Accumulate all you can lay your hands on: wire cutters and strippers, pliers, screwdrivers, clamps, magnifiers, assorted sprays and fluids, glues and lubricants, blades, hacksaws, drills and rotary tools, safety goggles, socket sets, spanners, files... The list is endless. A well stocked (and well organised) workbench makes the conduct of any job a much more straightforward affair.

Soldering and desoldering, and the associated tools are addressed separately in Chapter 10 – Constructing Circuits.

Bench Power Supply Units

Figure 8.1 HY3003-2 dual linear DC power supply (0–30V DC, 3A per channel).

The term bench PSU is used to refer to an adjustable power supply designed to provide versatile DC powering options on the workbench, as opposed to a standard PSU, which is designed to be supplied with or incorporated into a specific piece of equipment and provide just those power signals which are needed for that particular task.

AC outputs are possible from a PSU (bench or standard) but in the current context are much less generally useful and less commonly encountered than the standard DC supply. The variable autotransformer or variac® is the most common format for a variable AC power supply. A variable autotransformer allows for the AC voltage to be dialled up to any desired level, often including levels beyond the mains input level (remember transformers can step up as well as stepping down). These devices are often used for slowly ramping up to full mains voltage when servicing or fixing a piece of mains powered equipment such as a power amplifier. They are extremely useful in this application, but are not the kind of PSU which is of general interest here.

A typical bench PSU (Figure 8.1) will usually be able to provide a DC output voltage level between zero and some maximum, at a current up to some maximum level. Adjustment of the voltage, and often also the current limit, is possible through the frontend controls, with displays providing feedback as to the voltage and current levels during configuration and in operation.

Figure 8.2 Trio PR-602A linear DC power supply (0–25V DC, 3A), with two presets and one freely variable output control.

If the maximum current level is reached the device may switch from constant voltage mode into constant current mode. These two modes are often indicated by two LEDs on the front panel labelled CV (for constant voltage) and CC (for constant current). Normally it is the aim to operate in constant voltage mode, but if the load attached to the PSU is too heavy, and the current limit is reached, then the PSU will reduce the voltage level to whatever is needed in order to allow no more than the set maximum current to flow. This is constant current mode. By setting a maximum current just a little higher than the current which a circuit is expected to draw, it is possible to minimise the chances of damaged or destroyed components and equipment due to an error or a fault in the circuit being powered.

It can be useful to consider the operation of the device under four general headings: inputs, outputs, controls, and displays.

Inputs

A bench PSU almost always derives its power via a mains power connection. Advanced models can include remote control interfaces for automated operation but this is less common and not considered here. No other inputs are required.

Outputs

Terminals or connection points are provided in order to deliver power into the circuit. They are usually colour coded – black for negative and red for positive. Often a separate ground point is provided, typically coloured green or black and marked with an earth ground symbol (⏚). This ground terminal will be hardwired directly to the earth of the mains power lead.

The positive and negative outputs can either be floating or referenced to ground. The outputs on the devices in Figure 8.1 and Figure 8.2 are floating by default. The middle terminal in each cluster of three is the earth ground mentioned above, and can be connected to either positive or negative if specific ground referencing is desired (a jumper can be seen in Figure 8.2 bridging between the negative and ground terminals; this can be rotated to connect the positive rail to ground instead, if desired). The two outputs on the supply in Figure 8.1 are designed so that they can be wired in series to make a split or dual power supply if needed. A split supply provides three connection points: positive, negative, and ground.

Two supplies with their negatives hardwired to earth ground can not be wired in this way or one of the supplies' positive rails will be shorted to ground. Although most of the circuits in this book use a single supply, a split rail supply can be very convenient. The spring reverb drive and recovery circuit on p. 366 illustrates a split supply (notice the opamp negative power terminal is connected to −9V not ground).

Sometimes a bench power supply offers a number of different styles of output. Notice the selection switch on the supply in Figure 8.2. While the middle of the three options connects to the main control pot, the two others use the settings on the two recessed trimmers. These provide the same range of possible voltages as the main output, but must be adjusted with a small screwdriver. As such they are intended to be set at commonly used voltages (say 9V and 12V), and left there; a kind of preset facility.

Controls

Generally the controls on a standard bench PSU are very straightforward. Typically one knob sets the voltage while another sets the current limit (if available). There may be a second 'fine' control to tweak the voltage more precisely. Setting the current limit usually involves either entering a standby/setup mode or just shorting the outputs together with a good solid piece of wire so that current can flow in order to be measured. Shorting inputs is generally a thing not to be done lightly, so be sure to carefully check the procedure for the particular unit in question before proceeding.

For more elaborate units some extra controls may be found. For instance the two buttons labelled 'Tracking' at the centre of the front panel of the digimess unit in Figure 8.1 allow for a number of modes to be selected: independent (two separate outputs), series (a split supply as described above), and parallel (a single output with double the maximum current capability). The manual leaves the fourth possible switch combination a mystery – not ideal.

Both units illustrated here provide a switch which allows either the voltage or the current to be displayed and monitored at any given time. Many modern bench PSUs incorporate a microcontroller which expands the control possibilities considerably. Even very low cost units now often provide a menu based interface giving access to greater levels of control and feedback than is possible with more traditional units.

Displays

Either a moving coil style analog display as in Figure 8.2, or a digital readout as in Figure 8.1 will provide feedback as to the prevailing settings and operating conditions. In both units shown here switches allow either voltage or current to be displayed for each output; both cannot be viewed together. There are two displays on the digimess but one display is dedicated exclusively to each output. Even the more modest kind of microcontroller based units mentioned above will usually display voltage and current (and possible some other parameters) simultaneously.

FUNCTION GENERATORS

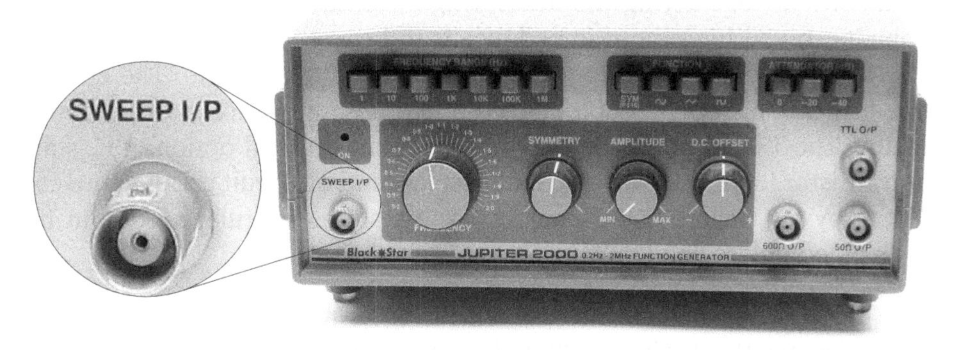

Figure 8.3 Black Star Jupiter 2000 0.2Hz–2MHz function generator. The inset illustrates the BNC connector type.

Function generators (or signal generators) provide a convenient source of controllable standard signals for use in the testing and characterisation of circuits being built, modified, or repaired. In addition to standalone units, other equipment often incorporates a function generator as a part of their advanced functionality. Oscilloscopes and signal analysers in particular are commonly found to include this option. Even the 72-7770

DMM (Figure 8.4a) includes a setting which provides a 50Hz square wave output at about $5V_{pk-pk}$. Not the most useful signal for audio work, and rather too limited to be classed as a function generator, but an interesting option on a low-end DMM. Similarly, even oscilloscopes which do not provide an actual function generator usually include a calibration output signal, typically a 1kHz square wave at $5V_{pk-pk}$, designed for adjusting the setup of $\times10$ probes (see the section on oscilloscopes).

Inputs

Function generators are usually mains powered, and advanced models can include remote control interfaces for automated operation. The Jupiter 2000 in Figure 8.3 has a BNC input on the left side of the front panel, marked 'SWEEP I/P' (see inset), which allows the output signal's frequency to be set or swept using an external signal applied to this input. This can be useful in more advanced applications, but in general can be ignored. By and large, as is the case with the bench PSU, function generators don't require or utilise much in the way of inputs.

Outputs

The core purpose of a function generator is to output a test signal. In common with much audio test equipment, BNC connectors are the most commonly encountered output connectors found on function generators. The Jupiter 2000 (Figure 8.3) sports three outputs to the righthand side of the front panel. These are labelled, '600Ω O/P', '50Ω O/P', and 'TTL O/P'.

As discussed in Chapter 9 – Basic Circuit Analysis, output and input impedances are important when considering the interfacing of electronic circuits. Most often low output impedance and high input impedance is to be expected in modern interfacing, and the 50Ω output is usually going to be the one to opt for here. The 600Ω output reflects the fact that historically 600 ohm to 600 ohm impedance matched interfacing was common. The TTL output is also somewhat of a historical option. TTL stands for transistor transistor logic, and refers to a particular kind of circuit architecture common in years gone by but of less importance today. As a general rule, an output like this should be avoided unless you know it is what is needed.

Controls

An analog device like the Jupiter 2000 makes for a very straightforward user interface. On the other hand, the FG085, which is a digital device, provides more functionality at the expense of higher complexity in operating it. In either case the most fundamental controls allow for the output's shape, frequency, and amplitude to be controlled.

The commonest wave shapes provided by function generators are sine, square, and triangle. Ramp and sawtooth waves can be seen as variants of the basic triangle wave shape. Other common options include DC offset, symmetry, and duty cycle.

A typical test procedure for working with audio circuits would be as follows. Configure the function generator by selecting the sine wave shape, setting the frequency to 1kHz, and dialing the signal amplitude down to zero. Then the generator output can be connected to the input of the circuit under test, and with the input and output of the circuit displayed on an oscilloscope, the amplitude is increased on the function generator to observe the operation of the circuit.

Displays

A simple analog device such as that shown in Figure 8.3 may have no information display at all (in this case a power LED is the only indicator present). Configuration settings are read from the legends surrounding controls, and exact values may be measured externally where they are required.

On the other hand, a device such as the FG085, which incorporates a simple graphical display, can be designed to provide direct feedback on the current device configuration. Even a small dot matrix display can be programmed to provide a wide range of feedback and information.

MULTIMETERS

The multimeter represents the commonest and most widely useful of all electronics test and measurement equipment. Multimeters come in a variety of types, ranging from cheap and basic to extremely expensive and highly functional. Figure 8.4 illustrates two hand-held DMMs (digital multimeters), while a benchtop unit is shown in Figure 8.5, and finally an analog multimeter (aka VOM, volt-ohm-milliammeter, and also sometimes called an Avometer®) can be seen in Figure 8.6. No matter the shape and size, and the variety of advanced functions notwithstanding, the basic functionality provided is more or less universal – the measurement of voltage, current, and resistance.

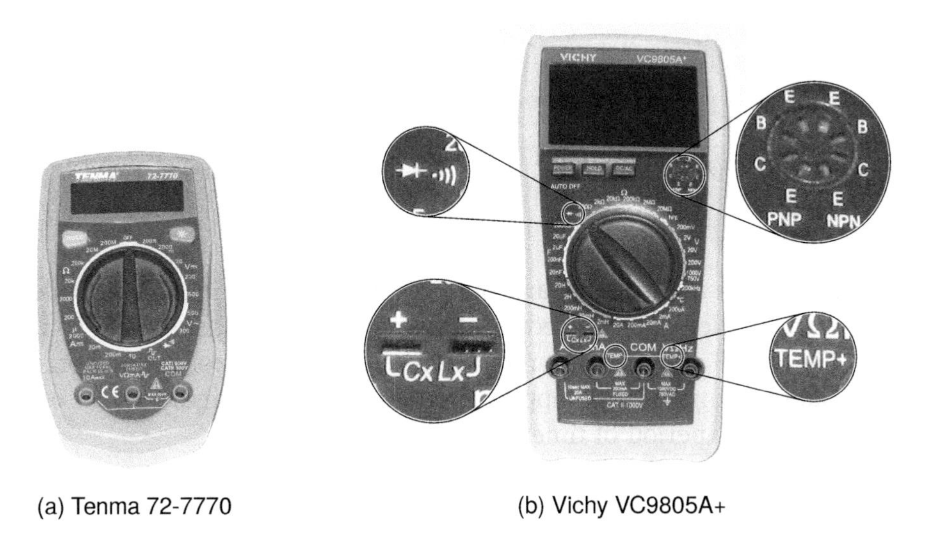

(a) Tenma 72-7770 (b) Vichy VC9805A+

Figure 8.4 Two examples of hand-held DMMs.

Inputs

Multimeters almost always use banana plugs for connecting test leads (see inset in Figure 8.6). There are typically either three or four banana sockets, two of which will be used for any given measurement. Usually one socket, labelled 'COM' – for common – is used in (almost) all measurements. Most measurements apart from current use one other, with the final one or two typically reserved for current measurements. Generally the labelling on the sockets makes clear which are to be used for any given measurement. A rare case where the common terminal is not used can be seen in the VC9805A+ unit in Figure 8.4b. In this unit temperature can be measured by attaching a special probe between the two terminals labelled 'TEMP−' and 'TEMP+'.

The VC9805A+ provides two other inputs worth mentioning. At the top right is a BJT tester (inset to the right), the three legs of a bipolar junction transistor can be inserted into the appropriate slots and a reading of the device's gain (or h_{fe} value) can be obtained. See Chapter 17 for more details on this important transistor parameter.

To the bottom left of the VC9805A+ front panel are two additional slots with '+' and '−' above them and '$C_X L_X$' below them (inset to the bottom left). These inputs allow for the measurement of capacitors and inductors. By selecting the appropriate range on the dial, capacitances in farads (up to 2000μF) or inductance in henries (up to 20H) can be measured.

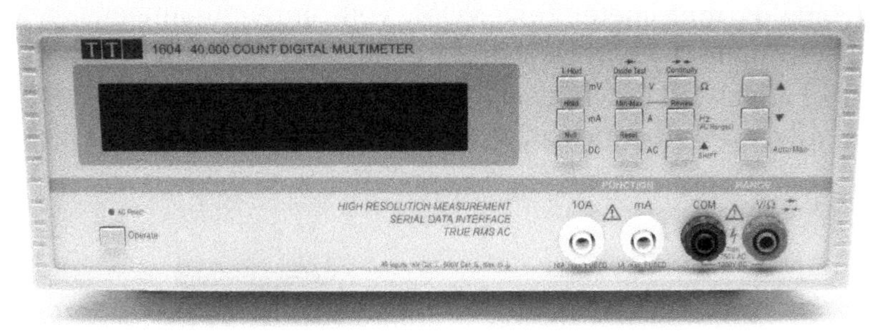

Figure 8.5 TTi 1604 benchtop DMM.

Outputs

Hand-held units rarely if ever have any outputs. Benchtop units may provide a serial or USB connection which can be used for data logging as well as remote control. The TTi 1604 in Figure 8.5 provides a serial interface on its rear panel and comes with a rudimentary data logging software application which allows a large number of readings to be taken over a long period of time.

Controls

On a hand-held device the main control is almost always a large rotary switch used to select between the often large number of measurement options available (the VC9805A+

features no fewer than 30 switch positions). The labelling usually makes it quite clear what can be measured in each position, although some ranges require a little more information to fully understand. As always, the user manual is a valuable resource.

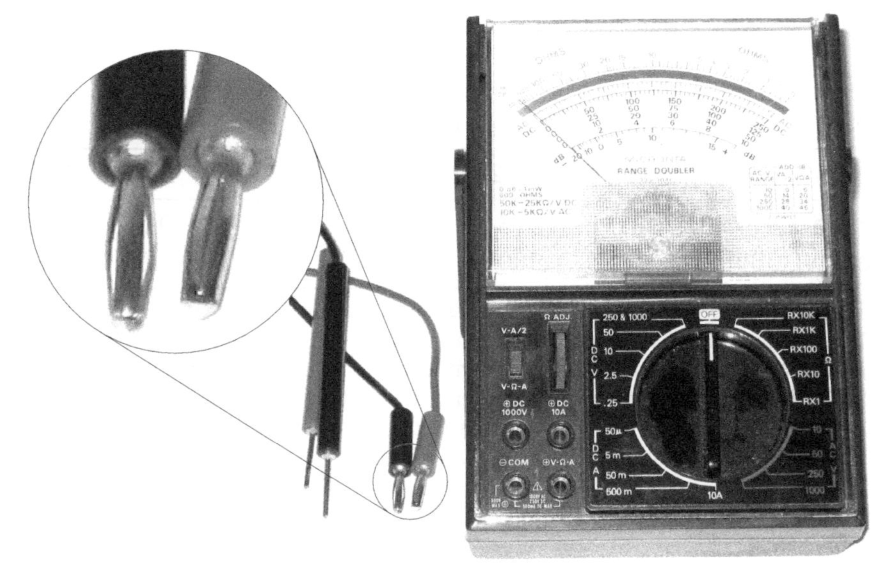

Figure 8.6 Micronta 22-204C analog multimeter (VOM), with banana plug connectors.

The hand-held units shown both feature fully manual range selection. Many multimeters (such as the 1604 benchtop unit from Figure 8.5) implement an auto-ranging facility where only the basic measurement type needs to be selected, and then the specific range to be used is determined by the device. This makes taking measurements an easier process but does inevitably take longer to get a result as the unit must assess the signal to be measured and determine the range to use before making and reporting the final measurement.

Whether a rotary switch or a keypad interface is being used, the selected range value (for a manual ranging device) usually indicates the maximum value which can be successfully measured in that range. What happens if that limit is exceeded can vary from range to range, and from meter to meter. Sometimes the display will flash, or display something other than a valid reading. A common out of range display for resistance measurements is a number '1' displayed on the leftmost position of the display. Other meters might display 'OFL', meaning overflow, or some other similar indication. When measuring voltage and current in particular, it is best to avoid selecting any range lower than that required, as this can potentially overload the meter and, at best, blow a fuse. For this reason it is best to start the measurement of unknown voltages and currents at the highest range available, and work down until an optimal reading is achieved.

Generally if a range higher than the optimal is selected, a valid result will be returned; it will just not be as precise as it could be. For instance, Table 8.1 illustrates the readings obtained testing the same resistor across every resistance range available on two DMMs. The results make it clear that the resistor in question is a nominal 1kΩ component. It is in fact a 1k carbon film device with a tolerance of 5%, and as such appears to be within spec. See Chapter 12 – Resistors, to expand on tolerances and resistor specifications.

Table 8.1 Resistance measurement results for two hand-held DMMs. The component under test is a 1kΩ, 5% carbon film resistor

	200	2000/2k[a]	20k	200k	2M	20M	200M
Tenma 72-7770	1 .	976	0.98	01.0	–	0.00	01.0[b]
Vichy VC9805A+	1 .	.980	0.98	00.9	.001	0.00	–

a. The two meters feature similar sets of resistance ranges, but for two points of note: 1) the Tenma uses 2000 where the Vichy uses 2k. The resolution is the same but notice that the Vichy indicates 0.98kΩ while the Tenma reports 976Ω.
2) the Tenma omits a 2M range but extends up to 200M, where the Vichy does not.

b. Testing indicates that the Tenma appears to have a bug in its 200M range which results in readings consistently 1M high.

Displays

A multimeter will either have a large moving needle display (for a VOM) or a numerical readout (for a DMM). These digital displays can incorporate all sorts of useful icons and glyphs in addition to the primary numerical readout. They also usually incorporate a minus sign. Most DMMs are designed to handle DC voltage and current measurements in either polarity. Analog meters on the other hand (where the needle's resting position is almost always at one extreme of the meters movement) should only be connected in their correct polarity, to avoid damage to the meter.

Analog meters are widely considered obsolete, as digital devices have taken over. Digital meters offer many advantages over their analog forebears, in terms of resolution, accuracy, repeatability, ease of use, flexibility, and robustness. There are still a couple of places where an analog movement may still have the edge over a digital readout. While single readings favour the digital format, slow moving signals and trends may be easier to detect and follow using a moving needle – a kind of halfway house on the road to oscilloscope functionality. A needle movement can also make it easier to spot and assess small glitches – for example in determining the polarity of an electric guitar pickup by observing the direction of needle movement when the pickup is touched with a metal rod (e.g. a screwdriver).

Common Measurement Tasks

Most measurements are performed by selecting the appropriate range on the multimeter, connecting the two test leads to the appropriate points, and reading the result off the multimeter display.

Resistance

Resistance measurements must not be performed with power applied to the device or component being tested. Most commonly a single resistor in isolation is the target of such a measurement, although it can also be useful to measure the resistance between two points in a circuit. When making in-circuit measurements it is important to assess the entire circuit region surrounding the test points. While the test probes may be connected on either side of a single resistor in a circuit, it should not be assumed that such a measurement will automatically return the value of this resistor. The circuit should be examined to see if other parallel resistance paths might exist, which will potentially affect the result. As indicated in Figure 8.7a, the best way to measure a resistor is to do so with it removed from any circuit it may be a part of. This is not always possible, but is the ideal.

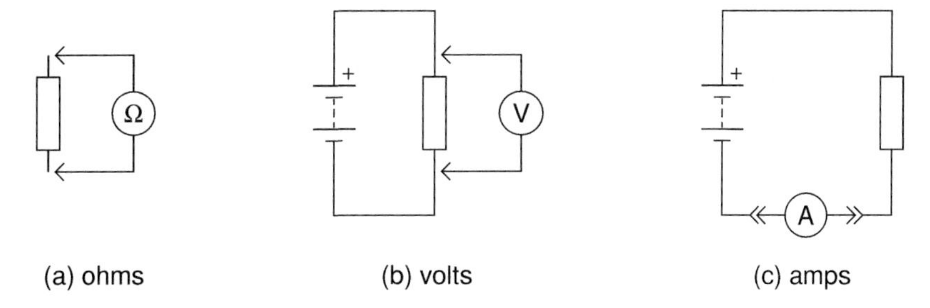

Figure 8.7 Connecting a multimeter to measure resistance, voltage, and current.

When using an auto-ranging multimeter it is sufficient to simply select the ohms range and let the meter do the rest. For a manual meter it will usually be necessary to switch through several range settings in order to get the best possible reading, as illustrated by the results presented in Table 8.1 above. Clearly in this case the 2k range gives the best result, maximising resolution and hence precision. Switching down a range takes the meters beyond their maximum readings, while going up a range, while producing a valid result, does loose a decimal point of precision.

Voltage

Voltage measurement, Figure 8.7b, involves selecting the appropriate range, and bridging across two points in a circuit with the probes of the multimeter. With voltage and current measurements it is very important to know whether it is a DC or an AC signal which is to be measured, and to set the meter correspondingly. While most meters provide for both AC and DC voltage measurements, AC current measurement is less commonly available (the Vichy in Figure 8.4 provides it but the Tenma does not).

Current

Current measurement is the most involved of the three primary measurement types. It is necessary to make a break in the circuit path to be analysed, and to insert the meter into the circuit, as shown in Figure 8.7c. The most common place to do this is in either of the two connections between circuit and power supply. This test can be used to determine the overall current draw of a circuit, which can be useful information, both in checking that there are no faults or errors in the circuit causing an unexpected current drain, and in assessing the powering requirements for a circuit – is it, for instance, suitable for battery powering or would it drain batteries too quickly?

Continuity

While volts, ohms, and amps measurements are the core functionality of any multimeter, one of the most useful functions provided by most meters is the continuity or buzz tester (Figure 8.8). In electrical terms, continuity means a low resistance electrical connection. The action of a buzz tester is very straightforward; if a sufficiently low resistance is detected between the measurement probes then a buzzer within the meter sounds. Typically the threshold for a continuity tester is somewhere in or around the 10Ω to 30Ω region. At resistances below this limit the buzzer activates; thus a buzz should always be heard when the test probes are touched together while the multimeter is in this mode.

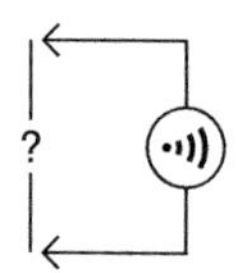

Figure 8.8 Multimeter continuity test mode. The meter beeps if a sufficiently low resistance is connected.

Continuity testers are particularly useful for testing cables and circuit boards. Any time solder has been applied, it can be a worthwhile exercise to 'buzz out' the work before deeming it complete. It is important to check both that all the required connections are present and solid, and that there are no unwanted connections due to solder going somewhere that it shouldn't have. The latter is particularly common when soldering tightly laid out stripboard circuits, as is discussed in Chapter 10.

Diode

The diode test range on many hand-held multimeters shares a switch position with the buzz tester, and is typically indicated by a combined symbol as in the upper left inset in Figure 8.4. Diodes are one of those components where the orientation of the device is significant, so it is important to observe the connection order when interpreting the results of a diode test, see Figure 8.9. For a diode in good working order, the reading indicates the forward voltage required to get a small amount of current (typically about 1mA) to flow through the device. This voltage varies for different types of diodes. See Chapter 16 – Diodes, for more details on diode behaviour and typical characteristics.

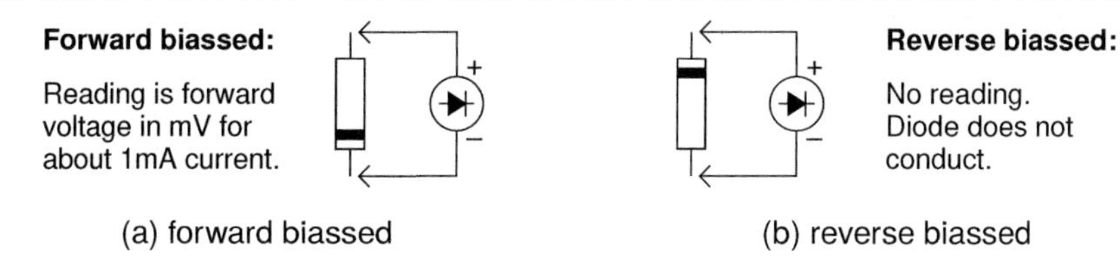

Figure 8.9 Multimeter diode test mode. In forward bias, the meter reports the voltage required to get a small current to flow.

Oscilloscopes

Oscilloscopes are one of the most variable pieces of standard lab equipment, both in terms of the functionality they provide, and in terms of the interface used to access and control that functionality. In this section, all the bells and whistles which any given device may provide are ignored. The fundamental functionality and how it is controlled stays fairly consistent across the gamut of oscilloscopes likely to be encountered. This involves displaying a steady picture with the optimum horizontal and vertical on screen sizing. While the multimeter presents a single reading, the oscilloscope allows a changing signal to be displayed graphically over a period of time.

Oscilloscopes come in a variety of form factors. Old analog scopes are large and heavy, with a very deep body in order to accommodate the long cathode ray tube used for the display screen. An analog scope is often referred to as a CRO or cathode ray oscilloscope. More modern digital scopes are usually called DSOs or digital storage oscilloscopes, because they can store and redisplay a signal plot after the signal is gone. USB scopes provide signal capture and A-to-D conversion hardware, but no screen. They rely on an attached computer for control and display purposes. Soft scopes on the other hand are software applications without any dedicated hardware. They take their input signals from standard audio interface hardware, and so are only usually of any use up to audio frequencies (below about 20kHz or so). Even a basic hardware scope will usually extend up to at least a couple of hundred kilohertz, and more likely to a couple of megahertz or more. Even when working exclusively with audio circuits this extended oscilloscope bandwidth can be useful. High frequency noise can be a problem which needs to be visualised in order to be addressed effectively.

Figure 8.10 illustrates the major controls found on most analog oscilloscopes. Most of this functionality is shared across digital scopes too, with just one or two exceptions. Both analog and digital types can exhibit a wide array of additional functionality, but the basic controls covered here allow the user to make all of the basic measurements typically required. There are five groups of controls shown in the figure: sweep, mode, trigger, channel 1, and channel 2. Most oscilloscopes have two channels, allowing two signals to be viewed simultaneously. This common configuration is assumed in the example provided here.

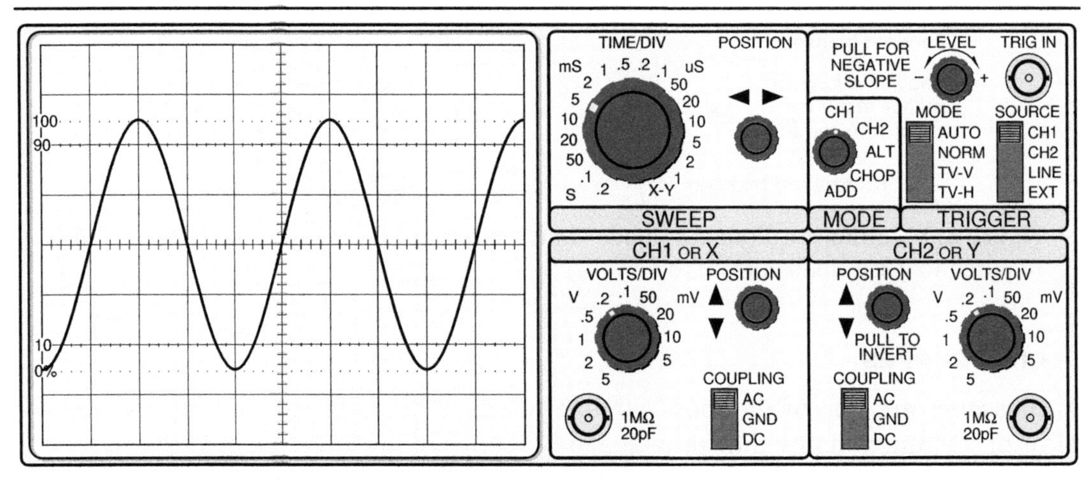

Figure 8.10 Typical oscilloscope front panel controls. Oscilloscope interfaces and functionality vary greatly, but most of these controls are always present in some form.

Mode

The Mode section just has a single five-way rotary switch, used to set the device's operating mode, selecting which channels to display and how to display them.

CH1: display channel one only
CH2: display channel two only
ALT: display both channels (default, best at higher frequencies)
CHOP: display both channels (prevents flicker at lower frequencies)
ADD: display the sum of the signals on the two channels

Trigger

The trigger section controls how and when a trace starts drawing on the screen. Getting the trigger setup wrong results in the trace continually running across or jumping around the screen (see Figure 8.11d). A well configured trigger gives a good solid trace without any movement. The trigger settings generally apply to both channels. Some more advanced oscilloscopes allow for offset or independent triggering of channels.

SOURCE: selects what signal is used in order to generate the trigger which initiates the trace. CH1 is the right choice in most situations.

MODE: controls the rule used to initiate triggering. AUTO is usually the right choice.

LEVEL: chooses the vertical position on the screen at which triggering will happen. The middle of the pot rotation is the best place to start. Adjust up and down if needed, until a steady trace is obtained.

SLOPE: can be negative or positive, controlled here by pulling out or pushing in the level knob. Usually best left positive (rising) unless it is specifically necessary to zoom in on the negative slope (falling edge) of the signal.

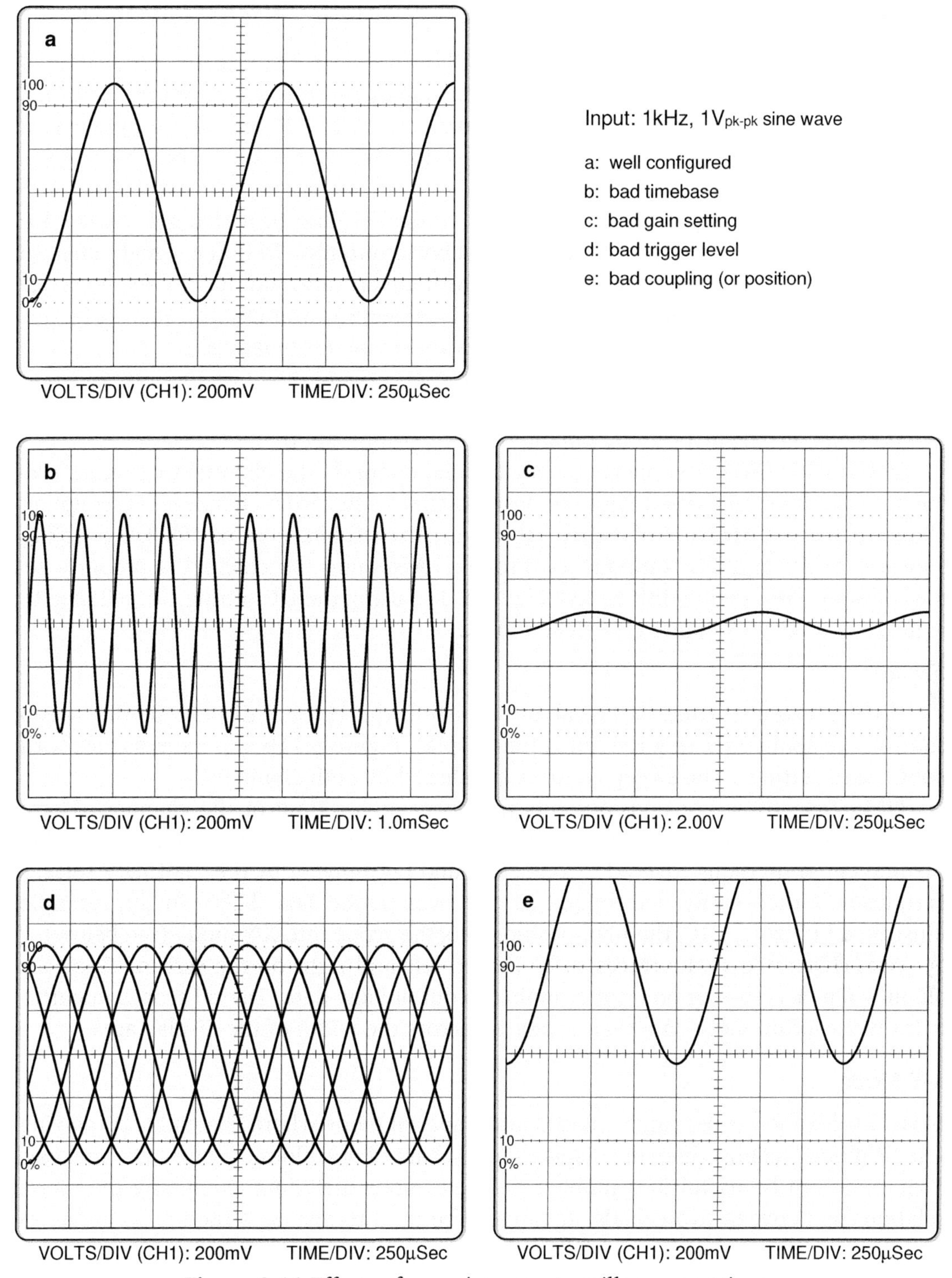

Figure 8.11 Effects of some important oscilloscope settings.

Channels 1 and 2

Each set of channel controls is usually more or less identical. In this case only a channel two invert option differentiates the pair. The channel controls are primarily about adjusting the vertical aspects of the display (Figure 8.11c). Three controls are shown here (plus the channel two invert option), as well as the BNC connectors where the probe is attached for each oscilloscope channel.

VOLTS/DIV: (volts per division) sets the vertical scaling of a channel. There are eight divisions marked out on the screen from bottom to top. To find a signal's peak to peak amplitude, count the number of divisions it covers, and multiply by the volts per division setting. If a times ten probe is used (see below), multiply the answer by ten. Digital scopes can usually be told to make this correction automatically.

POSITION: allows the whole trace to be moved up and down on the screen. Good for spacing the two channels to make the most of the available screen space. Channel two here can also be inverted by pulling out the position knob.

COUPLING: GND disconnects the signal, and connects the channel to ground, producing a flat line trace (used with the position control to zero a channel). AC coupling allows only the AC portion of a signal through. Any DC bias is removed. DC coupling allows both AC and DC components through. Assuming the channel is properly zeroed to start with, then a display like Figure 8.11e using the DC setting, will change to Figure 8.11a when coupling is switched to AC.

Sweep

Just as the 'channel' controls are about adjusting the vertical display, so the sweep controls are about adjusting the horizontal display. Figure 8.11b illustrates a trace with poor sweep settings. The sweep controls are shared by both channels.

TIME/DIV: (time per division) sets the horizontal scaling of the channel. There are ten divisions across the screen. The period of a signal is determined by counting the number of divisions covered by one cycle and multiplying by the TIME/DIV setting. Frequency is then easily determined as one over period (Eq. 5.2). In the example illustrated in Figure 8.10, the final position sets the scope into X-Y mode (see below).

POSITION: allows the traces to be moved left and right on the screen. Good for aligning a plot's crossing point or scanning to out of view data. Digital scopes in particular often capture more data than is plotted across the width of the display area.

X-Y Mode

Most two channel scopes can be placed into a special mode where, instead of displaying signals plotted against time, the display shows channel one plotted against channel two. This mode can be useful in a number of applications including examining the phase relationship between two signals. It can also be used to display a diode's i-v curve, as described in Learning by Doing 16.2 (p. 281).

Oscilloscope Probes

The standard oscilloscope probe consists of a spring loaded hook terminal, which can often be removed to reveal a pointed contact pin beneath (Figure 8.12). Coming off to the side is a clip lead. The main contact is connected to the point in a circuit which is to be examined while the clip lead must be connected to a voltage reference point, usually the circuit's signal ground.

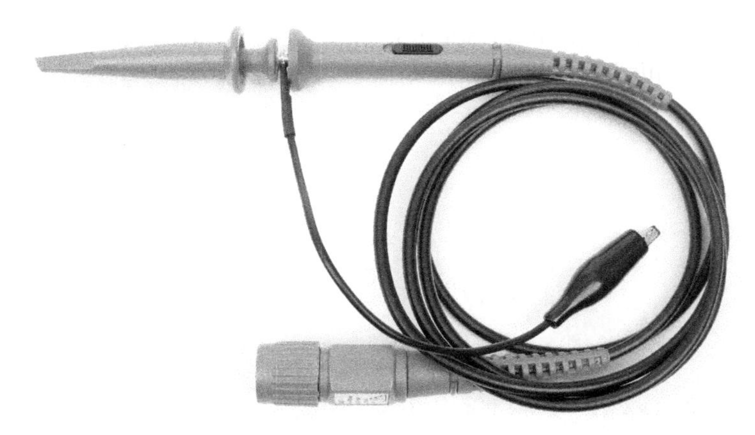

Figure 8.12 A standard ×10 oscilloscope probe. The black switch on the handle is the times one/times ten selector.

It is important to note that in a standard mains powered oscilloscope the ground clip lead is connected directly to earth ground through the mains power cord ground wire. This means that if a circuit also connected to earth ground is being examined the ground clip lead must only be connected through to this earth ground point in the circuit under test. There are various ways of overcoming this limitation (see Tektronix, 2011), the simplest of which is to use a battery powered oscilloscope. These issues are particularly mentioned in the same Learning by Doing 16.2 exercise mentioned above, which describes how to use an oscilloscope to visualise a diode i-v curve.

The ×10 probe (times ten probe) is the most common type of probe encountered in conjunction with a standard oscilloscope. A switch on the handle labelled '×1 ×10' switches in and out an attenuation network which in the ×10 position reduces the input signal by a factor of ten. The ×10 position also enables an adjustable compensation circuit which allows for the best possible accuracy of signal reproduction. Generally it is best to use this mode where possible, but it is important to remember that the signal amplitude readings taken must be re-scaled times ten (hence the name) to arrive at the true reading. In other words, a reading of 0.5V is actually 5.0V. As mentioned above, while this must be done manually when using an analog scope, a digital scope can usually be set up to perform the calculation automatically.

Audio Analysers and Advanced Test Gear

The systems from Audio Precision probably represent the de facto standard for audio test and measurement equipment. Prices for even the most basic systems start at several thousand euros. Many other manufacturers also offer audio analysers of various descriptions. The Keithley 2015-P Audio Analysing DMM shown in Figure 8.13 is one such device, providing measurement functions including harmonic distortion and spectrum analysis. Equipment such as this generally lies beyond the grasp of all but the specialist audio electronics practitioner, and details of its uses and utilisation fall well outside the scope of this book.

Figure 8.13 Keithley 2015-P Audio Analysing DMM.

References

G. Ballou, editor. *A Sound Engineers Guide to Audio Test and Measurement*. Focal Press, 2009.

E. Brixen. *Audio Metering: Measurements, Standards and Practice*. Focal Press, 2nd edition, 2011.

N. Crowhurst. *Audio Measurements*. Gernsback Library Inc., 1958.

J. Dunn. *Measurement Techniques for Digital Audio*. Audio Precision, 2004.

B. Metzler. *Audio Measurement Handbook*. Audio Precision, 2nd edition, 2005.

Tektronix. *Fundamentals of Floating Measurements and Isolated Input Oscilloscopes*. Tektronix, 2011.

9 | Basic Circuit Analysis

Before moving on to examine electronic components and how to use them it is worth taking some time to bring together key details of the most important primary electrical quantities which have been encountered so far, and to introduce four very simple but very important laws which facilitate a great deal of circuit analysis, providing the tools for many useful calculations to be performed on the circuits and systems which are encountered throughout the rest of this book. A good understanding of these quantities and these laws will help to develop an appreciation of what can be expected to happen in any given circuit and why things work the way they do in electronic circuits and systems.

Many other rules and equations, specific to particular components or particular circuit configurations are introduced at appropriate points throughout the rest of the text. These are gathered in Appendix B – Quantities and Equations. Together they provide the tools to analyse and design many useful circuits, and form the basis for a deeper understanding of the topics presented. As stated from the outset, this book aims first and foremost to provide an accessible introduction to audio electronics, avoiding too much in depth mathematical analysis. That being said, a firm grasp of these relationships can only be a good thing in assisting the reader to get the most out of their endeavours in practical audio electronics. While much of the maths can be skimmed on a first reading, returning to it and developing that deeper understanding is to be highly recommended.

Primary Quantities

Table 9.1 provides a reminder of the four most important fundamental quantities which have already been encountered: ohms, volts, amps, and watts. A firm grasp of what these concepts relate to and how they interact with each other makes all subsequent circuit analysis much easier to grasp.

Impedance is a vector quantity made up of two components, resistance and reactance. In this context a vector can be thought of as simply a quantity that has both magnitude and direction. For the purposes of this book only the magnitude (i.e. size, measured in ohms) is needed. The direction, or phase (measured in degrees) becomes important when more advanced and detailed analyses are being performed. Those

Table 9.1 The four fundamental electrical quantities

Quantity		Units		Notes
Impedance	(Z)	Ohm	(Ω)	also Resistance (R) and Reactance (X)
Voltage	(V)	Volt	(V)	aka EMF, PD
Current	(I)	Amp	(A)	aka Ampere
Power	(P)	Watt	(W)	also Apparent Power (VA)

familiar with audio theory will recognise phase as an important feature of how sounds interact with one another. No deeper understanding of the concept is needed here.

Impedance is a property which electronic components exhibit, representing their opposition to the flow of electricity. For some devices (e.g. resistors) it is a simple, constant value, while for others its magnitude and its behaviour depend on the circuit conditions at any given time. This more dynamic aspect of impedance is revisited when capacitors are introduced in Chapter 13.

Voltage is the electrical pressure or push which attempts to make electricity flow. A number of other terms can be used to refer to the same phenomenon. Electromotive force (EMF), potential difference (PD), and electrical tension all mean basically the same thing as voltage, although each is typically encountered in its own specific context.

Current is measured in amperes (often shortened to amps). The terms amperage or ampage are sometimes encountered, usually referring to the amount of current a particular device can be expected to draw under normal conditions or the amount of current which a device can handle before failing (e.g. What is the ampage of the fuse in that plug?).

Power in watts is used to measure the rate of energy usage in a circuit. The term wattage is sometimes used to refer to the amount of power a device can be expected to use (e.g. What's the wattage of that light bulb?). Apparent power, which has units of volt-amps (VA) rather than watts, includes both the power which is used up in a circuit and also the power which goes into but later comes back out of a circuit. This second element is called reactive power. Components like capacitors and inductors can store and later release energy, and as such can introduce an element of reactive power to the power usage of devices which involve them. Apparent power is worth being aware of but is of no real consequence in the context of this book.

Ohm's Law

Ohm's law states a very simple relationship between voltage, current, and resistance. It can be used to calculate the behaviour of very many circuits and is probably the single most useful calculation to know when it comes to working out what is going on in pretty much any electrical system. Some care needs to be taken when applying it to circuits which involve components other than simple resistances. This is addressed in more detail as these components are encountered along the way. The basic relationship represented by Ohm's law, as shown in Eq. 9.1, forms the bedrock of a solid understanding of electricity and electronics.

$$V = I \times R \tag{9.1}$$

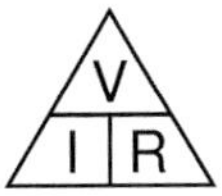

The triangle figure accompanying the equation is a useful way to remember how the law is used in different situations. As with any mathematical equation of this very simple form ($A = B \times C$) it can be used in three different variations depending on which two things are known and which one is required to be calculated. If current and resistance are known and voltage is to be calculated it is a simple matter of using the equation as presented. If however the quantities which are known are voltage and resistance, and it is the current which is to be calculated then multiplying the two known values does not result in the correct answer. In fact one needs to be divided by the other. In order to see which should be divided by which, the triangle can be used. In this case the required new equation can be read off from the triangle figure as $I = V/R$. Similarly if R is the unknown then $R = V/I$ can be arrived at from inspection of the triangle figure with equal ease. Thus the triangle figure allows the user to extract any of the three variations on Ohm's law with ease and in confidence that the correct relationship has been used.

Watt's Law

Watt's law, as shown in Eq. 9.2 provides for the calculation of power. Often different components can cope with different amounts of power, and so it can be important to calculate the power which will be dissipated in a circuit or in a particular component. In this way it is possible to ensure that components will not overheat and be damaged or destroyed if used in a particular circuit.

$$P = I \times V \left(= I^2 \times R = \frac{V^2}{R} \right) \tag{9.2}$$

The description presented above for the use of the triangle figure in Ohm's law holds equally well for Watt's law, which follows the same basic $A = B \times C$ pattern. It should therefore be obvious that the two variations not explicitly shown in Eq. 9.2 above are $I = P/V$ and $V = P/I$. Again, given any two of the three quantities the third can always be calculated.

Kirchhoff's Current Law (KCL)

Kirchhoff's current law, often abbreviated to KCL, states that the sum of the currents into any point in a circuit must equal the sum of the currents out of the point. What

9.1 – The Variants of Watt's Law

Eq. 9.2 shows in brackets two alternative variants of Watt's law. It is a simple matter to derive these alternate forms of the law by using the primary form in combination with the previously presented equation for Ohm's law (Eq. 9.1).

Given that Ohm's law states that $V = I \times R$, it is clear that the V term in Watt's law can be replaced with $I \times R$ in order to give the first of the two alternate versions.

$$\begin{aligned} P &= I \times V \\ &= I \times (I \times R) \\ &= I^2 \times R \end{aligned}$$

In order to arrive at the second variant, Ohm's law must be rearranged to give an expression for the current, $I = V/R$. Substituting this back into Watt's law generates the second alternate version.

$$\begin{aligned} P &= I \times V \\ &= \frac{V}{R} \times V \\ &= \frac{V^2}{R} \end{aligned}$$

this most commonly means in practice is that if a branch point is encountered in a circuit, where three or more components meet (a very common configuration), and the currents flowing in all but one of the branches are known or have been calculated then the current in the final branch is known by extension.

$$I_{in} = I_{out} \tag{9.3}$$

If the common image of water flowing in pipes is used to illustrate electrical current then Kirchhoff's current law becomes quite apparent. Whatever water flows into a junction in a network of pipes has to flow back out, there is nowhere else for it to go, and likewise no more water than that which has flowed in can flow out.

Kirchhoff's Voltage Law (KVL)

Kirchhoff's voltage law (KVL) states that the sum of all the voltage drops around any loop or closed path in an electric circuit is zero. In other words starting at any point in a circuit, moving from point to point along any path, and finishing up back at that same initial starting point, measuring the voltage changes as each new point is visited and adding them all up results in a final answer of zero volts. Imagine starting at any point on the side of a mountain and moving up and down to any number of other points, higher and lower on the side of that mountain, and finally arriving back at the starting point. The starting altitude is equivalent to the initial voltage. Each subsequent altitude corresponds to a new voltage. In order to arrive back at the same point, any movement up the hill must eventually be balanced out by an equal amount of movement down the hill, and vice versa, and so when arriving back at the starting point the sum of all the altitude changes equals zero.

$$V_o = 0 \tag{9.4}$$

When using Kirchhoff's voltage law the path taken around the circuit may or may not take in any power sources such as batteries which are present in the circuit, the law works for any path. Usually the loop of interest in any analysis is quite compact, but in theory the path considered can be as long and convoluted as required. The direction of travel around the loop is also irrelevant, all positive voltage steps simply become negative and likewise all negative voltage steps become positive, thus resulting in the same zero sum around the loop.

Kirchhoff's two laws are explicit statements of two underlying principles of electricity. More often than not they will be applied intuitively during a circuit analysis without ever making direct reference to the formal statement of the basic principles. Unlike Ohm's and Watt's laws, they tend not to be used for performing specific calculations, but rather help to guide the overall understanding of what particular circuit conditions will arise in any given situation. It is somewhat more common for them to be referenced (usually using the shorthand KCL and KVL) in a written description accompanying or replacing a more mathematical treatment of a problem.

Applying the Four Laws

The initial presentation of the four laws given above was kept deliberately brief, without examples or significant explanation. A few simple applications and instructive examples are presented here in order to clarify the use of the laws and to explore their implications.

Example One – Battery and Bulb

Consider the circuit shown in Figure 9.1. The battery is labelled as providing 1.5V. On its own this does not provide enough information to work anything else out. However if it is also given for example that the bulb has a working resistance of 15Ω it is now possible to calculate a) the amount of current which will flow in the circuit and b) the amount of power which it will dissipate.

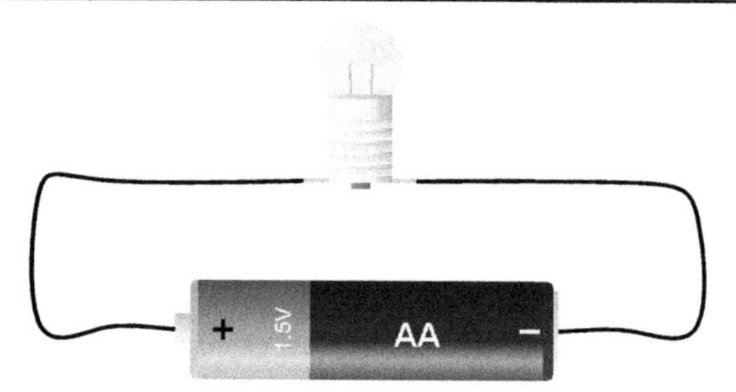

Figure 9.1 Battery and bulb diagram using nonstandard informal pictographic notation.

The first question to ask when addressing any problem of this kind is 'What is known and what is needed?'. In this case voltage and resistance are known, and current and power are needed. This leads to the question 'Is there a rule or law which involves only the known quantities and one of the unknowns?' In this case there are two possible routes which could be followed. Ohm's law ($V = I \times R$) fits, with current as the one unknown, and so does the second variant of Watt's law ($P = V^2/R$), with power as the one unknown. It makes sense to answer the first part of the question first.

In order to calculate the current, Ohm's law can be used. As stated above, the voltage provided by the battery, and the resistance of the bulb are known. This is a very simple circuit, with the bulb connected directly across the ends of the battery and no other components involved. As such Kirchhoff's voltage law can be applied to see that the full voltage of the battery must appear across the terminals of the bulb. (See the next example for a case where the available voltage gets broken up between different components.)

To see how KVL is applied here consider starting at the minus terminal of the battery and moving around the circuit in a clockwise direction. There is only one possible loop in this very simple circuit. First the battery itself is crossed. This results in a positive voltage step of 1.5V, the difference between the negative and positive terminals of the battery. Continuing around the circuit the next thing which is encountered is the left hand terminal of the bulb. Crossing over the bulb and continuing on leads straight back to the starting point with no further circuit elements encountered.

KVL tells that the sum of the voltage drops around this circuit must be zero. The voltage drops encountered around the loop were: +1.5V crossing the battery and an unknown voltage across the bulb. For a zero sum this unknown voltage drop must therefore be −1.5V.

Note that traversing the loop in the other direction will lead to the same results with the signs reversed. This just indicates the fact that the left hand terminal of the bulb is at a potential 1.5V above the right hand terminal, and likewise the right hand terminal is at a potential 1.5V below the left (and likewise for the battery). This is a trivially simple application of KVL but it does provide a straightforward example of how it works.

Using Ohm's law –

$$
\begin{aligned}
& V = I \times R \\
\Rightarrow \quad & I = \frac{V}{R} \\
& = \frac{1.5}{15} \\
& = 0.1\text{A} \\
& = 100\text{mA}
\end{aligned}
$$

Following this calculation the new situation can be summarised as, voltage, current, and resistance are known, and power is needed. The only law involving power is Watt's law, so this must be the one to use here. Since so many things are now known any of the three variants could be used (as was noted previously, variant two could have been used from the outset). Since it is available, it makes sense to use the primary form.

Using Watt's law –

$$
\begin{aligned}
P &= I \times V \\
&= 0.1 \times 1.5 \\
&= 0.15\text{W} \\
&= 150\text{mW}
\end{aligned}
$$

It is always worth doing a quick check where possible to make sure everything has been worked out correctly and no errors have crept into the calculation. In this case an obvious check would be to also calculate the power using Watt's law variant two as previously mentioned. A calculator could be used but, observing that $1.5 = 0.1 \times 15$, the calculation is easy enough by hand:

$$
P = \frac{V^2}{R} = \frac{1.5^2}{15} = \frac{1.5 \times 1.5}{15} = \frac{1.5 \times 0.1 \times 15}{15} = 1.5 \times 0.1 = 0.15\text{W}
$$

150mW as was calculated above.

The pictographic representation of a circuit as used in Figure 9.1 is not a standard nor a very useful method for drawing circuits. More usually circuit diagrams are constructed using a set of standardised circuit symbols to represent the components involved. A detailed description of this scheme and its interpretation is given in Chapter 7, and a comprehensive list of commonly encountered symbols is provided in Appendix C. The simple, symbol based schematic diagrams in the rest of this chapter should be easy enough to interpret without further explanation. First off a standardised version of the circuit examined above is presented in Figure 9.2. This represents exactly the same circuit as was shown in Figure 9.1 above (with the additional information on the bulb's working resistance added).

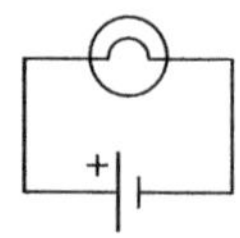

Figure 9.2 Battery and bulb schematic diagram.

Example Two – Resistors in Series

Figure 9.3 shows two resistors in series. The value of the first resistor is given, as is the current flowing through it, and the voltages at either end of the network are also provided. a) Calculate the value of the second resistor. b) Calculate the voltage at the point between the two resistors.

Note that in this case no complete circuit is actually shown. It is assumed that the circuit continues on from where the two voltages are specified eventually linking one side round to the other. This unknown section may be large or small, complex or simple, it makes no difference to the analysis of the shown section under investigation here.

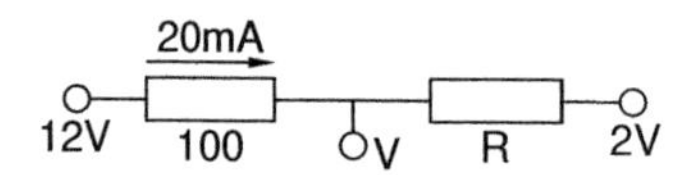

Figure 9.3 Resistors in series.

Consider a point in the circuit between the two resistors. By KCL current in equals current out. Therefore the current flowing through resistor R must be 20mA, since this is what is entering the point coming from the 100Ω resistor and there is only one route out, through resistor R. The voltage from end to end is $12 - 2 = 10\text{V}$, and the current from end to end is 0.02A, and so Ohm's law can be used to determine the resistance from end to end. Since resistors in series simply add their values (see Chapter 12), it is clear that the value of R can then be determined.

Using Ohm's law and the resistor series rule –

$$
\begin{aligned}
V &= I \times R && \text{– Ohm's law} \\
\Rightarrow R_{total} &= \frac{V}{I} \\
&= \frac{10}{0.02} \\
&= 500\Omega \\
R_{total} &= R_1 + R_2 && \text{– resistor series rule} \\
\Rightarrow \quad R_1 &= R_{total} - R_2 \\
\Rightarrow \quad R &= 500 - 100 \\
&= 400\Omega
\end{aligned}
$$

So the resistance of resistor R is 400Ω. Answering the second half of the question, what is the voltage between the two resistors, is now a simple matter based on all the information to hand. Applying Ohm's law once again gets most of the way there, but there is a trap to be avoided. Ohm's law will yield the voltage drop across either resistor, but this is not the same thing as the voltage at the midpoint.

When doing calculations always take great care in applying any multipliers which are present. In this case remember that milli- means divided by a thousand, and so in the above example 20mA becomes 0.02A. See Appendix A for details on all multipliers likely to be encountered.

Applying Ohm's law to the first resistor –

$$\begin{aligned} V &= I \times R \\ &= 0.02 \times 100 \\ &= 2V \end{aligned}$$

However this 2V is not the voltage at the midpoint, it is the voltage difference between the two ends of the resistor. The other end of the resistor is at a voltage of 12V, and so the voltage at the midpoint is $12 - 2 = 10V$.

To double check, the same calculation performed for the other resistor should lead to the same centre voltage since one point in a circuit can not be at two different voltages simultaneously.

Applying Ohm's law to the second resistor –

$$\begin{aligned} V &= I \times R \\ &= 0.02 \times 400 \\ &= 8V \end{aligned}$$

This means that there are eight volts across the second resistor. The far end of the second resistor is at 2V. Eight plus two equals ten, so again the midpoint voltage comes out as 10V. Note that in the first case the voltage drop (2V) was subtracted from 12V while in the second case the voltage drop (8V) was added to 2V.

The voltage at the midpoint must be between twelve and two, and so in order to find it the appropriate calculations are either to subtract from twelve (coming down from the upper limit) or to add to two (going up from the lower limit).

As further verification of the answer, eight volts across one resistor and two volts across the other yields ten volts from end to end. There is a potential of twelve volts at one end and two volts at the other, so the difference between the two ends is indeed ten volts. It looks like the calculation is correct.

Learning by Doing 9.1

Ohm's Law and Watt's Law

Materials

- Resistors (1k, 4k7, 22k)
- 9V PSU or battery and battery clip
- Clip leads
- Multimeter

Background

The idea of this exercise is to use the laws just introduced in order to calculate the expected behaviour of a very simple circuit involving no more than a single resistor connected across a PSU. Measurements are then made using a multimeter in order to confirm the expected behaviour and to assess any deviations observed.

Three resistor values (1k, 4k7, 22k), and a power supply level (9V) are suggested. Different values can be used for both but it is always a good idea to confirm that working limits will not be exceeded before actually connecting up these circuits. The basic questions for this simple circuit are how much current will flow, and as a result how much power will be dissipated. The key point is to ensure that the power dissipation capabilities of the resistors used will not be exceeded.

Practice

The steps below assess these issues as part of the practical exercise.

Step 1: Confirm the power ratings for the resistors to be used. This information will be provided in the manufacturer's data sheet, but if this is not readily to hand, common values for standard through hole resistors might be assumed. Very small through hole devices are usually rated about 125mW, with more typical sizes ranging from about 250mW to 600mW. If we aim for a conservative limit of 100mW in all the experiments below then we are unlikely to burn out any resistors.

Step 2: Once we have settled on the voltage to be used (let's stick to 9V here) it is possible to calculate a lower value for resistors which will not breach our 100mW limit. The final variant of Watt's law is the most useful in this case – $P = V^2/R$.

$$
\begin{aligned}
P &= \frac{V^2}{R} && \text{– Watts's law} \\
\Rightarrow R &= \frac{V^2}{P} && \text{– Rearrange} \\
&= \frac{9^2}{0.1} && \text{– Insert values} \\
&= 810\Omega && \text{– Calculate}
\end{aligned}
$$

From the equation ($P = V^2/R$) we can see that as R gets smaller, P gets larger, and vice versa, so clearly the calculated value is a lower limit. So long as we don't use anything less than 810Ω power dissipation will not rise above our 100mW limit. So the suggested values, starting at 1k, make sense. Now use the same equation to calculate the actual dissipation for each of the resistors to be tested. Clearly, all these results should come in under 100mW.

Step 3: The actual resistance values of the resistors you use will deviate from the exact nominal values. Use the multimeter to measure the actual values, and use these in subsequent calculations. The differences should not be more than a couple of percent at most (most resistors are either ±1% or ±5% tolerance), and would not normally be significant, but it is a good exercise to use the measured values here.

Step 4: Next Ohm's law yields the current which should flow for each resistor.

$$V = I \times R \quad \text{– Ohms's law}$$

$$\Rightarrow I = \frac{V}{R} \quad \text{– Rearrange}$$

$$= \frac{9}{1000} \quad \text{– Use your values here}$$

$$= 9\text{mA} \quad \text{– Calculate}$$

Step 5: Build the circuit, including the multimeter (set to measure milliamps), as illustrated below – always double check your build before connecting power. Record the current drawn by each resistor.

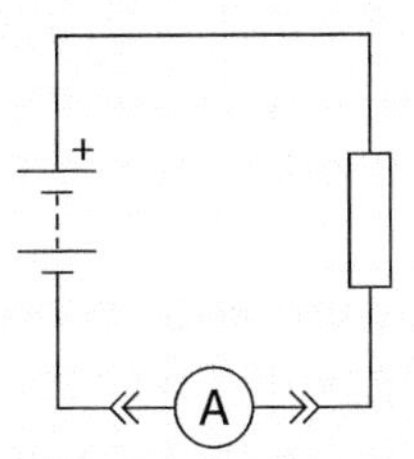

Step 6: Compare the calculated and measured values. They should be very close. Remember that your multimeter (especially the cheap hand-held types) may not be as accurate as you might hope, but should not be too far out.

Learning by Doing 9.2

Kirchhoff's Current and Voltage Laws (KCL & KVL)

Theory

In this exercise we use Kirchhoff's current law and Kirchhoff's voltage law to predict the behaviour of a circuit. Calculations and measurements are compared in order to confirm the expected behaviour. All the circled numbers around the circuit are just to allow for easy reference to different points – they will be referred to as nodes.

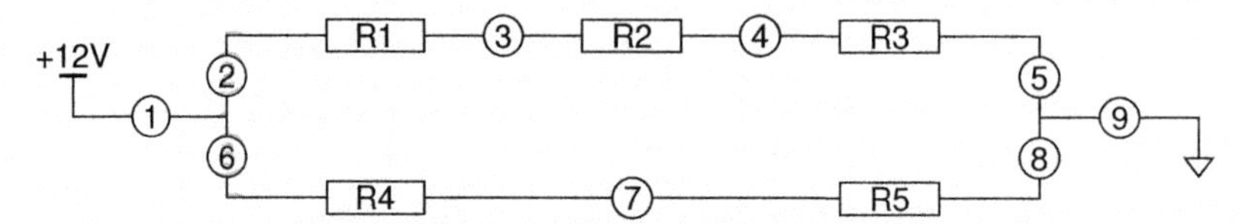

In the calculations which follow, two resistor values are used for the five resistors in the network (1k for R1–R3, and 1k5 for R4 and R5). These values have been chosen to make the calculations simple in this worked example. When doing this practical just keep all values between 1kΩ and 100kΩ, and the power supply between 5V and 15V, and you can't go far wrong. A power supply level of 12V is suggested here. Again this is for simplicity of example calculations.

The two things we typically measure in a circuit like this are voltage and current. Recall from Chapter 8 that voltage is measured between two points in a circuit, while current is measured by breaking into the circuit at a single point. This distinction has relevance for the selection of the nodes labelled in the circuit.

From a voltage point of view nodes (1), (2), and (6) are all the same point. On the other hand, from a current point of view, these three nodes are entirely separate locations in the circuit. This distinction is examined in the course of the exercise.

Practice

The resistor series and parallel rules (Eq. 12.1 (p. 198) and Eq. 12.2 (p. 199)) allow the total resistance across any part of a network such as this to be calculated. The resistance of each branch adds up to 3kΩ in our example, and the two branches in parallel thus equal 1k5Ω. You will need to use the equations referenced above to calculate your values.

Step 1: Select five resistors, measure and record their values, and build the resistor network shown – note carefully which resistor is placed in each location. Both calculate and measure the total resistance of each branch, and of the entire network. Record your results.

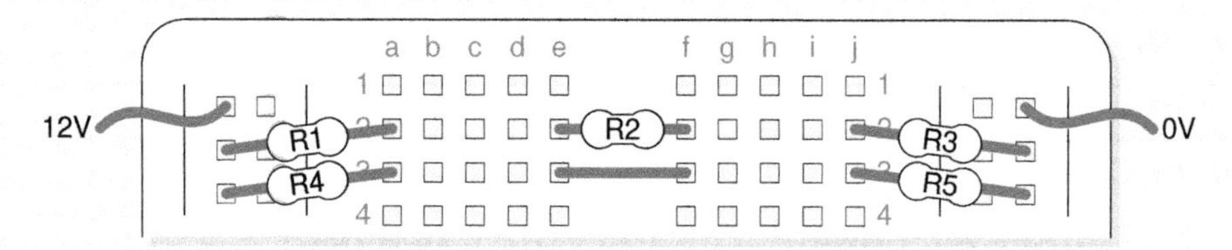

Steps 2 to 4 illustrate KCL: Currents into a point equal currents out.

Step 2: Use Ohm's law to calculate the current at nodes (1), (2), and (6).

Node (1): $I = \frac{V}{R} = \frac{12}{1k5} = 8mA$

Node (2): $I = \frac{V}{R} = \frac{12}{3k0} = 4mA$

Node (6): $I = \frac{V}{R} = \frac{12}{3k0} = 4mA$

Step 3: Use Kirchhoff's current law to confirm that the results calculated above look correct. Current flows from positive voltage to negative voltage, so in this circuit all current is flowing left to right.

$$
\begin{aligned}
I_{in} &= I_{out} && \text{– KCL} \\
\Rightarrow I_1 &= I_2 + I_6 && \text{– Expand KCL} \\
8mA &= 4mA + 4mA && \text{– Use your values here} \\
&= 8mA && \text{– Confirm matching currents in and out}
\end{aligned}
$$

Step 4: Compare calculated answers to measured results. Connect an ammeter into the breadboard at node (1), apply power, and record the current. Remove power, reposition the ammeter, and repeat measurements for nodes (2) and (6). The calculated and measured currents should match, and once again $I_1 = I_2 + I_6$, within some small margin of error.

Think carefully about where the ammeter must be connected in each case. Notice that the power bus on the breadboard acts as the 'T-junction' point between the three nodes in the circuit diagram. For each case, remove the corresponding connection from the bus (12V, R1, or R4), and connect the meter between the free end and the bus where it had been inserted.

Steps 5 and 6 illustrate KVL: Voltage drops around a closed loop equal zero.

When applying KVL, a loop can be traced in either direction, and can be started at any point around it. Obviously for a closed loop the end point must always return to be the same as the starting point.

When measuring voltages with a multimeter, it is important to maintain a consistent probe order around the loop, either black followed by red or red followed by black. Swapping the order will swap the sign of the measurements. It doesn't matter which you chose so long as you stay consistent. There should always be some positive and some negative voltage readings, or the sum can not possibly add up to zero. (With an analog VOM only positive readings can be made, so you have to watch the probe order and consistently add a minus sign where needed.)

Step 5: Remembering that the power supply is part of the circuit, there are three simple loops which can be traced in the circuit above.

Loop one: R1 – R2 – R3 – R5 – R4
Loop two: R1 – R2 – R3 – PSU
Loop three: R4 – R5 – PSU

Any number of more convoluted loops can also be traced, KVL still holds.

Loop four: R5 – R3 – R2 – R1 – PSU – R3 – R2 – R1 – R4

For each of the above loops, measure the voltage drop across each part of the loop in turn, and note down your readings. Adding the list of voltages for any loop should return a result of zero.

Step 6: Adding the extra jumper wire shown below changes the circuit and introduces some new loops. Map out a few loops and repeat the measurements described above. For instance, start at R1 and do a figure-of-eight through all five resistors. You should always get a zero sum on returning to your starting position.

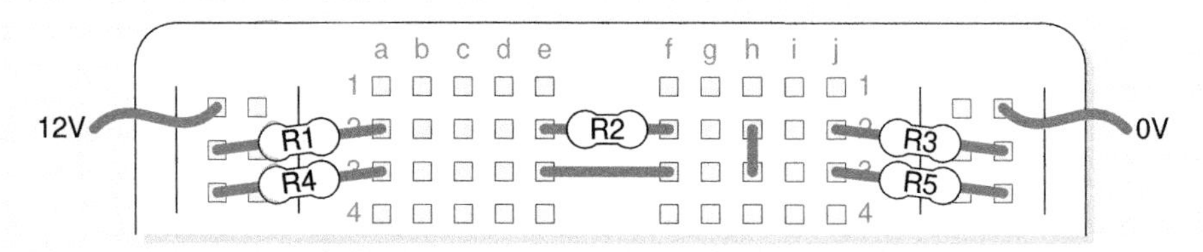

Step 7 (optional): You can also return to the KCL tests above, with this new layout. Which direction will current flow through the new link? The answer depends on the specific resistor values you have used in your circuit. See if you can analyse the situation first, and then make measurements to confirm your findings.

References

K. Brindley. *Starting Electronics*. Newnes, 4th edition, 2011.

A. Hackmann. *Electronics: Concepts, Labs, and Projects*. Hal Leonard, 2014.

P. Horowitz and W. Hill. *The Art of Electronics*. Cambridge University Press, 3rd edition, 2015.

P. Scherz and S. Monk. *Practical Electronics for Inventors*. McGraw-Hill, 3rd edition, 2013.

A. Sedra and K. Smith. *Microelectronic Circuits*. Oxford University Press, 7th edition, 2014.

10 | Constructing Circuits

The steps and stages involved in taking a project from an initial idea to a fully built, tested, and packaged end product are both numerous and variable. Some of these steps, such as circuit layout, and testing and analysis procedures, have already been examined in the preceding chapters. Here the entire process is addressed from end to end. The simple fuzz effect circuit first introduced in Chapter 7 is used as a case study in order to examine the steps involved in a fully developed audio electronics project.

THE PROJECT PIPELINE

The stages in any project can vary, but a typical sequence will include the following.

Initiation: The initial circuit idea is developed, either designed from scratch or identified in an existing circuit design.

Concept development and background research: Various questions can help to evolve and solidify the initial idea into a concrete design. Have others built this or similar circuits? What can be learned from their experiences? What worked and what didn't? What variations on the theme have been described? All of this is likely to mainly involve online research, and possibly reference to circuit cookbooks such as those by Penfold and others (see for example Anderton, 1992; Boscorelli, 1999; Penfold, 1992).

Procurement: A bill of materials (or BOM) is compiled based on the circuit to be built, and all the components and tools likely to be needed are assembled. A well stocked lab may provide all that is needed, but in the early days, building up supplies of useful kit and components can take some time.

Prototyping, testing, and development: It is most likely that a breadboard version of the circuit will be the logical first building phase. Based on the circuit developed over the research phase, a breadboard layout should be designed, and then building and testing can commence. Questions previously formulated can be examined, experimenting with possible circuit modifications until a final design is settled upon.

Final layout design: With the circuit finalised, a stripboard layout can be designed.

Circuit building and debugging: Once a detailed design is in place, the building process is a moderately straightforward affair.

Packaging: Questions of mounting and housing a project are beyond the scope of the material covered here.

Initial Idea, Concept Development, and Background Research

The following step by step project walk through uses as a case study the 'Bazz Fuss' fuzz effect circuit already encountered several times in earlier chapters. The Bazz Fuss is a circuit commonly encountered online, and often recommended as a good first project. Unfortunately, online resources are prone to move or disappear completely, as seems to be the case with the original version of this circuit. References to web sites are generally avoided in this book for this very reason, despite the fact that the internet is an invaluable source of information and ideas.

While the original Bazz Fuss appears to be long gone from the internet, many excellent examinations and modifications of this circuit can still be found, and diverging from the practice adopted in the rest of the book, some of these are highlighted below. The following online resources provide (at the time of writing) a wide range of excellent information and suggested modifications relating to this one modest circuit.

Beavis Audio Research presents the circuit in its most basic form, and directly references a now defunct original source. Some interesting modifications are also suggested (http://beavisaudio.com/beavisboard/projects/bbp_BazzFuss.pdf).

Escobedo's Circuit Snippets is most often encountered as a collection of simple projects, each presented as a circuit diagram with no more than a paragraph or two of notes. The original source once again seems to be long gone, and the contents of the collection can vary a bit between existing reproductions. One of the most complete versions of the collection currently accessible is at http://www.diale.org/escobedo.html. This includes an interesting Bazz Fuss variant, and brief but useful notes on the suggested modifications.

home-wrecker.com provides a much more expansive examination of the circuit (http://home-wrecker.com/bazz.html). It starts with the basic version and then proceeds to examine a number of possible routes to interesting modifications, suggesting various experiments for tailoring the tone of the circuit.

tonefiend.com takes a similar approach, again evolving the original circuit, adding various modifications to make the circuit more flexible and interesting
(https://www.tonefiend.com/wp-content/uploads/DIY-Club-Project-2-v02.pdf).

diystompboxes.com is one of many forums dedicated to building your own audio circuits. Forums threads can be long and rambling but often contain little gems of information. Chances are, if a circuit can be found online, then there are forum discussions ongoing into how to get the best out of it. Here is just one relating to the Bazz Fuss: https://www.diystompboxes.com/smfforum/index.php?topic=112626.0.

All of the above and many other sources besides, provide interesting ideas as to how this simple project can be expanded and experimented with, and as such provide the raw material of an excellent learning resource for anyone wishing to develop their own circuit project ideas. In this simple walk through, a bare bones version of the circuit, as shown overleaf in Learning by Doing 10.1, is taken as the starting point. Learning by Doing 17.3 (p. 302) examines the circuit in a little more detail.

PROCUREMENT

Hobbyist electronics retailers such as RadioShack and Maplin can prove good sources of materials, particularly if there is a physical store nearby (in 2018 Maplin closed all its stores and went online only). These sources can however prove a little more expensive than some of the larger online electronics suppliers such as Farnell, Radionics, Mouser, and Digikey. Budget online megastores such as Aliexpress can also be useful, but exercise care with these vendors, as quality control may not be as consistent as with some other retailers. Bargains can certainly be had, but it is perhaps best to steer clear until you have a good understanding of the tradeoffs involved.

Also of particular interest are the more specialist stores. Audio specific components like guitar pickups and reverb pans are not usually to be found at the general purpose electronics suppliers. Similarly, less common components such as vacuum tubes and germanium devices require a little more searching. Some good online stores to start with (amongst many others) might be: Banzai, Reichelt, Tayda, Thomann, and Tube Amp Doctor. It can be well worth maintaining a component inventory, updated with each new purchase, showing where and when components were bought, part numbers, price, and some basic notes.

Bill of Materials (BOM)

The BOM for this project is very modest, and includes no unusual parts. All are available from the Farnell online store. The part numbers and prices included here are taken from https://ie.farnell.com (accessed on 30th September 2019) All prices are quoted ex VAT. Minimum purchase quantities are noted, and high minimums have been avoided. The stripboard shown is probably enough for twenty or more projects (even given that most projects require a somewhat larger piece of stripboard than the tiny 8×3 section suggested for this project below in Learning by Doing 10.2). Also required are the basic tools of the trade: a breadboard, multimeter, soldering supplies, and sundry other tools.

Part		**Number**	**Unit price**	**Min purchase**
resistor	100k	2401807	€0.0316	×10 min
capacitor	220n	1141778	€0.17	×10 min
capacitor	4u7	9451234	€0.0465	×1 min
diode	1N4148	2675146	€0.035	×5 min
transistor	2N3904	1574370	€0.138	×5 min
jack (TS)		4169165	€0.721	×1 min
jack (TRS)		4169189	€0.885	×1 min
battery clip (PP3)		1183123	€0.339	pack of 10
stripboard		2503760	€2.95	100 × 160 mm
stranded hookup wire				
solder				
enclosure of your choice				

Prototyping, Testing, Development, and Final Layout Design

Learning by Doing 7.1 (p. 110) invites the reader to design a breadboard layout for this simple circuit, and as noted there, Appendix D offers one suggestion. The simplest possible modifications involve directly replacing components to see how different choices affect the sound of the circuit. While the version presented here starts with the original 2N3904 transistor, any NPN BJT might be tried. Most sources seem to favour a higher gain device, usually either a 2N5088 or even a Darlington part such as the MPSA13. Similarly diodes offer an easy option for a bit of variation, whether it be germanium parts, Schottky diodes, or LEDs.

Learning by Doing 10.1

Breadboarding Circuits

Materials

- Breadboard
- Jumper wires
- 100k resistor
- 220n capacitor
- 4u7 capacitor
- 1N4148 diode
- 2N3904 transistor
- TS jack (x2)
- 9V battery and clip (or 9V power supply)

For testing:
- Electric guitar
- Guitar amp
- Guitar lead (x2)

Theory

Building successful breadboard circuits takes a bit of patience and a light touch. Once you have a good layout and you know how to follow it, the single most important rule is, try not to bend the leads. Its not always possible, but it is worth making the extra effort to try. Breadboards vary in quality and some can prove very difficult to insert components into. The key is not to push harder, but rather to keep patiently adjusting the angle of attack until you get the lead lined up just right and it slides into the connection point with very little pressure.

Breadboard contacts consist of little spring loaded metal jaws. If a component lead catches on the lip and is forced, it will bend out of shape before it seats properly. While the component may (or may not) have been inserted successfully, these bent leads will inevitably result in unwanted contacts where adjacent leads touch. This can make breadboard circuits very frustrating and difficult to debug and get working. Much better to have the patience, maintain a neat build, and vastly increase your chances of ending up with a working circuit.

Rule number two is to make sure that component leads are clean before inserting them. Many components come in strips, all stuck together with tape. Sometimes when the tape is removed it leaves behind a layer of sticky gunk. Either cleaning or trimming the leads will avoid bad connections due to this gunk getting in the way.

Practice

For a first attempt at building a circuit on a breadboard, we return to the simple fuzz circuit introduced in Chapter 7, reproduced below. If you had a go at Learning by Doing 7.1 (p. 110) you should have a breadboard layout of your very own for this circuit. If not (or even if you have) below is an alternative you can use instead.

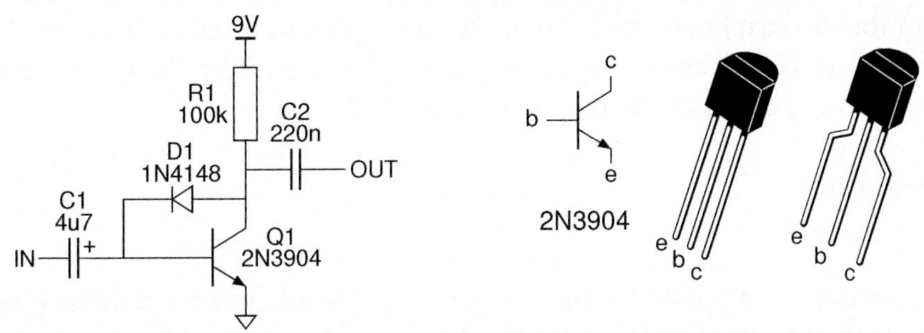

Following the layout of your choice, build the circuit carefully and check that no component leads are touching. Double check all connections, paying particular attention to the orientation of components: the flat side of the transistor, the stripe on the diode, and the minus signs on the polar capacitor. The smaller, nonpolar capacitor and the resistor do not have an orientation.

Be sure not to mix up the tip and sleeve of the input and output jacks, and finally double check that the battery clip is connected the right way round before connecting the battery. Be careful to work out the correct terminals on the battery before touching them to the clip. This circuit is fairly robust but some circuits will be damaged if the battery even briefly touches the contacts the wrong way round. For a standard nine volt clip remember 'squiggly to straight and straight to squiggly'. Never touch like to like. Power should always be the last thing connected, and only after everything has been double checked.

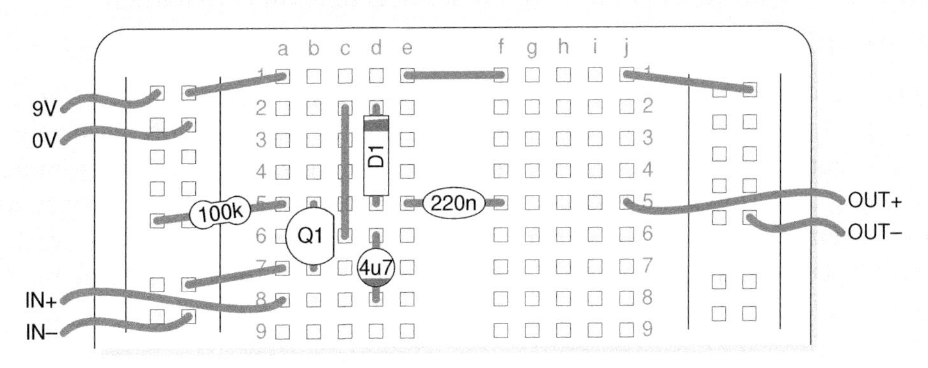

Once everything is connected up, all that remains is to test your circuit. There are no controls on this very simple fuzz, so it either works or it doesn't. Connect a guitar to the input and an amp to the output, turn it up, and have a listen. There is plenty which can be done to make this circuit more versatile but what we have here is a good start.

Construct a few variants on breadboard, spend some time playing and listening to the results, and see what you like the sound of. Switchable and swappable parts are an option too, but perhaps keep it simple for this first project. With experiments concluded and a design for the stripboard build finalised, a stripboard layout is needed. Learning by Doing 7.2 (p. 113) addresses this stage. Once again, be sure to have a go at a layout of your own before looking at the example in Appendix D.

Circuit Building

Soldering

With all the necessary components and supplies assembled, and a stripboard layout settled upon, it is time to start building the final soldered version of the circuit. Soldering is an important skill which requires practice above all. The best single piece of advice is to keep some solder on the working surface of a soldering iron at all times. In storage it protects the soldering surface and prevents oxidation, which is the death of any soldering iron tip. In use the solder on the tip vastly improves heat transfer from the iron into the joint being made, inproving the ease of soldering, and the quality of the results obtained.

There is much which can be said about soldering but for a good procedure to start:

Step 1: Secure the work to be soldered
Step 2: Heat the joint for a few seconds, contacting the solder on the tip
Step 3: Melt sufficient solder to make the joint
Step 4: Keep heating until the solder flows (you'll see it when it happens)
Step 5: Remove the iron as soon as the solder flows
Step 6: Do not move the joint for a few seconds to allow it to harden

Soldering and Desoldering Equipment

Soldering and desoldering equipment forms an important part of any electronics laboratory setup. Inexpensive soldering stations are readily available and perform well for basic work, but if much soldering is going to be undertaken, the cost of a more high quality unit is a worthwhile expense. Similarly, if only occasional desoldering and reworking is going to be attempted then a simple manual desoldering pump works well, but a powered desoldering gun makes the task much more straightforward, and can be a good investment if the task is to be performed often.

Learning by Doing 10.2

Stripboarding Circuits

Materials

- Stripboard (8 holes × 3 strips)
- Hookup wire
- 100k resistor
- 220n capacitor
- 4u7 capacitor
- 1N4148 diode
- 2N3904 transistor
- TS jack (x2)
- 9V battery clip

For testing:

- Electric guitar
- Guitar amp
- Guitar lead (x2)

Theory

Having mastered the soldering skills necessary, building stripboard circuits becomes a fairly straightforward business. As with breadboarding, preparation and patience are key. Double check your stripboard layout against the original circuit diagram before launching in. Check all your components and make sure components like resistors won't get mixed up. Most circuits will involve resistors of different values, which must not be swapped. Be sure you know how to identify the correct orientation for all components.

Practice

For a first attempt at building a circuit on stripboard, we return to the simple fuzz circuit introduced in Chapter 7, reproduced below. If you had a go at Learning by Doing 7.2 (p. 113) you should have a stripboard layout of your very own for this circuit. If not (or even if you have) below is an alternative you can use instead. Follow the procedure outlined in the main text and get building.

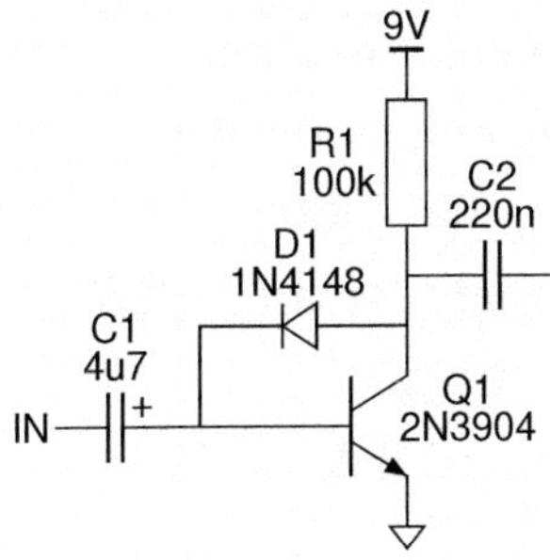

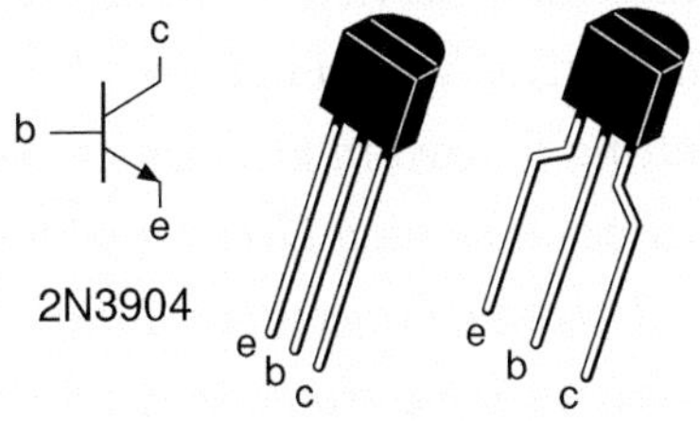

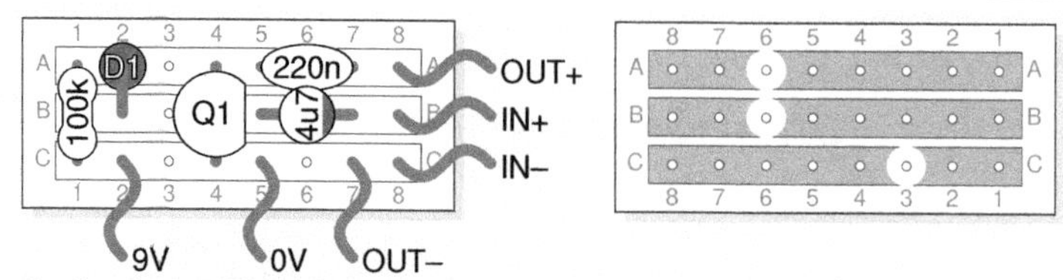

Just as with the breadboard version above, once built, test the circuit to see how it performs. As well as having a listen, it is always instructive to hook up a signal generator and an oscilloscope and visualise the performance of the circuit at different input signal levels. A 1kHz sine wave adjusted between zero and a couple of volts usually gives a pretty good idea of what's going on.

Packaging

The arrangement of off-board components such as pots, switches, power connections, and I/O jacks should always be considered with the final mounting and packaging in mind. Fly leads must be an appropriate length; too short and the will need to be rewired, too long or mismatched and the results will be messy at best. Is a tin, wooden, or plastic box to be used? If so, holes will require drilling. Labelling and graphic design are also beyond the topics considered here, but do need to be addressed to really finish a project.

References

C. Anderton. *Electronic Projects for Musicians*. Amsco Publications, 1992.

N. Boscorelli. *The Stomp Box Cookbook*. Guitar Project Books, 2nd edition, 1999.

N. Collins. *Handmade Electronic Music*. Routledge, 2nd edition, 2009.

R. Fliegler. *The Complete Guide to Guitar and Amp Maintenance*. Hal Leonard, 1998.

M. Geier. *How to Diagnose and Fix Everything Electronic*. McGraw-Hill, 2011.

R. Ghazala. *Circuit Bending: Build Your Own Alien Instruments*. Wiley, 2005.

A. Hackmann. *Electronics: Concepts, Labs, and Projects*. Hal Leonard, 2014.

T. Linsley. *Electronic Servicing and Repairs*. Newnes, 3rd edition, 2000.

F. Mims. *The Forrest Mims Engineer's Notebook*. LLH Technology Publishing, 1992.

F. Mims. *Getting Started in Electronics*. Master Publishing Inc., 2003.

R. Penfold. *More Advanced Electronic Music Projects*. Bernard Babani, 1986.

R. Penfold. *Electronic Projects for Guitar*. PC Publishing, 1992.

R. Penfold. *Practical Electronic Musical Effects Units*. Bernard Babani, 1994.

11 | Projects

There are a vast number and a wide range of useful and interesting audio circuits to be found in books and magazines, and on the internet varying in functionality and complexity from the almost trivial to the most elaborate imaginable. Typical audio projects might be broken down into three general types which can be broadly categorised as producers, modifiers, and amplifiers of sound. Within each category circuits will be encountered to suit every taste and level of experience.

The introductory projects presented here naturally stick close to the novice end of the spectrum. They should however provide sufficient insight and understanding to allow the reader to extend their explorations to whatever level and in whatever direction they so desire. Each of the three categories mentioned above covers a wide range of applications and approaches. Some of the key circuit types and classes are listed here under each of the three headings.

Sound producers include both sound and vibration detection devices such as microphone and pickup based projects, and signal generators such as oscillators, synthesisers, and general noise making circuits.

Sound modifiers range from simple filter and EQ circuits through to the vast array of audio effects which are available, including such things as distortion, phaser, flanger, vibrato, tremolo, echo, delay, and so on. This category also includes the wide range of ancillary circuits for controlling and modulating effects, such as LFOs and envelope generators.

Sound amplifiers are fairly self explanatory but there is still quite a range. Beyond the generic audio power amplifier (including home hifi, studio monitor, and PA and sound system style amps) there are also headphone amps, guitar and instrument amps, portable, battery powered, and practice amps, along with other specialist amplifiers employed in specific roles such as mic pre-amps, loud hailers, phono pre-amps, etc.

The three circuits presented as projects in this chapter include an example from each of these three categories. This gives a flavour of the range of audio projects which can easily be undertaken by the novice electronics experimenter. Guidance is offered on prototyping, modifying, testing, and building. A short introduction such as this can only scratch the surface of what is possible, and ultimately the limits of what can be achieved are set only by the imagination and ingenuity of the electronics project builder.

Project I – Low Power Audio Amplifier

One of the most ubiquitous integrated circuits encountered in audio electronics is a single chip low power audio amplifier called the 386. It comes from two main manufacturers who designate it respectively the LM386 and the NJM386 (aka JRC386). The consensus seems to be that the latter is usually more stable and on the whole to be preferred, and this has certainly been my experience. The former is however often more easily procured, and in general, with a little care, either will usually do the job in any given circuit.

This project is quite closely related to a design from runoffgroove.com called the Ruby, one of very many 386 based amplifier circuits to be found online. The main differences between the Ruby and the design presented here (see Figure 11.1) are:

1. The addition of a reverse voltage protection diode. (The 386 is easily damaged by applying a reverse voltage to its power supply.)

2. The JFET used has been changed from an MPF102 to a 2N5485. (This is mainly a question of availability – see Chapter 17 for a discussion on substituting transistors.)

3. The sizes of resistors R1 and R2 have been changed to more readily available values than those specified in the original Ruby design.

4. Where the Ruby uses a gain pot connected between pins one and eight of the 386, this is replaced with an spst switch and a ten microfarad capacitor. With the switch open the circuit operates at its lowest gain setting. Closing the switch raises the internal gain of the 386 from its default value of 20 up to its maximum level of 200.

As is always the case when working with an unfamiliar component it is highly recommended to review the data sheet for the 386 in order to fully understand these configuration options. A quick internet search will immediately yield this document. These data sheets are also well worth seeking out for the various integrated circuits encountered, as they often include useful example circuits. This is certainly the case with the 386 where a number of instructive circuit suggestions can be found. Figure 11.1 shows the circuit diagram for the particular design which is presented in this section.

Main Circuit Elements

The 368 low voltage audio power amplifier at the heart of this circuit is an eight pin DIL package IC (the type of chip pictured on the right of Figure 18.1 (p. 315)), designated IC1 in the circuit diagram. Similar to an opamp (opamp circuits are discussed in Chapter 18), the 386 has two inputs designated noninverting (+) and inverting (−), and the output is the amplified difference between these two inputs. Unlike opamps, which amplify voltage but do not have the current sourcing capability to drive a loudspeaker, power amplifiers like the 386 are designed to connect directly to a low impedance load (typically 8Ω for a standard moving coil loudspeaker).

Also in common with typical opamp practice one of the two inputs is usually wired to ground, in this case the noninverting input on pin three. The signal to be amplified is presented at pin two and the amplified (and inverted) result emerges from the chip

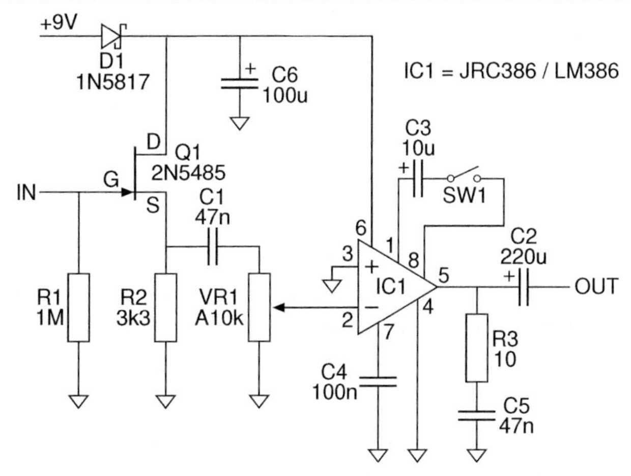

Figure 11.1 Circuit diagram for a simple 386 based low power (< 1W) audio power amplifier. Similar circuits are often described as guitar practice amps but they are ideal for use as a generic lo-fi utility audio amplifier well suited to experimentation and general audio amplification.

at pin five. Pins four and six provide the necessary connections to the power supply, in this case usually a standard 9V battery, although the chip can generally operate from a supply anywhere in the range 4V to 12V. The highest power variant, the LM386N-4, can be operated on a power supply ranging up to a maximum of 18V. Meanwhile pins one and eight allow for control over the amplifier's gain, and finally pin seven (usually labelled 'Bypass') is connected through a capacitor to ground in order to enhance the stability of the amplifiers operation, especially at high gain.

As mentioned above, the 386 comes in a few variants delivering different maximum output power levels ranging from 250mW up to about 1W. The upper end of this range is rather optimistic and if reasonably low levels of signal distortion are desired the maximum achievable output levels will be commensurately lower. On the other hand in its common application as an electric guitar practice amp, driving the chip to higher levels of distortion can result in quite a satisfyingly dirty sound. Surprisingly high sound levels can be coaxed out of this diminutive little chip, and it is quite capable of driving even moderately large loudspeaker drivers (or multi speaker cabinets) as well as making for an excellent compact utility amplifier when used in combination with a two and a half inch, half watt speaker.

The input buffer is comprised of a JFET (Q1) and two associated resistors (R1 and R2). It significantly improves the sound of this amp as compared to that of more rudimentary unbuffered 386 designs, especially when used connected directly to an electric guitar with passive pickup circuitry. This improvement is achieved primarily due to the increased input impedance which this input circuitry provides. Connecting directly to the input pins of the 386 presents an input impedance of about 50kΩ whereas here the input impedance is 1MΩ, set by the value of the resistor R1 connected from the JFET's gate to ground. The JFET itself presents an extremely high input impedance typically in the GΩ range (Evans, 1972, p. 73) and often considered to be infinite for all practical purposes (Sedra and Smith, 1987, p. 295). As such it does not have much effect on the overall input impedance as it appears in parallel with the 1MΩ resistor to ground.

A high input impedance is of particular value here because a guitar pickup has an output impedance which is also very high, typically in the hundreds of kilohms range. For optimum interfacing the input impedance typically wants to be about an order of magnitude larger than the output impedance which it is connected to. This provides for efficient transfer of signal from output to input and best preserves the tone of the guitar or other sound source. As such this JFET based input buffer also operates very well when interfacing to a piezo pickup. As mentioned in Chapter 20, piezos have an extremely high output impedance and as such benefit from connection to as high an input impedance as possible.

The observations in this section should serve to illustrate and re-enforce the idea that a solid grasp of impedance and the role it plays in interfacing audio circuitry is vitally important to developing a good understanding of the ins and out of signal flow and audio circuit design and analysis.

Working with discrete transistors it is always necessary to pay attention to the pinout of the particular device used. Pinouts are not consistent across transistors, even of the same type. Reference to the device's data sheet should be made in order to confirm the arrangement of the three terminals. In the case of the 2N5485 used here the pinout is shown in Figure 11.2. Transistor data sheets will include this information, usually in the form of a diagram similar to the one included here.

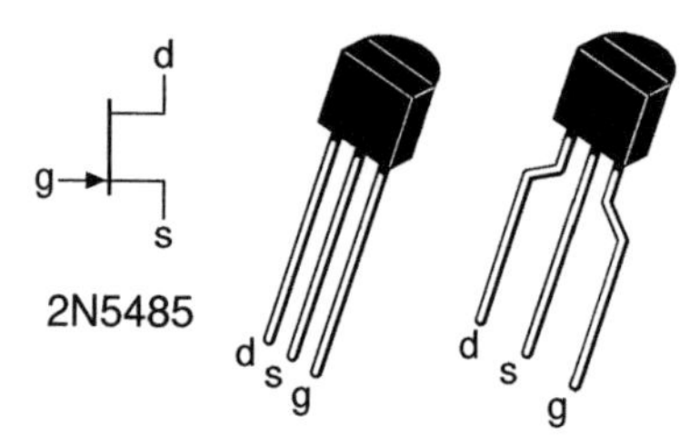

Figure 11.2 The pinout for the input buffer 2N5485 JFET.

Input and output coupling capacitors are common elements found in many audio circuits. They will often be referred to as DC blocking capacitors as one of their main jobs is to make sure that signals moving from one section of a system to the next have the correct DC bias applied to them. A guitar pickup generates an AC signal which varies up and down either side of ground or zero volts, but an oscillator circuit for instance is quite likely to present a signal whose midpoint lies at a positive voltage most probably halfway between ground and the positive power rail voltage powering the circuit. A general purpose amplifier must be able to cope with both of these signals, contending with these two very different signal bias conditions seamlessly.

The inputs on the 386 are internally biased and are designed to take ground referenced input signals, so while the guitar signal might couple directly into the chip without problem, attempting to do the same with the oscillator signal would result in massive distortion at best and a blown amp at worst. The solution is however very straightforward, an input coupling capacitor is placed between the incoming signal and the internally biased 386 input pin. Recalling that capacitors block low frequencies and pass high frequencies it should come as no surprise that any DC offset between signal and input will be soaked up in charging the capacitor while the AC component of the signal continues on into the input pin. In the circuit in Figure 11.1 this job is accomplished by the 47nF capacitor labelled C1 which feeds into the volume pot VR1.

Similarly at the output of the 386 the signal produced will include an offset bias equal to half the supply voltage. Thus if the amp is powered from a 9V battery then the signal will vary either side of 4.5V. However the loudspeaker which is likely to be connected to the amp's output wants to see a signal with zero bias voltage, otherwise the speaker driver's voice coil is likely to be burned out before long. Again a simple capacitor between output and speaker resolves the difficulty, in this case the 220μF capacitor C2 between pin 5 and OUT.

The input and output coupling capacitors will also tend to affect the tone of the amplifier. Each will combine with their respective following resistances to ground, to form high pass filters (see Chapter 13). Using Eq. 13.13, the cutoff frequencies can be estimated. The large 220μF capacitor on the output, in combination with the 8Ω loudspeaker, will pass all but the lowest audio frequencies unaltered.

$$f_c = \frac{1}{2\pi RC} = \frac{1{,}000{,}000}{2\pi \times 8 \times 220} \approx 90\text{Hz}$$

The smaller input capacitor C1 (in combination with the 10kΩ volume pot VR1, and ignoring the effect of the larger input impedance of the 386) will have a more significant effect, rolling off the level within a significant range of the lower band of audio frequencies delivered to the input of the 386.

$$f_c = \frac{1}{2\pi RC} = \frac{1{,}000{,}000{,}000}{2\pi \times 10{,}000 \times 47} \approx 340\text{Hz}$$

These calculations are somewhat simplistic in this context, but give a good sense of the general behaviour to be expected. Different sized capacitors can thus be used in this way to alter the overall tone of the amplifier to suit personal taste or specific applications. It makes sense to roll off any undesired low frequencies before amplification, especially since the 386, being a very modest amplifier, is easily overwhelmed especially by low frequencies.

The volume pot VR1 feeding pin two of the 386 operates as a voltage divider exactly as described in Chapter 12 in the section on Attenuators and Volume Controls. When the pot is turned fully down pin 2 of the 386 is connected directly to ground and as such sees no input signal at all. As the pot is turned up the level of signal reaching pin 2 steadily increases until finally the full signal voltage from the input buffer section is presented at pin 2 for amplification. Using a pot with an audio or log taper in this position is likely to give the best volume control characteristic. Using a linear taper pot will work fine but the volume control will bunch most of the level change into the first half of the pot's rotation, with the latter half having little further effect on the output volume. As discussed in Chapter 12, using a log pot will mitigate this problem and provide a more user friendly behaviour for the amplifier's volume control. The A in A10k specifying the value for VR1 indicates that an audio taper is preferred.

The gain switch and capacitor (SW1 and C3) which can be found bridging pins one and eight of the 386 provide a simple lo/hi gain option. Reference should be made to the data sheet for the 386 IC in order to see the effect of these components. With an open circuit between pins 1 and 8 the 386 amplifier has a native voltage gain of 20. That is to say, whatever input signal is presented at pin 2 the amplifier will endeavour to output a signal identical in shape but twenty times as large. Of course this being a rather simple amplifier some distortion is to be expected and as always the output is also limited by the power supply used, so with a single nine volt battery the output signal can certainly not extend beyond the range zero volts to nine volts.

Once the gain switch is closed the gain jumps up to 200. As is detailed in the data sheet intermediate gains can be achieved by placing a resistor between pins 1 and 8, and so a smoothly varying gain can be achieved using a pot. Given that the input volume pot already provides continuous control over the signal level, and in the interests of both simplicity and variation, a switch has been employed in this design. The capacitor is as recommended in the data sheet and is intended to assist in maintaining stability at high gains.

The stabilisation capacitor C4 connected to pin 7 also helps prevent stability and oscillation problems, again especially when in high gain mode. The addition of a capacitor on pin 7 is once more simply a case of following the advice given in the chip's data sheet. A common issue which can be encountered when amplifiers with high gain are employed is what are called parasitic oscillations where some frequencies can result in runaway amplification causing the amplifier to squeal or screech even when there is no input signal present. Often a judiciously deployed capacitor is the best solution to eradicating such unwanted oscillations, and in this case the connection on pin 7 is, according to the data sheet, the place to connect just such a capacitor.

The zobel network composed of R3 and C5 connected to the output in this circuit is a standard method of interfacing to a loudspeaker. A typical moving coil loudspeaker is an inductive load as the presence of the word 'coil' in its name might suggest. Recall that an inductor is just a coil of wire and so it should come as no surprise that a component which includes a coil as the primary electrical element used in its construction exhibits significant inductance. Remember that for an inductor impedance increases with frequency (inductors pass low frequencies and block high frequencies). An amplifier prefers to see a steady load across all frequencies in order to operate at its best. For a typical 8Ω loudspeaker (remember most speakers are rated 4, 8, or 16Ω, with 8Ω being by far the most common) the series resistor-capacitor network shown will do a pretty good job of flattening the overall load impedance which the amplifier sees, allowing it to perform to its best.

With rising frequency the coil impedance rises as the capacitor's impedance falls. Combined with the resistive impedances of both R3 and the wire in the coil of the loudspeaker, the overall impedance loading the amplifier is much more stable across frequencies than would be the case were the loudspeaker connected without any compensating network in parallel. This can have a very substantial effect on the operation of the circuit as can be seen from the oscilloscope traces in Figure 11.3. In each case the top trace is the input signal while below it is shown the amplifier's output. In situation (a) the zobel network is present exactly as shown in the circuit diagram. The only change made in order to obtain the traces in (b) is that the zobel network has been removed from the circuit. A large amount of high frequency instability has been introduced.

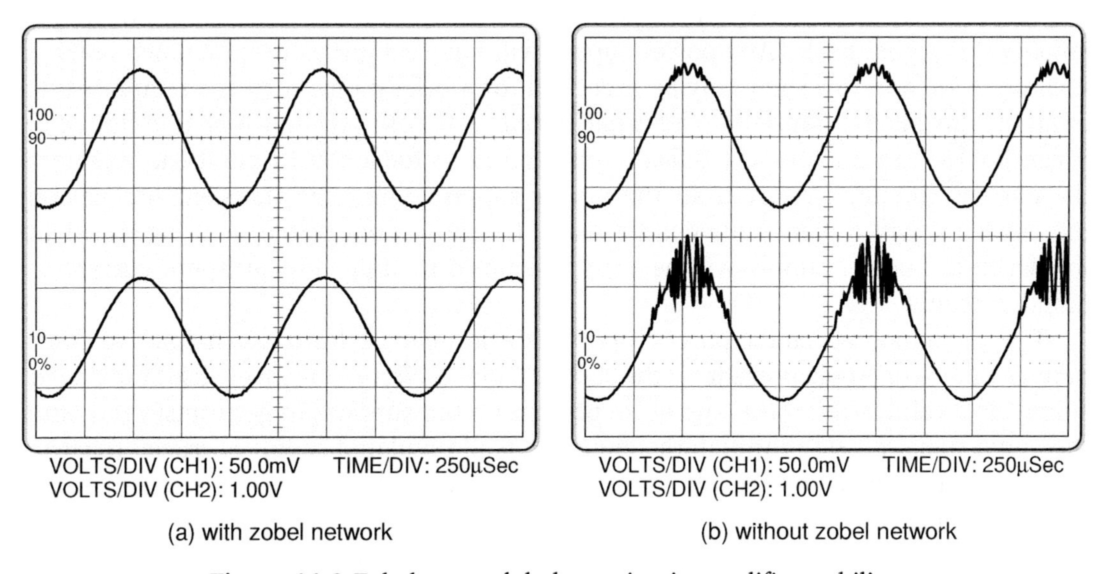

Figure 11.3 Zobel network helps maintain amplifier stability.

Both plots were obtained with the amplifier running at low gain. Notice that the legend states that channel 1 was set to 50mV per division and channel 2 was set to 1V per division. From this it is a simple matter to confirm that a gain of approximately twenty was indeed applied.

It is also easy enough to estimate the frequency of the test signal used given the 250μSec per division label also present. A 1kHz test tone is a very commonly used audio test signal, and so it should come as no surprise to arrive at this result here also. This is a very common but very simple approach to loudspeaker compensation. More advanced approaches (e.g. Leach, 2004) can achieve superior results in high quality audio systems but would certainly be overkill in a project such as this.

The reverse voltage protection diode D1 connected from the +9V supply line into the circuit illustrates one common method of implementing reverse polarity protection. In general it is important not to reverse the power connections to a circuit. If a standard nine volt battery is being used to power a circuit, it is very easy to accidentally touch the terminals to the battery clip the wrong way round when changing the battery. Without the protection diode to block the reverse voltage, if the terminals were to touch the wrong contacts on the battery clip even for a moment the 386 would likely be destroyed.

The reservoir/smoothing capacitor C6 connected between the +9V and 0V rails is another common circuit element encountered in many circuits. Even when not shown explicitly on a circuit diagram it can be a good addition depending on the characteristics and function both of the circuit and the power supply being used to run it. A large capacitor across the power supply can serve two useful functions.

Firstly it can act as a charge reservoir. When a high level transient arrives at the amplifier the circuit will require a short burst of high current in order to effectively amplify this signal peak. Any power supply will have a limit as to how much current it can provide and this can easily be exceeded for a brief moment by the requirements of amplifying such a transient. The capacitor can not hold very much charge but what it can hold it can release very quickly indeed and so for a short time it can augment the current coming directly from the power supply. Once the transient has passed the capacitor will replenish its charge ready for the arrival of the next big transient for which the power supply will need its assistance in supplying sufficient current to amplify cleanly.

The second job which the power supply capacitor can perform is in helping to eliminate any noise or ripple present on the supply. Since a circuit effectively uses the power supply as a voltage reference any noise present on the supply can get transferred onto the audio signal passing through the circuit. A well regulated supply is crucial to clean audio, and the capacitor can play a large role in making sure this is achieved. Low frequency ripple often appears in a voltage derived from the mains. Mains power is a large AC signal with a frequency of 50Hz or 60Hz, and when this gets into the audio signal it results in an annoying and all too familiar hum. Higher frequency noise can originate from many other places but one way or another it is highly desirable to keep it from getting into the audio.

The exact size of the smoothing capacitor is not crucial. Typically 100μF or larger is common. It needs to be big so that it can hold enough charge to make for an effective reservoir. The only problem with large capacitors is that they are usually a little less effective at working with high frequencies and so if high frequency interference is a potential issue then a second much smaller capacitor is often placed in parallel with the reservoir cap in order to deal with smaller but higher frequency noise on the power line. Sometimes theses smaller decoupling capacitors can be found placed throughout a circuit. Most commonly one can be found right next to the power pin of every IC on many commercially produced circuit boards. Often this is overkill but if a noise problem is encountered which cannot be eliminated at source, it may turn out to be a simple solution.

Breadboard Layout, Circuit Testing, and Modification

When a new circuit is being investigated, a breadboard build is almost always the best first step in the process. Because components can be easily plugged into and out of a breadboard they allow for simple testing and experimentation prior to soldering up a permanent version of the circuit on stripboard or PCB. Preparing a good, well laid out design prior to starting work on the board itself is to be highly recommended. The situation can quickly become messy and confusing when attempting to build directly from the circuit diagram. Much better to take some time to design a good layout first. This can be done easily by drawing on squared paper, but it is probably much simpler to use a software package designed for the purpose. There are many options available but for a good free solution a program called DIYLC ('Do-It-Yourself Layout Creator') serves very well. The layouts reproduced in this book were specifically generated in order to obtain optimum printing clarity and resolution but a similar on screen result is possible using DIYLC.

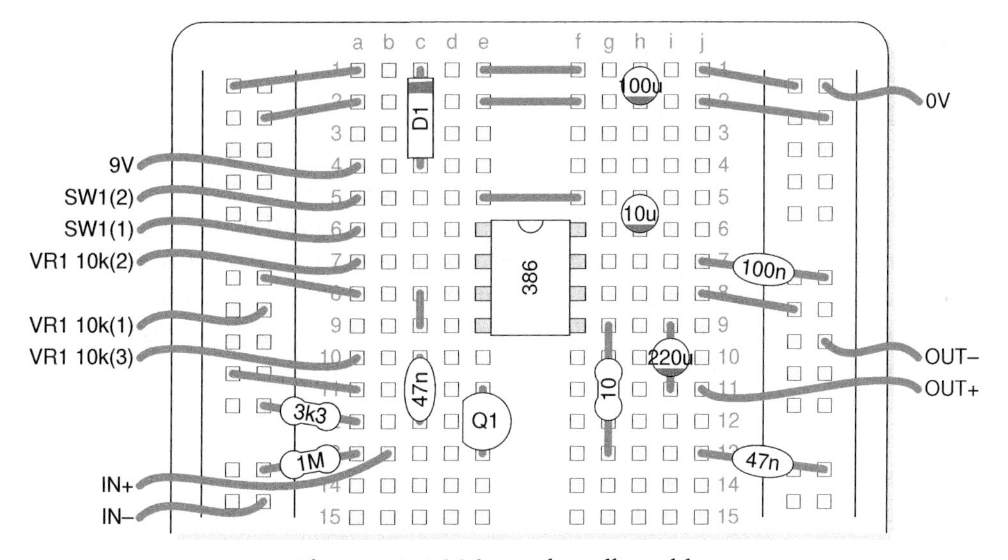

Figure 11.4 386 amp breadboard layout.

Figure 11.4 shows a possible breadboard layout for the 386 amplifier circuit. The two rows of jumper wires across the top of the board are a common way of distributing 0V and +9V up and down both sides of the board. This is often convenient as both power and ground connections are often required at various points around the circuit built on the middle section of the breadboard. These outer buses do not have to be used for power connections but they most often are. Very occasionally on more complex circuit layouts it can be convenient to use one of these buses to make a connection which would otherwise prove messy. Maintaining a clean layout is a great benefit when it comes to debugging a circuit, so it is well worth spending the time at the outset to get everything positioned, connected, and labelled in a clear and consistent fashion.

Symbols are often provided by layout software to indicate off-board components such as jacks, switches, pots, and power connectors. The style adopted in this book has been to exclusively use text labels for all off-board components (in this circuit these include power, in/out jacks, gain switch, and volume pot). This makes everything a little easier to lay out and avoids big looping connection wires snaking around the figure. Such graphical conventions are simply a matter of individual taste. I/O jacks are labelled with a (+) for the signal connection, usually the tip of a TS phone jack, and a (−) for the ground connection, usually the sleeve of a TS phone jack. Other components have their terminals named and numbered in a logical fashion.

The OUT connections may either go to a jack for connection to a speaker cab, or they may be connected directly to the terminals of an appropriate loudspeaker if this is desired. Both options can even be provided by using a switching jack. The switch SW1 is a simple spst device, and as such the order of connections does not actually matter. Pots have three terminals and throughout this book these terminals are numbered 1–3, with pin 1 being the left or counterclockwise connection point, pin 2 being the middle, variable connection, and pin 3 being the right or clockwise terminal. Thus turning the pot counterclockwise (conventionally considered down) moves 2 towards 1, and turning it clockwise (conventionally considered up) moves 2 towards 3.

Often it is not immediately obvious from a circuit diagram which way round terminals 1 and 3 should be connected in order to achieve the desired effect. If the wrong orientation is initially selected no harm will ensue, the operation of the circuit will simply be counter intuitive with 'up' becoming 'down' and vice versa. Confirming and carefully noting the correct connection orientation for expected operation is one task easily performed during the breadboarding phase of circuit test and development. Once it has been confirmed, correct labelling on the diagram will avoid a mixup later on down the line.

With a layout such as that shown in Figure 11.4 the process of breadboard construction is a simple matter, and with all components to hand should take no time at all. Always carefully check and recheck all connections before connecting power to the circuit. Pay particular attention to power and ground connections, making sure that none have been accidentally inverted. It is worth bearing in mind that the 386 will not tolerate reverse power connection and this is one sure way of burning out the chip.

Once the breadboard circuit is up and running, modifications can be tried and different component values can be compared. Try smaller coupling capacitors for a more trebly sound. Experiment with different JFETs in the input buffer. See what difference a bigger volume pot makes, or compare linear and log pot tapers. Revert to a gain pot between pins 1 and 8. Run at lower or higher voltages (within the limits the 386 can handle). Listen without the zobel network. See how well it works into different loudspeaker loads. Compare operation with and without the stabalising capacitor on pin 7, with particular attention to high gain operation. After all these variations have been examined a much better feel for how the circuit works, and what it is capable of will have been gained, and so the final design to be soldered up as a permanent circuit will provide the best performance possible.

Along with all these modifications there are a myriad tests and measurements which can be performed in order to quantify and visualise the consequences of each change. One invaluable tool at this stage is the oscilloscope. An example of the kind of information which can be obtained was shown above when discussing the role of the zobel network. Figure 11.5 below offers another snapshot into the way in which this circuit works. Here the low and high gain settings are being compared. As before the top trace in each case is the input signal and the bottom trace is the output, and again the settings used to take the measurements are noted below the figures.

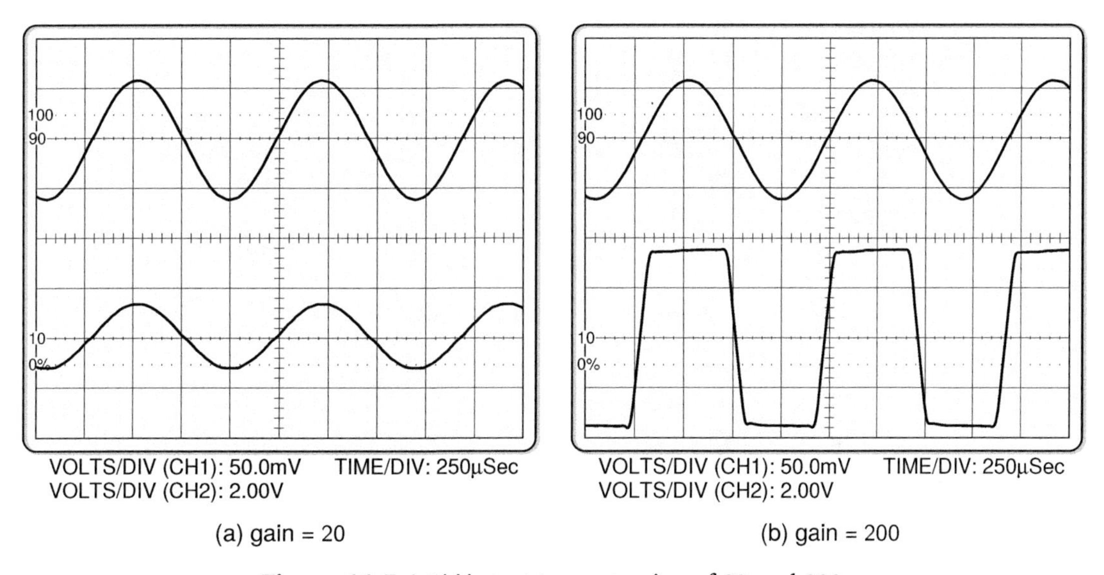

Figure 11.5 A 1kHz test tone at gains of 20 and 200.

Notice that the output has been displayed using the same settings in both instances. Clearly with the gain at 200 the output is not ten times bigger, and furthermore the clean sine wave input has been distorted badly into something more closely approximating a square wave. Close to the centreline the amplification appears to have been performed quite cleanly. However the amplifier quickly runs out of headroom as it attempts to extend the output signal beyond the range facilitated by the available power supply voltages, and so the signal is clipped at round about +1V at the bottom and +8V at the top, leading to the squared off signal seen in Figure 11.5b. This is classic signal clipping resulting from overdriving the amplifier. In listening to these two signals the second is louder certainly but not 20dB louder. It is however vastly more distorted. In this instance, this can actually work quite well as a basic electric guitar distortion effect.

Stripboard Layout and Final Project Build

With the design finalised on breadboard it is time to draft a stripboard layout ready for building. Figure 11.6 illustrates a good compact stripboard design. Notice that two copies of the board are shown. Above is the component side of the board while below it the board has been flipped over to show the copper side where all the soldering happens. The only things shown on the copper side are holes and breaks. Holes (the black rings on the diagram) just identify locations where mounting holes might be drilled through the board to allow standoffs to attach the board to whatever enclosure it is going to be built into.

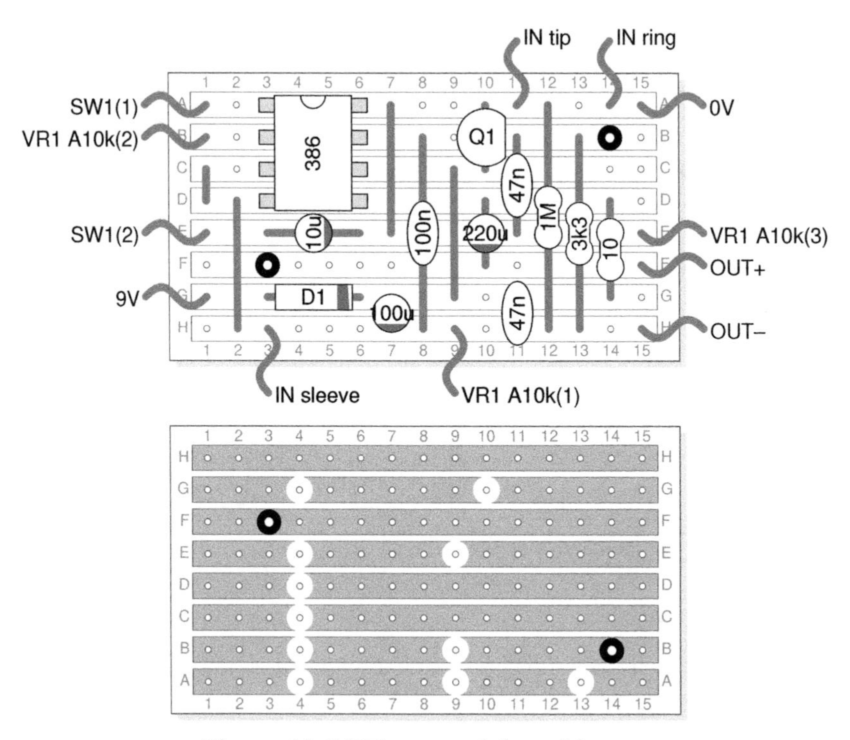

Figure 11.6 386 amp stripboard layout.

Breaks (the white rings on the diagram) are places where the horizontal copper strips of the stripboard need to be broken so that different sections of the same strip can be used to make different connections in the circuit. Notice that a DIL package IC will always need to have a column of breaks running under it to isolate the legs on either side of the chip (in this case breaks at locations 4A through 4D). Notice also that the strip indexing has been reversed on the copper side since when the board is flipped over the top row ends up at the bottom and vice versa. Great care must be taken to make sure that everything gets correctly located on the board as once soldering has happened it can be very difficult to undo any mistakes. Breaks in the wrong place are even more difficult to remedy and it is all too easy to mirror the required positions if sufficient care is not taken.

The step by step build procedures described in Chapter 10 should be followed in order to complete the project. Notice one small tweak which has been introduced in this final circuit when compared to the (otherwise identical) design provided for the breadboard layout. There will often be one or two housekeeping details which are not included until the final version. In this case automatic on/off switching has been implemented as described in Chapter 15.

Instead of a TS input jack a TRS jack has been used. The ring terminal is connected to the negative of the power supply while all other ground connections in the circuit are connected with the input jack's sleeve in the usual fashion. This means that even after the battery or other power supply has been attached, the circuit is not powered up until a TS jack plug (i.e. a guitar lead) is inserted into the input jack. At this point the sleeve of the plug shorts together the ring and sleeve of the jack and the circuit is live. This is an arrangement commonly encountered in guitar effects pedals. They do not tend to have a dedicated on/off switch but rather remain off until a lead is plugged into them.

The final circuit can be built up into any desired enclosure either with or without a built in loudspeaker. Internal/external speaker switching can be implemented as described earlier if desired, and similarly the automatic battery/PSU selection arrangement also described in Chapter 15 could be added allowing the amp to be run either from an internal PP3 9V battery or an external power supply.

Project II – Stepped-Tone Generator

The Stepped-Tone Generator (or Sound Synthesizer) was a circuit originally developed by Forrest Mims and published in a number of circuit cookbooks under the two names above, including Mims (1996), one of many circuit collections released by companies like Maplin and Radio Shack over the years, by various authors (Robert Penfold is another author well worth keeping an eye out for, see for instance Penfold (1986, 1992, 1994)). The circuit presented in this project has become popularly known as the Atari Punk Console or APC on account of the interesting noises it can produce which are not unlike early eight bit video game sounds like those heard on various Atari computer games. The circuit utilises a very popular integrated circuit, the 555 timer, which is often one of the first chips encountered by electronics builders, as it lends itself to the design of lots of interesting, and often fairly simple, circuits. The APC requires two 555 timers or a single 556, which is a chip which just packages two 555s into one IC.

As is always the case with such popular designs, variations and modifications abound. One well known and interesting variant called the Vibrati Punk Console adds a third 555 timer to the original complement of two, which acts as an LFO to provide a further level of modulation to the sounds available. The project presented here sticks fairly closely to the original Forrest Mims design but as always, development, modification, and experimentation is to be encouraged. In this case, just the sizes of the three pots have been adjusted, with the output connection left unspecified. The 555 is often used to drive a small loudspeaker directly (as in Mims' designs), but this loads the chip excessively and is probably not a good idea long term. Ideally the output here would be connected to the higher impedance input of a power amplifier, avoiding the heavy load on the 555 or 556 of driving a speaker directly.

Circuit Operation

The circuit diagram for this project can be seen in Figure 11.7. As stated, it differs from Mims' original design in a couple of respects. The two control pots have been changed from 500k to 100k (either works well), and where the original has a 5k pot directly driving an 8Ω loudspeaker, here a much larger 100k pot is specified with the output sent to a jack for connection to an external amplifier (such as the one described in Project I) or to some other signal input.

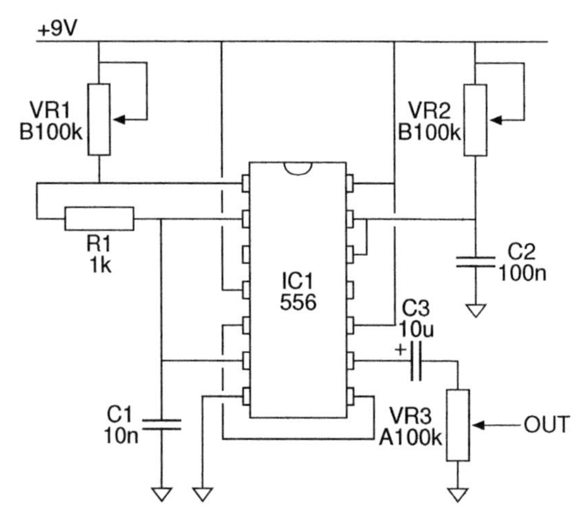

Figure 11.7 APC circuit diagram.

Notice that the frequency pots are specified as linear taper pots by way of the B designator in the circuit diagram. The output level pot, with an A designator, is indicated to have an audio or log taper. These niceties are not important to the operation of the circuit but can make for a more user friendly interaction characteristics as explained in the section on pot tapers in Chapter 12. It is always worth trying different pot variants in any circuit if the opportunity presents itself.

The final difference from the original design is that the orientation of the output capacitor is reversed. The original Forrest Mims design had the speaker referenced to the positive power rail (a somewhat unusual configuration) whereas the output jack here exhibits the more traditional ground referenced sleeve connection. It would be interesting to investigate if the symmetry of the output signal means that the originally specified mode of operation has any benefits or sonic differences.

As explained, the circuit is built around a commonly available IC called the 555 timer. The 555 timer itself is a very well known and much loved device popular with electronics hobbyists for the vast array of simple projects (audio and otherwise) which have been designed around it. The 556 chip specified in this project is just two 555 timers on one chip. The same project can also be found laid out for two individual 555 chips – the two alternatives are functionally identical.

The 555 timer itself is advertised as a high precision, high stability timing module which can be used to construct a variety of useful timer and oscillator circuits. Versatile circuits can be developed which can include such features as variable duty cycle and pulse width modulated square wave output signals. As is often the case, the chip's data sheets provide guidance on how to design circuits for a wide range of applications, and as mentioned above the popularity of the device means that a vast range of other resources have been developed using the 555 timer in many and varied applications.

As mentioned already, the circuit here is built using a 556 chip which contains two individual timer blocks. The circuit uses the first timer to construct an oscillator. The output of this is then fed into the second timer which is configured to act as a variable frequency divider. The result is some very interesting sounds emanating from the circuit. It is difficult to convey visually the nature of the sounds available, which derive their primary character from the dynamic variations achieved as the controls are adjusted. The basic idea of the operation, and in particular the frequency dividing nature of the operation of timer two, can however be glimpsed in the traces shown in Figure 11.8. Channel 1 (top) shown the output of the first oscillator, whose frequency is controlled by pot VR1. This feeds into the second timer, whose output is displayed in the second trace. The frequency division aspect can be seen in the lower number of transitions present in trace two. The division factor changes in jumps as pot VR2 is adjusted giving surprisingly musical runs of ascending and descending tones.

The first of the two control pots varies the frequency of the initial oscillator while the second pot adjusts the frequency divider's operation. The final control pot (VR3) is just a simple output level control. The waveform coming out of the chip will be too large for most signal inputs that it might be connected to, and so this pot provides whatever degree of attenuation might be required. Notice from the legend in Figure 11.8 that the display on each channel is set to five volts per channel and so it is clear that the output has an amplitude of about $7V_{pk-pk}$.

Consider that the 386 amp in Project I has a maximum output swing of about 7V when run from a 9V battery, and that its minimum gain is 20. From this it is an easy

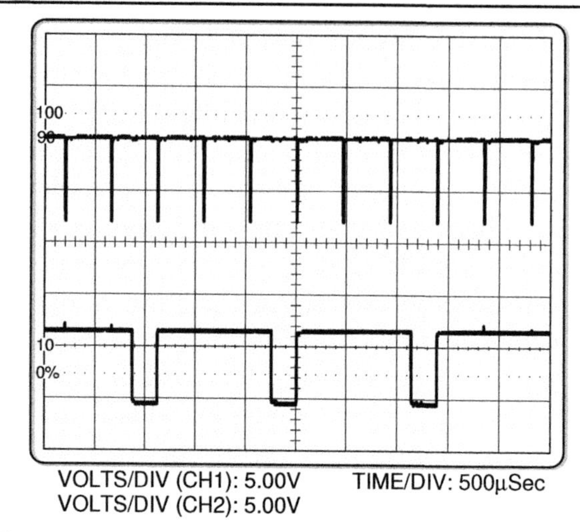

Figure 11.8 Output of the two oscillators in the APC.

matter to calculate the maximum peak-to-peak input signal level which the amplifier can handle without major clipping – $V_{pk-pk(max)} = 7/20 = 0.35V$. Clipping a signal which is primarily bi-level in nature such as that illustrated in Figure 11.8 makes little difference to its shape and thus its sound, but clearly the output of the APC is what could be called a hot signal, and an effective output attenuator is going to be a good idea.

Layout and Construction

As always the next step in the development of a project once a new circuit has been designed or otherwise procured is to come up with a good breadboard layout in order to test the basic design and also to test any modifications which might want to be considered. Figure 11.9 provides a layout for this project. The circuit itself is quite simple and as such an easy to follow design layout is not too taxing to arrive at.

It would be good practice and most instructive for the reader to develop a layout from scratch. It need not turn out the same as the one provided here, indeed it is most unlikely to do so. The important thing is that it is (of course) correct but also organised and easy to follow. The designs included in this book sometimes include connections which might not seem the most obvious but are usually chosen in order to keep a neat and easy to read diagram, avoiding where possible crossing wires and component leads, and keeping connections short and in as much as it is possible, running vertically and horizontally. These details are not important to the operation of the unit, but can make the processes of building, testing, and debugging the actual circuit a far less painful experience.

Once satisfied with the results of the breadboard circuit the project can proceed on to the design of a suitable stripboard layout as shown in Figure 11.10, in order to allow for a permanent soldered circuit to be constructed. Often some of the physical layout

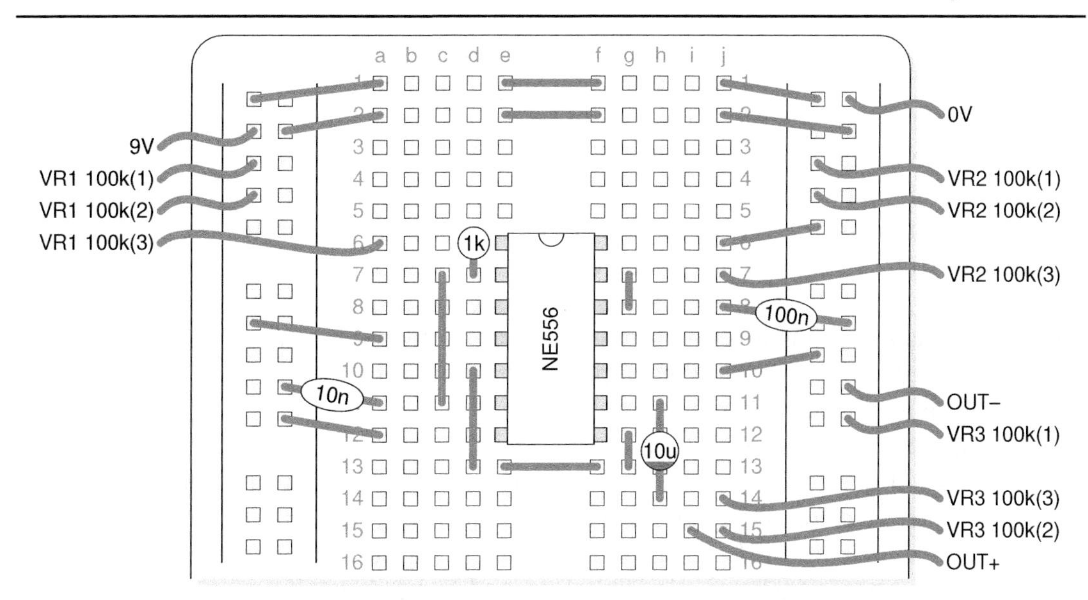

Figure 11.9 APC breadboard layout utilising an NE556 dual timer IC.

can be derived from the existing breadboard layout, but the differing nature of the two constructions means that much of the layout will best be considered from scratch. It will often be the case that a couple of strips above or below an IC might best be designated as the primary positive and negative power rails for instance.

In this case with the two capacitors to ground emerging from pins close to the top of the chip it makes sense to have a ground strip at the top. Further consideration of the layout places the +9V bus at the bottom, as it becomes too crowded to get those connections up to the top as well. The second strip which has been used below the chip is perhaps a little wasteful of space. It is only used to make the connection between pins five and eight. This could have been achieved using a short jumper wire under (or indeed over) the chip. For the sake of clarity the extra strip was employed and the overall design remains compact.

Again the copper side of the board is shown in order to identify the correct locations for all the necessary breaks in the copper strips. This time it has been fitted to the side rather than underneath as in the previous project, and as such it must be remembered that positions are mirrored left to right rather than top to bottom as in the previous case. To reflect this difference notice that now the column numbers are inverted (1 to 12 becomes 12 to 1) while row letters (A to J) remain unchanged. Once again this maintains consistency when referencing locations on the board from either the component side or the copper side. Note for instance that column 6 runs directly under the chip on the components side, and on the copper side it hosts a column of breaks in the copper strips in order to remove the unwanted connections between the pins on each side of the chip.

It is always a worthwhile exercise to retrace the connections on the stripboard, remembering to take the breaks into account, and confirm that every connection corresponds to what is happening in the original circuit diagram. For instance, the break at 10J allows the negative terminal of the 10μF capacitor to connect to pin three of the output pot while separating this connection from the five +9V connections which come together on the rest of strip J. Similarly the break at 10A separates the final output connection from the ground bus which occupies the rest of this strip. Each point in the circuit can be similarly examined to confirm that the required connections, and only the required connections are being made. A useful technique when dealing with an integrated circuit is to go around each pin on the chip in turn and check that it goes to all the right places (and no wrong places).

The connection between pin two of VR3 and OUT+ at the top righthand corner of the board could be left out, and wired directly between the pot and the output jack. It has been included for clarity as this is intended to be a project suitable for a novice circuit builder. It is often the case that the wiring harness (all those connections to off-board components such as pots, switches, and jacks) is itself a complex affair and will regularly merit a separate set of build instructions dedicated specifically to getting it right.

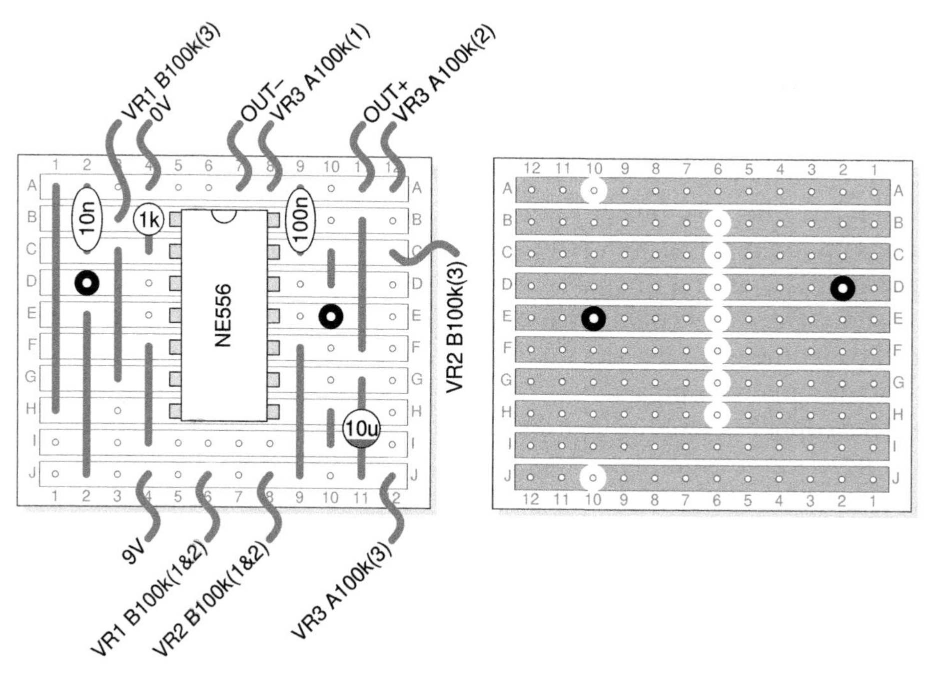

Figure 11.10 APC stripboard layout utilising an NE556 dual timer IC.

Project III – Modulation Effect

Having built an amplifier and a noisemaker, project three presents an audio effect circuit. Effects take an audio signal as their input and they output an altered version of that signal. In the case of the modulation effect, how this signal alteration is achieved can in broad terms be described as follows. An oscillator modulates the state of a transistor in the audio path, and this produces a pulsating effect in the audio, which is heard as the 'wobble' which gave the original circuit its name – this particular project is based on a design called the 'Wobbletron', taken from a collection of interesting audio circuits generally referred to as 'Escobedo's Circuit Snippets'. The circuits in this collection were designed and annotated by Tim Escobedo and the collection has become one of the most popular sources of interesting circuit ideas for audio electronics hobbyists and experimenters. It is an excellent place to start for anyone interested in getting more deeply into practical audio electronics.

The circuit diagram for the modulation effect is shown in Figure 11.11. The only differences from the original design are that a value has been specified for C1, and the three transistors have been substituted (based on availability and using the substitution procedures outlined in Chapter 17). The circuit separates conveniently into two parts. To the bottom left can be seen the oscillator, while to the top right is the section which represents the audio path through the circuit. A single wire from the 500kΩ pot VR2 to the gate of the JFET Q3 connects the two. This wire carries the oscillators output into the JFET, turning it off and on and thus wobbling the audio as it passes through this point in the circuit, producing the effect. It is worth noticing the circuit symbol used for the variable resistor VR1. This is a shorthand for the configuration seen for VR1 and VR2 in the APC circuit diagram shown in Figure 11.7. In the case of the APC the three connections to the pot terminals are shown explicitly. Here the diagonal arrow indicates a variable resistor but does not show exactly how to connect it. It is assumed in this case that the circuit builder will understand how to implement the appropriate wiring.

Circuit Operation

In electronics there are many ways to build an oscillator. This particular one is what is called a phase shift oscillator. When a signal passes through a capacitor it experiences a frequency dependant shift in its phase due to the lag introduced as the capacitor is charged and discharged. The amount of phase shift introduced is in fact a function of three things: the frequency of the signal, the capacitor's size, and the values of surrounding resistors in the circuit. In this circuit a ring of three capacitors (C4, C5, and C6) can be seen to run from Q2's collector to its base. Each capacitor introduces a little more phase shift to any signal passing along this route.

The transistor Q2 is connected in a common emitter configuration which is to say that its output is taken from the signal at the collector, which is an inverted version of the signal at the base. So if a particular frequency within this signal experiences

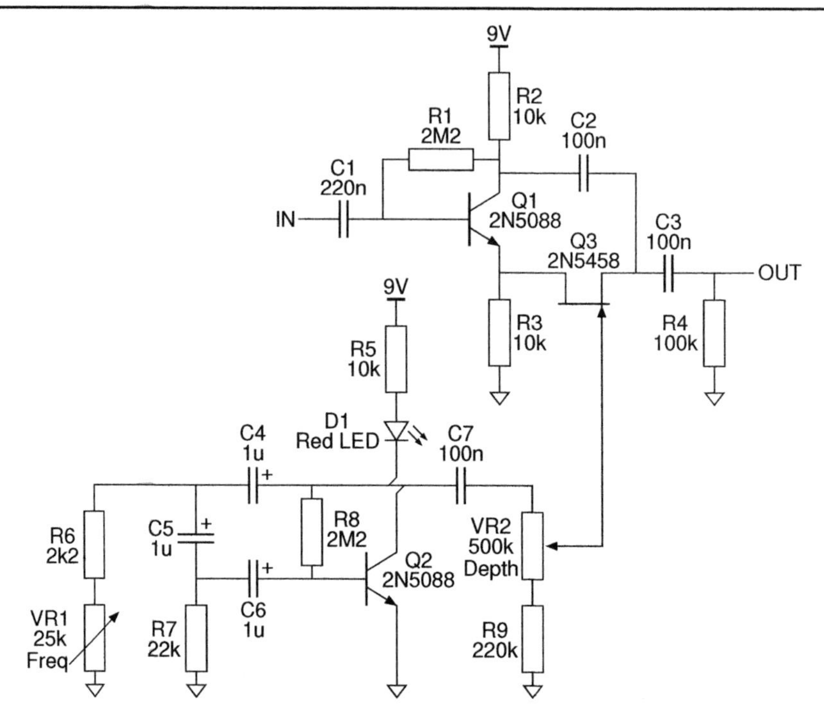

Figure 11.11 Modulation effect circuit diagram.

180° of phase shift in travelling through the three capacitors between the collector and base then when this is added to the natural 180° phase shift introduced through the transistor's action that frequency will end up back in phase as it emerges from the transistor and as such will be amplified. Meanwhile all other frequencies which might have been present in the signal end up out of phase after travelling around this loop and therefore rather than reinforcing, these frequency components tend to cancel themselves out. In this fashion the initial random noise of the circuit kicks off self sustaining oscillations at a particular frequency while damping down all other frequencies.

Changing the setting of the 25kΩ pot VR1 alters the circuit conditions in the phase shift network and produces a different frequency with just the correct amount of phase shift to be reinforced. In this way a variable frequency LFO signal emerges from the oscillator ready to be applied to the gate of the JFET, driving the operation of the effect. With the component values shown the frequency range of the LFO typically runs from round about two hertz up to something in the vicinity of five hertz. This range can be quite variable from circuit to circuit due to component variability and can be adjusted in any particular case by selecting different values for the capacitors and resistors making up the phase shift network.

The LED provides visual feedback of the operation of the LFO as it flashes in time with the oscillator. This can prove to be quite a convenient first check when building and debugging this circuit as it immediately confirms whether or not this section of the circuit is operating. The flashing is likely to be quite dim as the LED is not driven hard, but it should be sufficient to be clearly visible in most cases.

Moving up to the audio section of the circuit, the audio signal which is to be affected enters the circuit at the point labelled IN. It then passes through the familiar coupling capacitor C1 and enters the base of transistor Q1. The three resistors R1, R2, and R3 set up the transistor's operating point ensuring that suitable bias voltages appear at the various terminals of the device. Following the audio path as it continues on, signal is tapped off from both the collector and the emitter of Q1, the upper path via C2, a 100nF capacitor while the lower path is through the JFET Q3 whose state is being modulated under the control of the LFO. These two independently phase shifted signal components are recombined to yield the final modified signal which emerges from the circuit through the final output coupling capacitor C3.

Figure 11.12 illustrates the effect which this circuit has on an audio signal. Panel (a) shows the control signal emerging from the LFO on channel 1 while channel 2 shows the circuit's modulating effect on the amplitude of an audio signal passing through it.

From the figures given in panel (a) the frequency of the LFO can be read off the plot as about five hertz, the top of the LFO's range. The audio in the lower trace is at a frequency of approximately 1kHz which translates to about 500 cycles across the sweep of the plot. This is why the signal shows up as a solid black area – it is only the varying amplitude envelope which can be seen. On the other hand, panels (b) to (d), with a much quicker time base (seconds per division) allow the shape of the test signal to be seen. These three panels show the input and output signals at three different positions along the progress of the LFO modulation, labelled as points (I), (II), and (III) on the plot in panel (a). As can be seen, the input signal remains steady as it should, while the output signal goes from unaltered when the LFO level is sufficiently low to successively more heavily distorted moving into the centre of the wobble.

These plots were taken with the depth pot VR2 set close to maximum. As the depth setting is dialed back the amplitude of the signal going to the gate of the FET Q3 is progressively reduced. With the voltage swing driving the FET's gate reduced the amount of modulation is reduced until finally there is not enough variation to modulate the signal at all. Thus the depth pot VR2 allows for control over the strength of the effect while the frequency pot VR1 alters the nature of the effect from a slow sweeping effect at low LFO rates to a strongly pulsating sound as the rate of the LFO increases.

Layout and Construction

The modulation effect is certainly the trickiest of the three project layouts presented here, but with a little effort a pair of breadboard and stripboard layouts which are

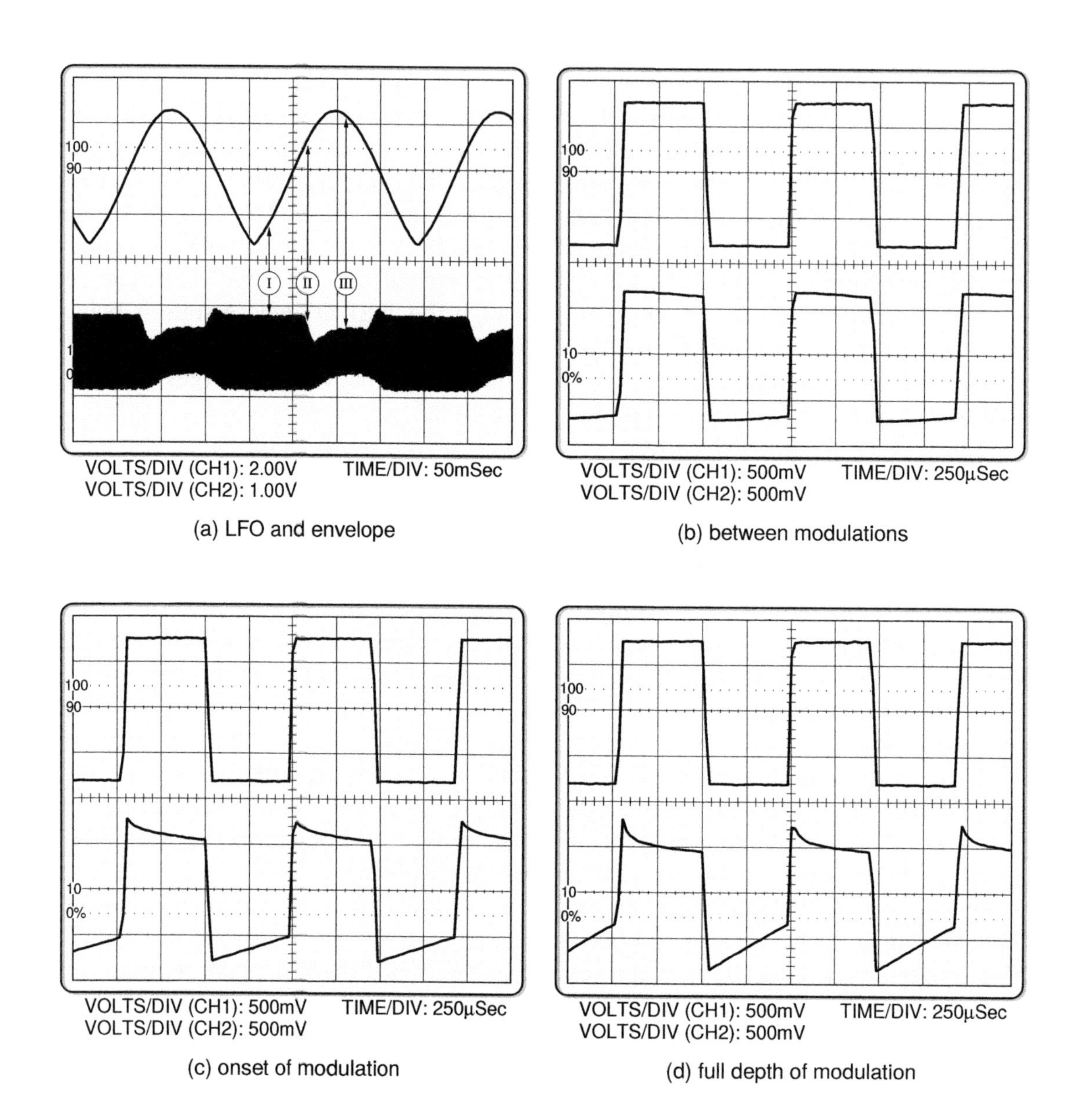

(a) LFO and envelope

(b) between modulations

(c) onset of modulation

(d) full depth of modulation

Figure 11.12 Operation of the modulation effect.

both clear and compact can be achieved. When laying out transistors it is essential to know the exact part number which is to be used. A quick search online will offer up a data sheet for the device, and the ordering of the three pins can be confirmed. The pinout can differ from one part type to another, so it is important to confirm the correct ordering for the particular devices to be used.

Figure 11.13 provides the necessary information on the parts used in this case, and shows the correct pin designations. For the 2N5088s which are bipolar junction transistors Q1 and Q2 the pinout is shown as EBC, indicating that the pins are emitter, base, collector from left to right looking at the flat face of the device with legs down. For the 2N5458 JFET Q3, the DSG similarly stands for drain, source, gate, thus identifying the arrangement of pins in this device. Once all this is known, an appropriate circuit layout can be designed.

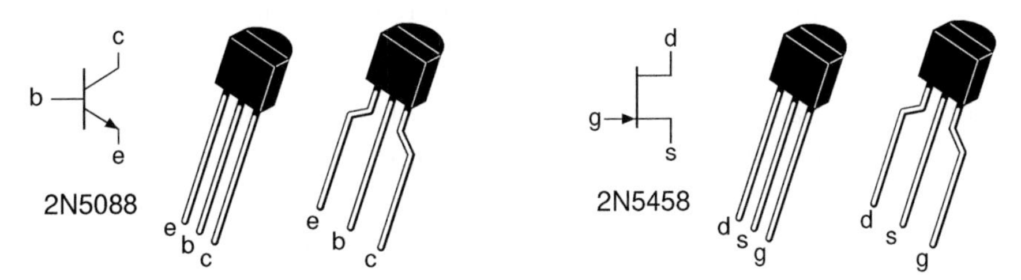

Figure 11.13 Pinouts for the transistors used in the modulation effect.

The separation of the circuit into two parts described above can be seen in the breadboard layout also around row 7 on the left and 10 on the right. Above these rows the audio section is laid out while below them is the LFO. With three transistors in the circuit, great care must be taken not to mix up the identical looking devices as they are inserted into the board, and it is of course also important to get the orientation correct. In this design the BJTs Q1 and Q2 on the left side of the board have their flat faces to the right, while JFET Q3 has its flat face to the left.

Notice also that the standard power rail connections have been included distributing 0V and +9V to both sides of the board. This has been done here as a standard step even though the +9V rail is not needed on the right side of the board and these three jumper wires across the top could be omitted.

Once again, having built a working circuit, it can be worth considering if any tests or modifications might be worth investigating. Reviewing the notes included in the original 'Circuit Snippets' document throws up a suggestion that the overall tone of the effect might be altered by trying different values for the input coupling capacitor. In fact the original circuit diagram does not specify any value at all. The 220nF size suggested here is simply a fairly typical value for this standard component. Larger values allow the full frequency range of the incoming audio signal to enter the circuit while smaller values will roll off progressively more and more of the low frequency components. As such, in common with many audio circuits, a lighter more trebly sound can be achieved

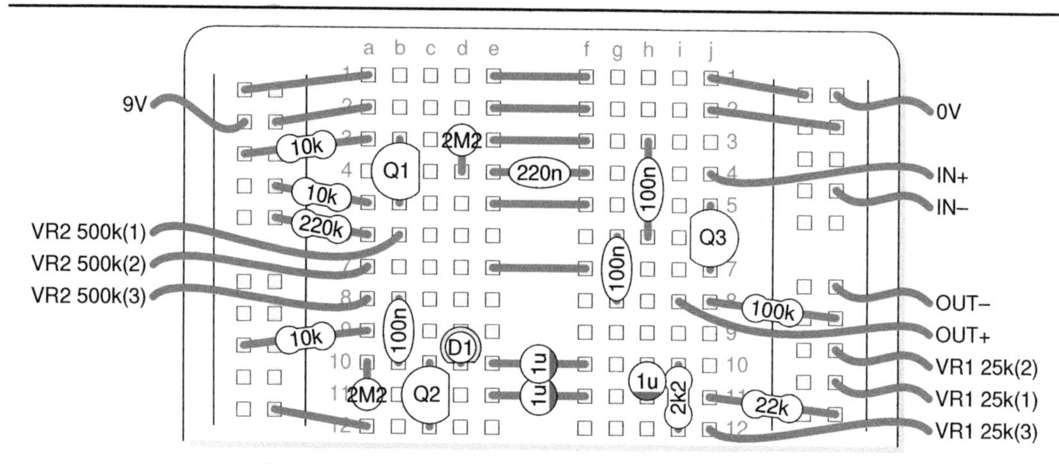

Figure 11.14 Modulation effect breadboard layout.

using smaller input capacitor values while a fuller or darker tone can be achieved with larger values (assuming the original audio has significant low frequency components in the first place).

The characteristics of the LFO which is used to drive the modulation at the heart of this effect can also be altered. Changing the values of the three 1μF capacitors will alter the base frequency of the oscillations, while using a different size pot for VR1 will allows for a larger or smaller range of frequencies to be produced. As always, experiment, find the setup that gives the best results, and go with that.

With testing completed and design decisions finalised it is time to move on to the drafting of a stripboard layout. Again an example layout is presented here, in Figure 11.15 but it is always good practice to attempt to draft one from scratch. The layout given here is particularly busy and a less complicated one could certainly be developed if a larger stripboard matrix were used. While developing the smallest possible footprint can be a rewarding challenge, a more spread out stripboard design can make the building process much easier. This can be especially important until ones soldering skills have developed to the point where such tight quarters do not pose a problem in keeping solder joints safely separated.

The relatively large number and scattered nature of the breaks on the copper side of the board could also cause problems for a novice builder, in addition to the soldering of so many components at tight quarters. Given the regular 0.1 inch hole pitch of standard stripboard, it can be seen that this really will be a very small board, and aiming for a larger footprint might not be a bad thing for a first attempt at this project. With rows and columns of only six and sixteen respectively the board shown here will only be about half an inch high and one and a half inches wide. Not much room to play with but perhaps a good test of soldering skills.

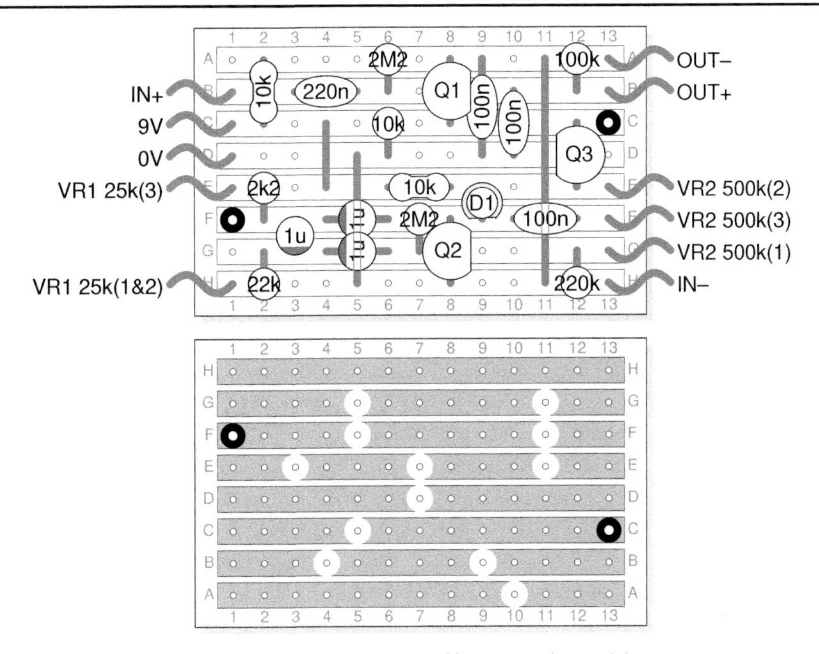

Figure 11.15 Modulation effect stripboard layout.

References

C. Anderton. *Electronic Projects for Musicians*. Amsco Publications, 1992.

N. Boscorelli. *The Stomp Box Cookbook*. Guitar Project Books, 2nd edition, 1999.

B. Duncan. *High Performance Audio Power Amplifiers*. Newnes, 1996.

E. Evans, editor. *Field Effect Transistors*. Mullard, 1972.

W. Leach. Impedance compenstion networks for the lossy voice-coil inductance of loudspeaker drivers. *Journal of the Audio Engineering Society*, 52(4):358–365, 2004.

F. Mims. *Engineer's Mini-Notebook: 555 Timer IC Circuits*. Radio Shack, 3rd edition, 1996.

R. Penfold. *More Advanced Electronic Music Projects*. Bernard Babani, 1986.

R. Penfold. *Electronic Projects for Guitar*. PC Publishing, 1992.

R. Penfold. *Practical Electronic Musical Effects Units*. Bernard Babani, 1994.

A. Sedra and K. Smith. *Microelectronic Circuits*. HRW, 2nd edition, 1987.

D. Self. *Audio Power Amplifier Design*. Focal Press, 6th edition, 2010.

G. Slone. *High-Power Audio Amplifier Construction Manual*. McGraw-Hill, 1999.

R. Wilson. *Make: Analog Synthesizers*. Maker Media, 2013.

Part III

Component Reference

12 | Resistors

Resistors drop voltage and limit current

The resistor is the simplest and also the most fundamental of all electronic components. It is designed to have a constant resistance across all frequencies and under all normal circuit conditions (within reason). Resistors limit the amount of current flowing through them and they drop voltage across their terminals. Voltage drop is the term used to talk about the difference in voltage between two points in a circuit, also commonly referred to as the potential difference between the two points.

Resistors come in many shapes and sizes, but the most common format used for handmade electronics is the little dumbbell with coloured stripes on its body and leads sticking out from either end as in the image in Figure 12.1a. These are called through hole resistors and they are designed to be used by bending the legs, inserting them through holes in a circuit board, and then soldering them in place on the boards reverse side.

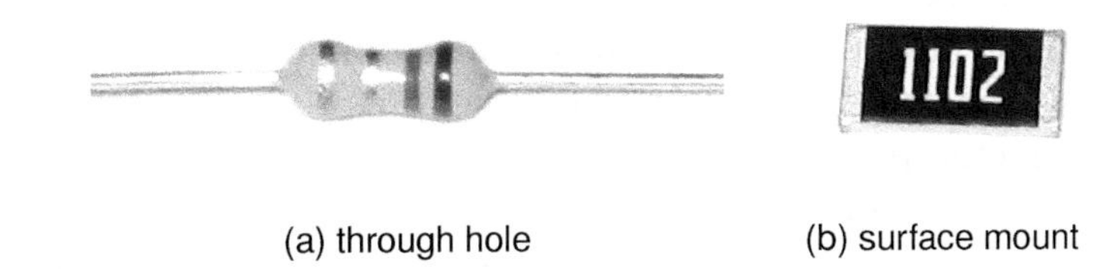

(a) through hole (b) surface mount

Figure 12.1 A through hole resistor and a surface mount (SMD) resistor.

For comparison, Figure 12.1b shows the kind of resistor typically used in modern machine assembled circuits. These resistors look like small chips of black plastic, and are called surface mount devices or SMDs. They come in different sizes but tend to be very much smaller than typical through hole resistors. Some people do hand solder the larger of these surface mount devices but it is a tricky skill to master and is not addressed in this book. The shiny regions at either end of the device are the solder pads where it is soldered into the circuit, and the number indicates the value of the resistor in ohms. The first two digits are the start of the number and the third digit tells how many zeros to add. So in the case of Figure 12.1b the value is thirty followed by three zeros, i.e. thirty thousand ohms or 30kΩ. By contrast the resistance of a through hole device is determined by decoding the coloured stripes on its body. The interpretation of resistor colour codes is examined shortly.

Two different symbols are commonly used to represent resistors when drawing circuit diagrams. In this book the convention of a rectangle is used exclusively as in Figure 12.2a. The zigzag symbol in Figure 12.2b is used widely elsewhere and indeed both are encountered regularly. There is no difference between the meanings of these two alternate forms of the resistor circuit symbol.

Figure 12.2 Circuit diagram symbols for a resistor.

The resistance value measured in ohms is the most important quantity describing any given resistor. There are however quite a number of characteristics which together define the precise nature and performance of any resistor. In addition to the resistance value, these include such things as its manufacturing tolerance, power handling capability, composition, and temperature coefficient. Self (2015, pp. 44–63) provides an examination of the effects of tolerance, along with consideration of the noise and distortion contributions of resistors in audio circuits. The following sections examine the most important facets of each of the above characteristics in turn.

E-Series Preferred Values

Certain particular resistance values turn up again and again, such as 47 and 82. There is good logic behind the use of such commonly encountered and yet strangely irregular numbers. Resistors come in values from a fraction of an ohm up to several million ohms (the most commonly used values are generally in the range 100Ω to 1MΩ). It would be impractical for manufacturers to offer every conceivable resistor value in these ranges, and so this has led to the use of the E-series preferred values. These series specify sets of standard values. Each series uses a different number of values to cover a decade (a multiple of 10). So for instance the E3 series includes three values while E12 consists of twelve. The standard series extend up to E192 for very precise resistance values, but E12 and E24 are probably the two most commonly used sets of values.

Table 12.1 shows the values in the first three series: E3, E6, and E12. There is certainly no need to remember these values but the commonest, like 10, 22, 33, and 47 will quickly become familiar to anyone regularly working with electronic components. Clearly it makes sense for smaller values to be spaced more closely and larger values to be more spread out. While a gap of twenty ohms might be reasonable when for instance going from 100Ω to 120Ω, providing the option of say both one million ohms and one million and twenty ohms makes less sense. The most logical way to choose what values to make readily available is to use a multiplication factor to go from one value to the next. In this way the step size slowly and steadily increases as the resistor values increase.

Table 12.1 The first three resistor E-series preferred value series

E-series	Preferred values											
E3	10				22				47			
E6	10		15		22		33		47		68	
E12	10	12	15	18	22	27	33	39	47	56	68	82

So in order to work out what values should be included in a given series the first thing to do is to decide on the number of values to be available in each decade (i.e. the E3, E6, E12, E24 thing). Next a bit of maths gives the necessary factor, and then any value can be multiplied by this factor and the answer rounded to the nearest sensible number to give the next value in the series (the maths will give answers like 14.677992... – that one is the third in the E12 series, so let's just call it 15). Notice how the gaps between values increase going up through the E12 series shown in Table 12.1, and remember of course that the next step up will be from 82 to 100, a step size of eighteen, the biggest yet. Similarly, the next values heading down from 10 will be 8.2, 6.8, and so on. Thus standard values in any range can be specified, going up as high or down as low as is needed.

Tolerance

Every resistor will have an actual value which is slightly different from its specified value due to the limitations of the manufacturing process. Resistors are manufactured to a certain tolerance, the commonest being $\pm 5\%$ and $\pm 1\%$. The tolerance indicates the guaranteed maximum deviation from the labelled value. Five percent of two hundred and twenty is eleven, so a 220Ω resistor with a $\pm 5\%$ tolerance might actually have a value anywhere in the range 220 ± 11 (i.e. 209Ω on up to 231Ω). Resistors with tight tolerances will generally cost more. In most of the circuits considered in this book (and in many circuits in general) precise values are not crucial to the operation of the circuits, so tolerances are not usually of any great concern, and $\pm 1\%$ or even $\pm 5\%$ will usually be quite sufficient. It is still worth being aware of their existence and their values for the components used.

There is a logical relationship between the tolerance and the E-series. Consider two resistors, one 220Ω and the other 230Ω, both with $\pm 10\%$ tolerance. Ten percent of 220 is 22 so the 220Ω resistor might actually have a value anywhere between 198 and 242. Similarly with ten percent of 230 being 23, the 230Ω resistor can fall anywhere in the range 207 to 253. Now clearly the 220Ω resistor is supposed to be the smaller of the two but its largest allowed value (242Ω) is significantly bigger than the smallest allowed value of the nominally larger resistor (207Ω). It would make little sense to have these two resistors in the same set, so maximum sensible tolerances for each E-series can be defined which avoid large overlaps in the possible values of adjacent nominal values. These maximum tolerances are shown in Table 12.2.

Table 12.2 E-series and their maximum standard tolerances

E-series	Maximum typical tolerance
E3	±50% tolerance (no longer used)
E6	±20% tolerance (seldom used)
E12	±10% tolerance
E24	±5% tolerance
E48	±2% tolerance
E96	±1% tolerance
E192	±0.5% and tighter tolerances

Notice that Table 12.2 indicates a maximum standard tolerance of ±10% for the E12 series. From Table 12.1 the next value up from 22 in the E12 series is seen to be 27. Applying the ±10% bands to these two standard values gives allowed ranges of 19.8–24.2, and 24.3–29.7 respectively. So even in the very worst case scenario (which is itself extremely unlikely) a nominal 22Ω resistor with a ±10% tolerance will at least remain slightly smaller than a nominal 27Ω resistor with a ±10% tolerance. Most resistors these days are actually ±5% or ±1%, even when sold as E12 series components, so generally the actual values of resistors match their nominal values quite closely.

Resistor Colour Codes

The coloured stripes on a standard through hole resistor tell two things: the resistance value and the manufacturing tolerance. It is not necessary to remember the colour codes, they can always be looked up as and when they are needed. It is however useful to know how to interpret them. The code on a standard resistor will consist of either four or five coloured stripes. The colours can be interpreted using Table 12.3.

The last two stripes always have the same meaning – the second last stripe is the multiplier and the last stripe is the tolerance. This leaves either two or three colours which get turned into the first two or three digits of the value. So a five stripe code allows for the greater precision needed in E48 series and above, where numbers like 105 and 226 are encountered.

So for instance the 220Ω ± 5% resistor mentioned above might have either a four or five colour code depending on the E-series it has been manufactured as a part of. The four colour code would be: Red, Red, Brown, Gold, while the five band code would be: Red, Red, Black, Black, Gold. However it is worth noting that a five band code is more

Table 12.3 Resistor colour codes

Colour	Digit	Multiplier	Tolerance
Black	0	×1	–
Brown	1	×10	±1%
Red	2	×100	±2%
Orange	3	×1k	±3%
Yellow	4	×10k	±4%
Green	5	×100k	±0.50%
Blue	6	×1M	±0.25%
Violet	7	×10M	±0.10%
Grey	8	×100M	±0.05%
White	9	×1G	–
Gold	–	×0.1	±5%
Silver	–	×0.01	±10%
None	–	–	±20%

likely to be found on a ±1% tolerance resistor, so the last stripe would be brown instead of gold.

In theory there should be a slightly wider gap between the last two stripes so that it is possible to tell which way round to read the code but in practice it is often difficult to differentiate this larger space. On ±5% resistors this isn't a problem because Gold does not represent a digit and so can not appear as the first stripe. With ±1% resistors it is more problematic as Brown is also a valid digit.

It is worth saying however that the sometimes tricky task of deciphering colour codes is often unnecessary. With loose resistors it will usually be easier to use an ohm meter to measure the value. However with resistors which are soldered into a circuit, using a meter is not a reliable guide, as other components in the circuit may contribute to the answer which the meter provides. Sometimes a little bit of investigative work is needed to confirm the actual value.

Power Rating

As with all components, when current flows through a resistor it heats up. The power rating indicates how much heat it can dissipate before it runs the risk of being damaged. Standard resistors can usually handle between about 125mW and 0.6W of power safely. The power ratings of standard through hole resistors are relatively modest. The three components shown in Figure 12.3 have, going from top to bottom, power ratings of 600mW, 250mW, and 125mW. The smallest of the three, at barely 3.5mm long, is a slightly less common format, and the upper two exemplify by far the most usual range of through hole resistors encountered.

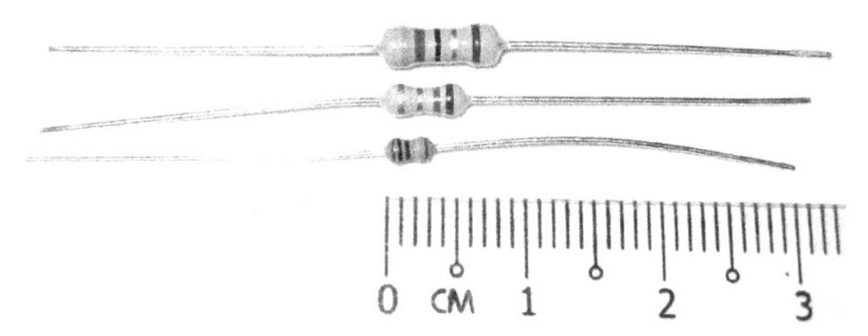

Figure 12.3 Standard through hole resistor power ratings correlate closely with the size of the components. The three resistors shown are rated at, from top to bottom: 600mW, 250mW, and 125mW.

For higher dissipations physically bigger components are needed, possibly with integrated heat sinks to help get rid of excess heat. Figure 12.4 shows examples of three types of power resistor with different power ratings.

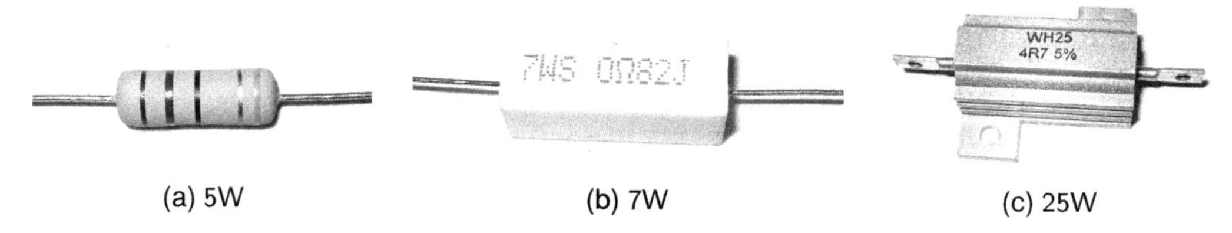

(a) 5W (b) 7W (c) 25W

Figure 12.4 Power resistors.

Power resistors are usually of wire wound construction (see the next section), often encased in a ceramic material which can handle high temperatures. For the highest power handling capabilities they can also be encased in an aluminium housing and, like the 25W component shown in Figure 12.4c, mounting holes may be provided to allow the component to be bolted onto an additional external heat sink. This provides extra heat conduction away from the resistor keeping it cooler and avoiding failure. High power components such as these can get very hot indeed when working hard and extra care must be taken when investigating such a circuit. The three high power components shown in Figure 12.4 are capable of dissipating five, eight, and twenty five watts respectively.

Material and Temperature Coefficient

The commonest constructions for standard resistors are carbon film and metal film but there are others. High power resistors will often be of wire wound construction for instance. Other makeups have also been employed. Carbon composition used to be common but has poor characteristics in many respects and is rare these days. More exotic materials are also used to provide better performance in various characteristics but these types tend to be very much more expensive and are only really of interest to manufacturers of high end gear and precision measurement equipment.

While an understanding of what these various materials and methods of construction entail is by and large unimportant, it can be useful to recognise the terms and have an appreciation of some of the more important implications of the particular types of resistors used in a project. Of the two most commonly used materials carbon film tend to be cheaper than metal film, but also poorer on most counts, with looser tolerances (typically ±5% vs. ±1% for metal film) etc. Thick film and thin film variants also exist. As a general guide, avoid carbon composition and thick film components, and choose metal film over carbon film unless the small cost savings of using carbon film are important.

One last component characteristic worth mentioning briefly is the temperature coefficient. A resistor's temperature coefficient indicates how its value changes as its temperature changes. A typical temperature coefficient may be quoted as ±50ppm/°C, where ppm stands for 'parts per million'. This value thus indicates that for every °C change in temperature, the resistance may change by up to 50 millionths of the total value, e.g. a nominally 1MΩ resistor might change in value by up to ±50Ω for each °C change in its temperature. Typically with resistors as the temperature goes up the resistance goes up, and as the temperature goes down the resistance goes down. This is called a positive temperature coefficient. There are also components which exhibit a negative temperature coefficient (as temperature goes up, resistance goes down, and vice versa). Both these property are sometimes used to good effect in specialised components, but on the whole, for standard resistors, the smaller the temperature coefficient the better. Ideally the resistance should not change at all as the temperature changes, but in reality it will always vary to some extent.

Electrical Behaviour

The details of how resistors behave and how they are analysed in particular circuits are addressed in more detail in Part II. Here just a brief examination of their electrical characteristics is presented. In particular, two important aspects of a resistors behaviour are touched upon, described here under the headings 'Transfer Function' and 'Frequency Response'.

Transfer Function

A transfer function is a general description of the behaviour of a device or system, presented in terms of how the output of the system changes as the input to the system is altered. This kind of analysis can take on many specific forms depending on the nature of the system under consideration. In addressing the electrical behaviour of a simple resistor, the input is considered as a voltage applied across the ends of the resistor, and the resulting output is quantified in terms of the current which flows through the resistor as a result of the applied voltage. (This particular relationship, plotting current against voltage is often also referred to as an i-v characteristic.) As discussed in Chapter 2, voltage can be thought of as the electrical push or pressure, and current is the flow of electricity which comes about as a result of that push.

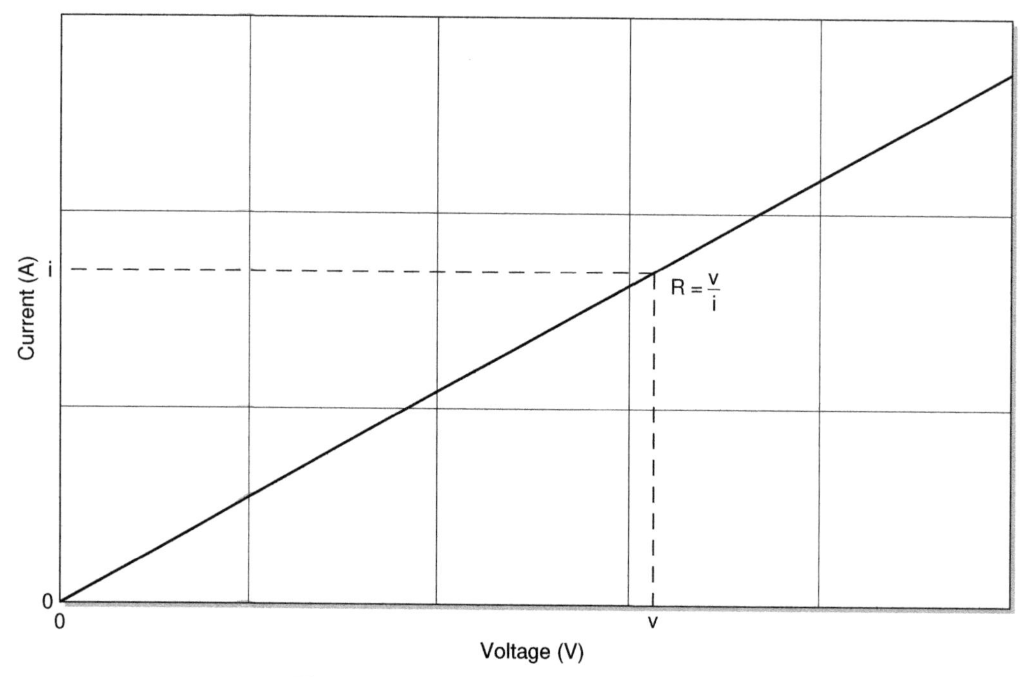

Figure 12.5 Resistor transfer function.

So the transfer function of a resistor can be presented in terms of current against voltage. This relationship is illustrated by the graph in Figure 12.5 which shows that the current (measured in amps) rises as the voltage (measured in volts) rises, and falls as the voltage falls. The graph is a straight line, which represents a linear relationship between the current flowing through the resistor and the voltage which is causing it to flow. The line passes through the origin; when the voltage is zero, zero current flows. The slope of the line represents the size of the resistor (measured in Ohms). A bigger resistor is represented by a shallower slope, and a smaller resistor is represented by a steeper slope. Later on, in Chapter 9, this relationship is formalised with Ohm's Law, one of the most important and useful rules for analysing circuits. For now it is more important to develop a general understanding of the behaviour, and to be able to visualise in general terms how a changing voltage across a resistor will alter the amount of current flowing through it. While this is a general rule in electronics, more voltage equals more current, more electrical components do not exhibit the linear relationship which is illustrated here. Generally more complex behaviour is to be expected.

Frequency Response

As the focus of this book is audio electronics, frequency response is an aspect of the behaviour of any circuit or component which is of great interest. Music, and audio in general is all about the frequency content of the signals being encountered. An amplifier wants to preserve the frequency content of a signal, an EQ aims to alter the frequency balance in a controlled way, and a distortion effect looks to introduce new frequencies related to those already present.

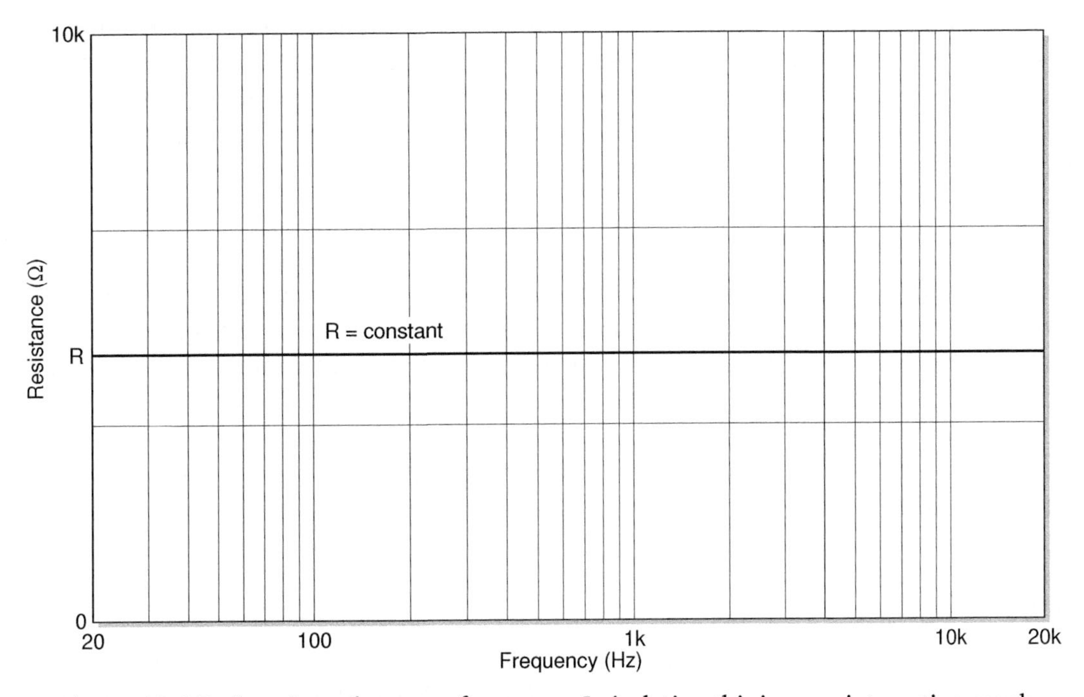

Figure 12.6 Resistor impedance vs. frequency. In isolation this is an uninteresting graph. The impedance-frequency relationship becomes very important once capacitors and inductors are added to the mix as shall be seen in the next chapter. An appreciation of the differences in behaviour between these three components is vital to understanding the operation of many key audio circuits.

As such it is clear that an understanding of how various components affect the frequency content of a signal is of great import. It should be of little surprise that the resistor, being the simplest of all components, presents very straightforward frequency behaviour, as shown in the frequency response graph presented in Figure 12.6.

The frequency response of a device or system represents how that system behaves when presented with signals of different frequencies. It can be thought of as a particular kind of transfer function where the question is how the level of the output changes as the frequency of the input signal is changed. As can be seen from the graph, the answer in this instance is that the output level remains constant regardless of the frequency of the input. This is in contrast to the behaviour of the next two components which are examined: capacitors and inductors. In that case very clear and predictable variations are observed, which lead to the conclusion that capacitors and inductors are ideal components for building filters – circuits which modify the frequency balance of a signal.

The frequency response in Figure 12.6 is shown in terms of how the resistance (here called impedance, don't worry about the difference for now) measured in Ohms changes as the frequency of the input signal changes. As has already been stated, in this case there is no change. In Chapter 13 on capacitors and inductors the same frequency response graph reappears with two new plots added, in order to compare the behaviours of these three important components. This leads naturally to an understanding of how some very simple filter circuits can be designed.

Series Rule and Parallel Rule

When analysing circuits it is often necessary to work out the total resistance of a number of resistors wired in various configurations. Even the most complex resistor network can be reduced to a single resistance value through the successive combination of pairs of resistors wired either in series or in parallel.

Two simple rules are used to achieve this, usually referred to as the resistor series and parallel rules. What these two rules are saying is that the two resistors in series or parallel could be replaced by a single resistor whose value is given by the appropriate rule without changing the operation of the circuit.

Series Rule

The series rule for resistors just says that the combined resistance of two resistors wired one after the other (see Figure 12.7) is the sum of the two individual resistor values.

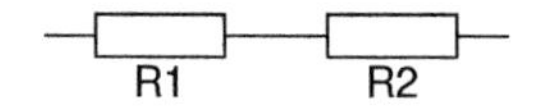

Figure 12.7 Resistors in series.

For two resistors in a circuit to be candidates for an application of the resistor series rule a number of conditions must hold. The two resistors must be directly connected at one end and must not be directly connected at their other ends (or they would be in parallel). It also must be the case that there are no other connections into the point where the two resistors join. If these conditions hold then the two resistors may be replaced by a single resistor without altering the operation of the circuit.

$$R_{series} = R_1 + R_2 \tag{12.1}$$

Parallel Rule

The parallel rule allows a combined value to be calculated for two resistors wired up side by side as in Figure 12.8. Here the calculation is a little more involved though still not what might be call complex.

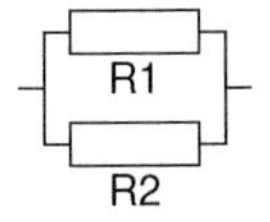

Figure 12.8 Resistors in parallel.

There is another formulation of the parallel rule which allows for the combination of more than two resistors in parallel simultaneously but there is no need to confuse the issue by introducing that equation here. On the rare occasions when three or more parallel resistors require summing, multiple applications of Eq. 12.2 yields the correct result. The form of this alternative version, along with a proof of its equivalence, is examined in the highlighted Box 12.1 (p. 212).

$$R_{parallel} = \frac{R_1 \times R_2}{R_1 + R_2} \tag{12.2}$$

Example

As a simple example of the application of these rules consider the circuit shown in Figure 12.9. Equipment A is to be connected via a cable to equipment B. However the output and input impedances do not match and so resistors R_1 and R_2 are inserted at the output of equipment A as shown. The task is to ensure that the modified output impedance seen by equipment B has the correct value.

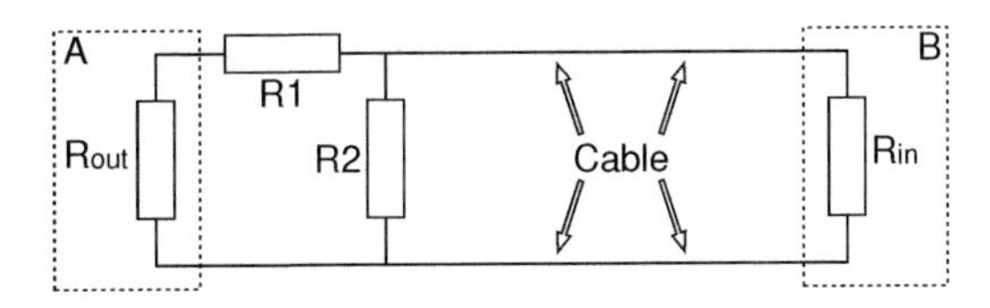

Figure 12.9 Series and parallel law example.

Equipment B looking back down the cable sees the combined impedance of R_1, R_2, and R_{out} the output impedance of equipment A, but how do these three impedances combine? First it is necessary to add R_1 and R_{out} in series.

$$R_a = R_1 + R_{out}$$

The resulting resistance is in parallel with R_2, and so they need then to be added using the parallel rule to give the final answer.

$$R_{total} = \frac{R_a \times R_2}{R_a + R_2}$$

Combining these two equations into a single calculation gives

$$R_{total} = \frac{(R_1 + R_{out}) \times R_2}{R_1 + R_{out} + R_2}$$

This particular example describes a simple circuit used to connect two pieces of equipment with different digital interfaces, as described in Bohn (2009). Equipment A has an AES digital output and equipment B has an S/PDIF digital input. The former has an output impedance of 110Ω while the latter has an input impedance of 75Ω. Digital connections like this prefer to have matched impedances (i.e. the same output impedance at one end as the input impedance at the other).

If R_1 is set to 330Ω and R_2 is set to 91Ω a suitable match is achieved,

$$R_{total} = \frac{(330 + 110) \times 91}{330 + 110 + 91} = \frac{440 \times 91}{531} \approx 75.40\Omega$$

and so it is clear that the series and parallel rules have here been successfully applied to confirm that the suggested values for R_1 and R_2 will indeed result in a suitable resistance at the AES digital output of equipment A, as seen from the other end of the cable at the S/PDIF input of equipment B.

Voltage Divider Rule (VDR)

The next step in the simple analysis of resistor circuits involves calculating the voltage which will appear at the midpoint of two resistors in series given the voltages at either end. This question can be considered in two parts. First the more straightforward case is examined where the voltage at one end equals zero volts, with an arbitrary voltage applied to the other end. This is then generalised to the case where arbitrary voltages may appear at both ends.

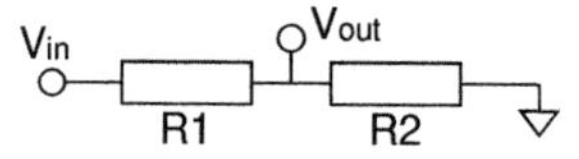

Figure 12.10 Simple voltage divider where the voltage at one end equals zero volts.

The circuit for the first case is illustrated in Figure 12.10. With the voltage V_{in} applied at the top of R_1, and the bottom of R_2 connected to ground (zero volts), the question which is to be answered is what is the voltage V_{out} which will appear at the junction of R_1 and R_2? There is a simple calculation needed to find the answer, and the equation is given in Eq. 12.3.

$$V_{out} = V_{in} \times \frac{R_2}{R_1 + R_2} \tag{12.3}$$

This basic configuration appears very often is real circuits. An attenuator or 'pad' circuit used to reduce the level of a signal by a given amount as it passes from one device to another might be no more than the simple voltage divider circuit shown here.

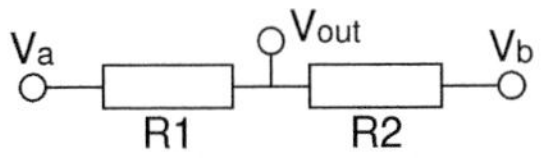

Figure 12.11 Generalised voltage divider configuration, where $V_b \neq 0$.

An obvious generalisation of this arrangement is when the voltage at the bottom of R_2 is not zero, in which case the calculation becomes just slightly more complicated. This variation is shown in Figure 12.11, and the modified calculation is given in Eq. 12.4.

$$V_{out} = (V_a - V_b) \times \frac{R_2}{R_1 + R_2} + V_b \tag{12.4}$$

It is always worth developing a sense for roughly what the answer should be. This helps greatly in gaining a deepened understanding for the behaviour of ever more complex circuits as they are encountered. The basic rules remain unchanged no matter how elaborate the circuit. It is also well worthwhile getting into the habit of performing what is sometimes referred to as an 'idiot check' whenever a calculation is performed in order to confirm the answer arrived at is at least in the right ballpark. Familiarity with expected behaviour helps greatly in this regard.

In order to utilise an idiot check it is necessary to be able to apply some general rules. The first thing to remember is that the voltage at the mid point will always lie somewhere between the voltages at either end, and the second useful rule is that bigger resistors have larger voltages across them and smaller resistors have smaller voltages across them. If the two resistors are equal then the voltage at the middle is halfway between the voltages at either end. Likewise if R_1 is twice the size of R_2 then it has twice the voltage across it, and so V_{out} will be two thirds of the way from V_a to V_b.

LEARNING BY DOING 12.1

THE VOLTAGE DIVIDER RULE

Materials

- A collection of resistors (between 1kΩ and 47kΩ)
- Clip leads
- PSU or 9V battery
- Multimeter

Practice

This exercise uses the voltage divider rule to calculate the expected voltage at points in a resistor network. A more elaborate network can be used for a more challenging analysis.

Step 1: Select three resistors at random from the collection, measure their values, and record the results as R_x, R_y, and R_z

Step 2: Using the clip leads (or a breadboard) connect the three resistors in series, with R_x at the top and R_z at the bottom

Step 3: Assuming a 9V power supply, use the series rule and the VDR to calculate the expected voltages at the midpoints between R_x and R_y, and between R_y and R_z, and record these values as V_1 and V_2

Step 4: Connect the 9V power supply, measure the voltage across each resistor, and record these values as V_x, V_y, and V_z

Step 5: Confirm that the following three relationships hold true (to within some small margin of error)

$$V_1 = V_y + V_z$$
$$V_2 = V_z$$
$$V_x + V_y + V_z = 9V$$

Variable Resistors

Variable resistors are extremely useful. They consist of a strip of resistive material with a terminal attached at each end of the strip, and a third terminal attached to a moving contact which connects to the strip. This contact can be positioned and repositioned at different locations along the strip to provide a continuously variable resistance between the third terminal and each of the two end terminals.

Variable resistors come in many shapes and sizes. Figure 12.12 shows some examples: a) is what is called a potentiometer or pot, and is the kind of device which might

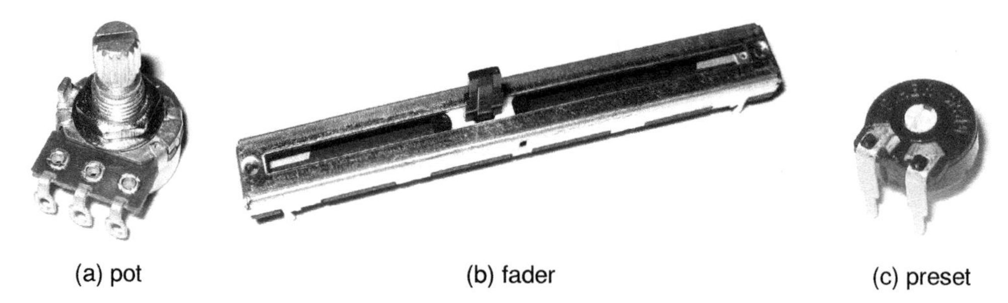

Figure 12.12 Examples of three types of variable resistor.

be found under a volume control for instance, b) is called a fader or linear fader, because it is moved in a straight line instead of rotating as in a pot, c) is variously referred to as a preset, trimmer, or trim pot, and while a) and b) usually provide user controls, presets are used for calibration of a circuit. They are usually intended to be adjusted when the circuit is first built and then left alone unless a recalibration of the circuit is ever required.

There are a number of variations on the circuit symbols which may be used in drawing variable resistors in circuit diagrams. Figure 12.13 shows four possibilities, two for pots and faders, and two for trimmers, although the first pair will quite often be used when a trimmer is intended in a circuit. Sometimes it is necessary to assess the context and the function being performed by the variable resistor in order to decide what style to use.

Figure 12.13 Variable resistor symbols.

The reason there are two symbols for each type is that sometimes a variable resistor is connected into a circuit using only two of its terminals, and sometimes all three are used. The symbols may be a little confusing at first since when only two terminals are used it will always be the variable one plus one of the two others, but in the symbol it looks like the two end terminals are used and the variable one is left open. This would of course make no sense as the device would then be acting as a fixed value resistor and not utilising its variable function at all.

A common practice when only two terminals are needed is to attach the unused terminal directly to the variable terminal. This makes no functional difference to the circuit but can be beneficial in reducing the noise sometimes generated when adjusting a pot (especially older pots with accumulated dirt). It can also be beneficial in the event that the moving contact in the pot breaks. The circuit will then still operate, it will just behave as if the pot has been rotated all the way to the end and left there permanently – not great but better than the alternative of an open connection.

Generally a pot can be thought of as being used to perform one of two different functions based on the way it is wired, either a potential divider or a variable resistance. In the case of a potential divider (or voltage divider) the three terminals are wired into three different points in the circuit in which case the middle (variable) contact acts as a voltage tap point providing a voltage somewhere between the voltages found at the two outer terminals. When used as a variable resistance only two terminals are actually needed, the variable one and one of the outer ones. The second outer terminal is often wired together with the variable terminal for the reasons explained above but this connection plays no role in the actual functioning of the circuit.

Pot Tapers

The final question which needs to be addressed is that of how the resistance changes as the variable resistor is adjusted. This characteristic is called the device's taper. The logical and most common behaviour is that the amount of movement and the amount of change in resistance are directly proportional. This is called a linear taper but it is not the only possibility, as illustrated in Figure 12.14.

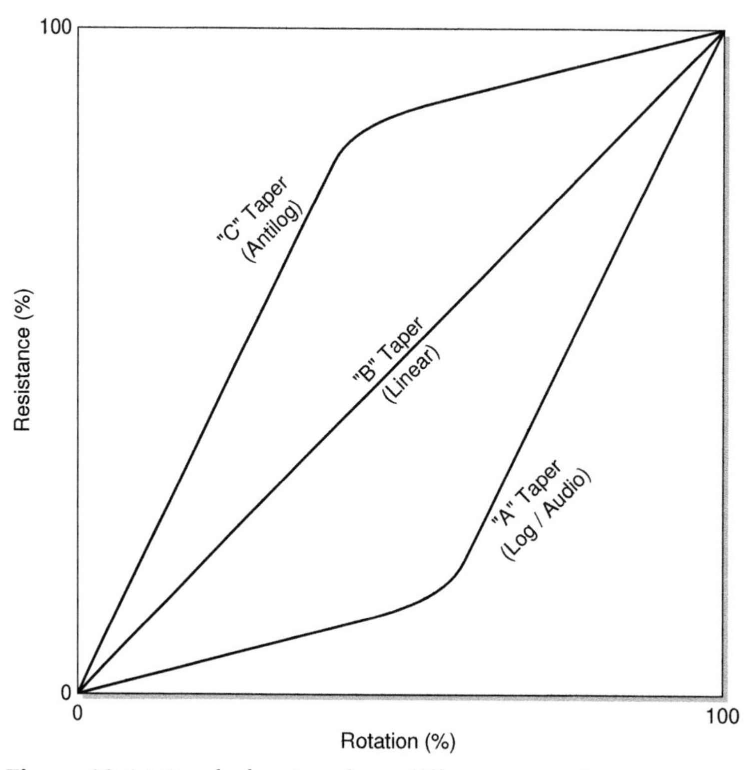

Figure 12.14 Graph showing three different potentiometer tapers.

The linear taper or 'B' taper says that the percentage rotation and the percentage variation in resistance between the terminals track in a linear fashion. However sometimes this behaviour does not provide satisfactory results. A common situation encountered especially in less expensive equipment is where a volume control seems to have most effect in the first third of its movement while it seems to do little or nothing in the last third of its movement. This is a natural consequence of the complex relationship between a signal's level and the perceived volume of the sound that it can produce.

One way to mitigate this problem is to use what is called an 'A', log, or audio taper volume pot. Now the first third of the rotational range corresponds to a much smaller percentage of the total resistance change. Thus the volume is changed less in this first portion of the pots movement, reserving more of the available volume change for the latter portion of the pots movement.

The third taper illustrated on the graph is called a 'C' taper or anti-log taper. This is less common but is sometimes used in conjunction with an log taper to implement the left and right channels of a pan pot. This kind of dual gang pot will be wired so that as one signal is turned up the other is turned down – a standard left-right panning function.

Electrical Characteristics

When it comes to the various manufacturing characteristics previously discussed for standard resistors, the values typical of pots, faders, and trimmers tend to demonstrate larger tolerances (typically of the order of ±20%) and lower power ratings (typically 100mW to 200mW). The range of values commonly available is more limited, with pots below about 1kΩ and above about 1MΩ being uncommon. Likewise the number of values encountered within this range also tends to be limited. Probably 90% of the pots encountered in audio circuits will come from the E3 series: 10, 22, 47 (often rounded to 10, 20, 50 for simplicity).

Variable Resistors and Rotary Encoders

Rotary encoders look very similar to potentiometers (Figure 12.15) but they operate on an entirely different basis. They are generally used in digital systems and are not of much utility here (they are great for working with Arduino, Raspberry Pi, or other similar microcontroller style platforms – a topic for another day). They are mentioned here mainly to avoid confusion if they are encountered along the way.

Rotary encoders rotate all the way around continuously, and are divided up into segments. They often have a clicky feeling to them when rotated, and each movement clockwise or counter-clockwise causes two switches to open and close indicating the movement. There is no way to know what the current position of a rotary encoder is – instead a microcontroller is typically used to count how many clicks left or right it has been turned.

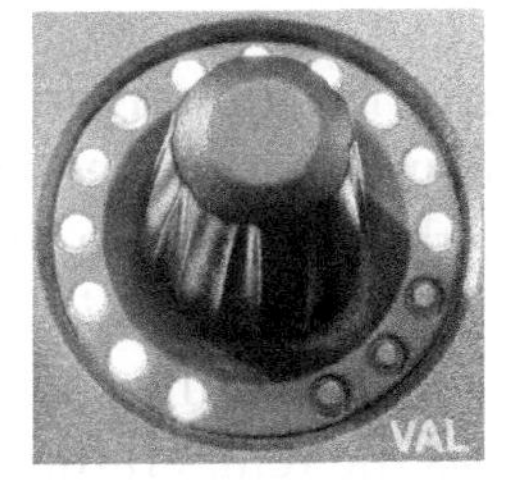

Figure 12.15 Two rotary encoders. The second is surrounded by an LED ring to indicate the current setting or position.

They are useful as rotation sensors or selectors in digital systems, and are particularly handy because effectively they can be automatically reset and adjusted by the system. Many rotary encoders will include a ring of LEDs which are used to give a visual indication of their current setting. Rotary encoders will very often be encountered on digital sound desks and control surfaces.

Photoresistors

A component which can be used to great creative effect in audio electronics is the photoresistor, light dependant resistor, or LDR. Figure 12.16 shows what they look like. The characteristic pattern snaking across the top of the device is the light sensitive region. As more light shines on the device this region becomes more and more conductive, reducing the resistance present between the two leads.

Figure 12.16 A light dependant resistor or photoresistor.

As with a number of circuit symbols, arrows at an angle indicate light. Here arrows arriving at a standard resistor symbol (Figure 12.17) indicate light sensitivity. Similarly, arrows departing at an angle can indicate light production, as is seen with the light emitting diode later.

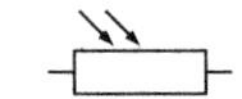

Figure 12.17 Photoresistor symbol.

Thermistors

Less common, but worth mentioning in this chapter about resistors is the thermistor or temperature dependant resistor, see Figure 12.18. Thermistors come in two types depending on whether their resistance rises or falls as their temperature rises. Positive temperature coefficient or PTC thermistors exhibit a rise in resistance with rising temperature. With negative temperature coefficient (NTC) thermistors the effect is reversed with resistance falling as the temperature rises.

Thermistors are often used in power amplifiers and other high power circuits to monitor the operating temperature of critical parts and activate some kind of safety circuitry if the temperature rises above a given threshold. They can also be used in so called soft start circuitry to limit current flow until the circuit has had a chance to get

(a) NTC (b) PTC

Figure 12.18 A negative temperature coefficient (NTC) thermistor and a positive temperature coefficient (PTC) thermistor.

going. They are indicated on circuit diagrams using the symbols shown in Figure 12.19. The type (NTC or PTC) will often be indicated by two little arrows. Both arrows up indicates PTC, while one up and one down indicates NTC.

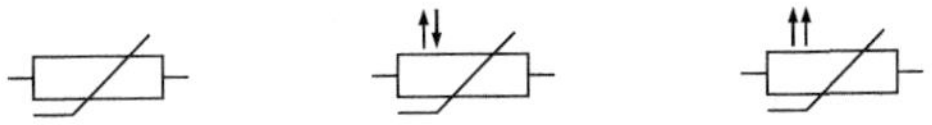

Figure 12.19 Thermistor symbols (unspecified, NTC, and PTC versions).

Instructive Examples

To help cement an understanding of the role resistance plays in electronic circuits, the following sections describe the structure and function of a number of commonly encountered subcircuits in which resistors play a pivotal role.

Generating a Reference Voltage

Audio circuits often operate from a power supply providing two rails. For instance the two terminals of a standard 9V battery might be connected to provide two power rails, labelled plus and minus, power and ground, +9V and 0V, or some similar designations. It is often necessary to derive an intermediate reference voltage at for instance 4.5V to supply key points in the circuit. This can be achieved using a simple voltage divider composed of two equal value resistors (Figure 12.20). Using the Voltage Divider Rule described earlier in this chapter, it is easily confirmed that the point marked V_{ref} in this figure will sit at a voltage of four point five volts. In fact any pair of equal size resistors will, as can be seen from the equation, yield the same result.

$$V_{ref} = V_{in} \times \frac{R_2}{R_1 + R_2} = 9 \times \frac{100k}{100k + 100k} = 9 \times \frac{100k}{200k} = 9 \times \frac{1}{2} = 4.5V$$

This approach to generating a voltage reference while useful does have some limitations. In isolation the subcircuit shown in Figure 12.20 will indeed provide a voltage at the point marked V_{ref} which is half way between the voltages at its top and bottom. However when this reference voltage is used by connecting it to some point in a larger

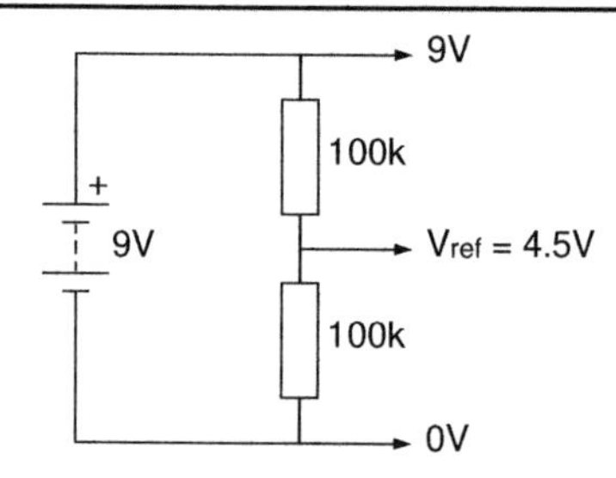

Figure 12.20 Voltage reference derived from a pair of equal resistors.

circuit the circuit to which it is connected can easily pull the voltage up or down or even cause it to swing back and forth constantly as the circuit operates.

In order to use this kind of simple voltage reference to provide a stable and well defined bias voltage in a circuit it is essential that the place or places to which it is connected in the circuit present a high impedance. This will make sure that they have a minimal effect on the level of the voltage reference which they are using. The commonest place this kind of simple voltage reference is encountered is connected to the input of an opamp as illustrated in Figure 12.21. Opamp inputs are generally designed to have a very high input impedance. As such they will draw minimal current and thus have as little an effect on the reference voltage as possible.

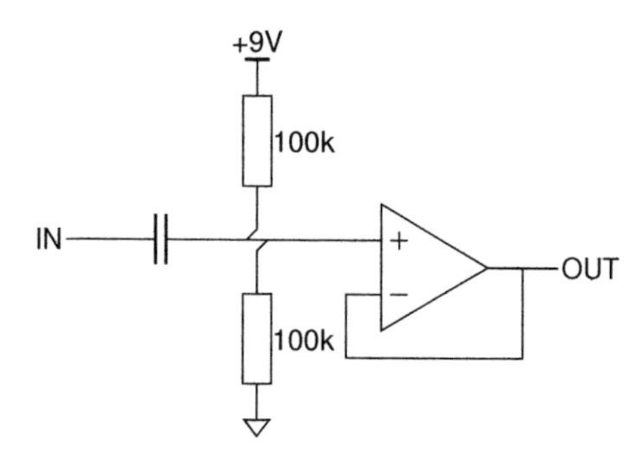

Figure 12.21 Typical circuit employing a voltage divider based reference voltage.

As stated above, in order to provide a voltage which is half way between the top and bottom voltages, the two resistors in the voltage divider must be of equal size. (Other voltage references can be generated using two non-equal sized resistors.) In Figure 12.20 and Figure 12.21 the resistors are shown as being 100kΩ each, and this is a value commonly used in this application. The value chosen is a compromise. The two resistors provide a direct path from power to ground and as such current will flow through them continually. Smaller values (less than perhaps 10kΩ) will place an unacceptable load on the circuit's power supply (the 9V battery would be drained too quickly and would need to be regularly replaced). Conversely too large a value would result in a less stable reference as the circuit to which it is connected will have a greater impact on its value. These two opposing factors lead to the commonly encountered values between about 10kΩ and 100kΩ.

The other two components can be ignored for now. They are a capacitor and an opamp, and their roles in the circuit fragment shown here becomes clear in later chapters. Another capacitor will often be seen connected between V_{ref} and ground in this kind of voltage reference subcircuit. One of the roles of a capacitor is to store charge and in this case a large (perhaps 10μF or larger) capacitor provides a little current reservoir which will help to stabilise the reference voltage and prevent the action of the circuit to which it is connected from moving it about too much. This capacitor while not shown here is a fairly common feature in this kind of voltage reference subcircuit and helps to minimise the unwanted voltage swings described above.

Attenuators and Volume Controls

If an audio signal is applied to the top of a voltage divider then what appears at the midpoint of the voltage divider is a smaller version of the same signal (Figure 12.22). How much smaller this signal is than the original is controlled by the relative sizes of the two resistors forming the voltage divider. This arrangement as illustrated in Figure 12.22a is often referred to as an attenuator or pad and it simply reduces the level of the incoming signal by a set amount. Once again, the Voltage Divider Rule can be used to determine the amount of attenuation in any given case.

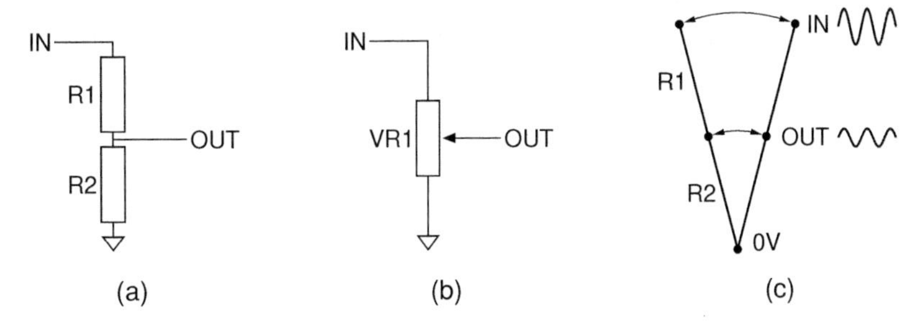

Figure 12.22 Voltage divider reducing the level of an incoming signal.

What is shown in Figure 12.22b is not a pair of fixed resistors as found in a pad but rather it is a pot or variable resistor. This can however be analysed in exactly the same way. A pot can be thought of as a simple voltage divider where the relative sizes of the two resistors vary as the pot knob is turned. As one gets larger the other gets smaller but the sum of the two always remains the same.

Imagine the arrow of the pot symbol moving up and down along the length of the rectangle as the pot knob is turned. The rectangle remains the same length but sometimes more of it is above the arrow and sometimes more of it is below the arrow. So for example, when the arrow is at the bottom, R_1 in the voltage divider equals the total pot resistance and R_2 equals zero. Putting these values into the Voltage Divider Rule indicates that $V_{out} = 0$.

$$V_{out} = V_{in} \times \frac{R_2}{R_1 + R_2} = V_{in} \times \frac{0}{R_{pot} + 0} = 0$$

Conversely when the arrow is at the top, R_1 in the voltage divider equals zero and R_2 equals the total pot resistance. Putting these values into the Voltage Divider Rule indicates that in this case $V_{out} = V_{in}$, maximum volume.

$$V_{out} = V_{in} \times \frac{R_2}{R_1 + R_2} = V_{in} \times \frac{R_{pot}}{0 + R_{pot}} = V_{in}$$

As the pot knob is turned the arrow can be thought of as moving between top and bottom, and V_{out} the output voltage or volume changes correspondingly between maximum (V_{in}) and minimum (0V).

Limiting Current Draw

By definition limiting current is what a resistor does. There are a couple of commonly encountered situations where this particular aspect of its action come to the fore. Sometimes a resistor is placed between the power supply and the rest of the circuit (Figure 12.23). This has the effect of reducing the overall current drawn by the circuit and thus extending the battery life. If the circuit in question does not need the full voltage available from the power supply this is a simple way of limiting current draw and extending battery life. In fact, this technique is sometimes employed more for the effect it has on the sound than for simple power control. An audio circuit can behave in different ways as the power supply is altered. Voltage droop is a recognised effect in some circumstances.

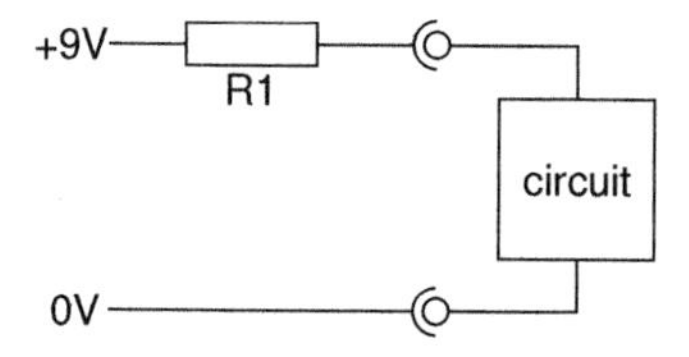

Figure 12.23 Resistor limiting current drawn from power supply.

This is certainly a crude approach to limiting the amount of current which is drawn by a circuit but can be quite useful. Some good examples of this technique in action can be found in the circuit ideas compiled by Tim Escobedo collected under the title 'Circuit Snippets' and readily available on the internet. (A number of the circuit ideas explored in this book can trace their origins to the 'Circuit Snippets' collection and anyone interested in exploring audio electronics will find something of interest in it.)

The current limiting action of a resistor is also employed in a very direct way when adding an LED to a circuit. LEDs require a specific amount of current in order to light up to their intended brightness. This current is usually controlled by a resistor as shown in Figure 12.24. Not enough current and the LED will be very dim while too much current will burn it out. The correct size of resistor must be chosen in order to allow the LED to operate correctly. This application is examined in more detail when LEDs are discussed in Chapter 16.

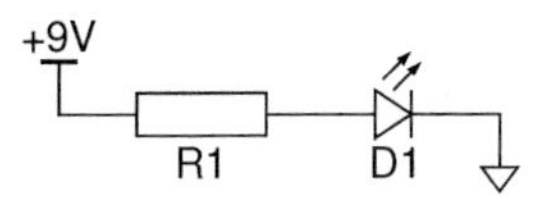

Figure 12.24 Resistor controlling current to LED.

In the examples above, the details of how much the resistor will limit the current depend on the rest of the circuit as well as the current limiting resistor itself. It is however a simple matter to determine the maximum current draw of the system, based on the value of the resistor used, along with the supply voltage. Suppose R1 is 330Ω in either of the above examples. Recalling Ohm's law from Chapter 9, the required calculation becomes clear.

$$
\begin{aligned}
V &= I \times R \qquad \text{– Ohm's Law} \\
\Rightarrow I &= \frac{V}{R} \\
&= \frac{9}{330} \\
&\approx 27.3\text{mA}
\end{aligned}
$$

Similarly, if R1 is increased to 1kΩ, the maximum possible current draw is reduced to $9/1{,}000 = 9\text{mA}$. If the impedance of the rest of the circuit is low then this limit set by the resistor will be approached. As the impedance of the rest of the circuit rises, the actual current draw drops off accordingly.

References

G. Ballou. Resistors, capacitors, and inductors. In G. Ballou, editor, *Handbook for Sound Engineers*, ch. 10, pp. 241–272. Focal Press, 4th edition, 2008.

D. Bohn. *RaneNote 149: Interfacing AES3 & S/PDIF*. Rane Corporation, 2009.

D. Self. *Small Signal Audio Design*. Focal Press, 2nd edition, 2015.

12.1 – The General Form of the Parallel Rule

It should be obvious that the series rule can be extended in order to combine any number of resistors in series.

$$R_{series} = R_1 + R_2 + \cdots + R_n$$

The form of the parallel rule given in the main text does not lend itself to such an extension. The equation given can however be rearranged into a form which follows a similar pattern to the series rule, and which can be extended in a similar fashion.

$$\frac{1}{R_{parallel}} = \frac{1}{R_1} + \frac{1}{R_2} + \cdots + \frac{1}{R_n}$$

It is not immediately obvious that the form quoted as Eq. 12.2 can be derived from this more general form. The derivation is given below.

$$\frac{1}{R_{parallel}} = \frac{1}{R_1} + \frac{1}{R_2} \quad \text{– parallel rule for two resistors}$$

$$= \frac{R_2}{R_1 \times R_2} + \frac{R_1}{R_1 \times R_2} \quad \text{– normalise denominators}$$

$$= \frac{R_1 + R_2}{R_1 \times R_2} \quad \text{– add fractions}$$

$$\Rightarrow R_{parallel} = \frac{R_1 \times R_2}{R_1 + R_2} \quad \text{– invert both sides}$$

The form of Eq. 12.2 is preferred because it has as its subject the final resistance value $R_{parallel}$, rather than $1/R_{parallel}$. The alternate version can also be rearranged in order to make $R_{parallel}$ the subject.

$$\frac{1}{R_{parallel}} = \frac{1}{R_1} + \frac{1}{R_2} + \cdots + \frac{1}{R_n}$$

$$\Rightarrow R_{parallel} = \frac{1}{\frac{1}{R_1} + \frac{1}{R_2} + \cdots + \frac{1}{R_n}}$$

However this makes the resulting equation somewhat unwieldy, so since it is usually only required to add together two resistors in parallel, and if necessary multiple applications of the rule can be used, the more compact special form is often the one which is quoted, as in Eq. 12.2. The two formulations are entirely equivalent.

13 | Capacitors and Inductors

Capacitors block low frequencies and pass high frequencies
Inductors pass low frequencies and block high frequencies

Capacitors and inductors are examined together because these two components are very closely related, despite being markedly different in structure and operation. At the very simplest level a capacitor is composed of two sheets of conductive material separated by a layer of insulating material, while an inductor is nothing more than a length of wire wound into a coil. The relationship between capacitors and inductors stems from their very differences. While not holding in all cases, there is a symmetry between these two components. As a general rule whatever one does the other is likely to do the exact opposite. This is illustrated in the one line descriptions for each component, highlighted in the box above; capacitors block low frequencies and pass high frequencies, while inductors exhibit the opposite behaviour passing low frequencies and blocking high frequencies. Often this is as much information as is needed in order to see what role a capacitor or inductor plays in a particular circuit, although capacitors in particular are used to do a great many useful things in audio electronics.

Capacitors

After resistors, capacitors are the most commonly encountered component in typical audio electronic circuitry. Capacitors can also be said to exhibit what from many perspectives is the widest variety of different types and subtypes of any electronic component. Where for instance resistors are most often of either carbon film, metal film, or sometimes wire wound construction, and transistors usually appear in the form of either BJTs or FETs with a few variations of each, capacitors come in a wide array of forms and formats the full range of which is only touched upon in this chapter. The range of shapes, sizes, and types provides for a rich area of experimentation in audio circuits as the sometimes subtle differences in the performance of different types can result in slight but sometimes audibly significant variations in their effect.

A variety of circuit diagram symbols can also be encountered representing capacitors, the most standard of which are illustrated in Figure 13.1. Quite a few variations on these symbols are found but they are all pretty similar, and on the whole fairly self explanatory. Watch out for polar or polarised capacitors (most often electrolytic or sometimes less common tantalum types). Polarised capacitors must be inserted into a circuit the correct way round. They should not be used in positions where the polarity of the applied voltage is expected to reverse in normal operation. Most often polarised capacitors are indicated in a circuit diagram by a symbol with a plus sign on one side of the component as in Figure 13.1b. Occasionally a different symbol with one straight and one curved line is used, and other variations are also encountered, the most common of which are included in the schematic symbol reference chart in Appendix C.

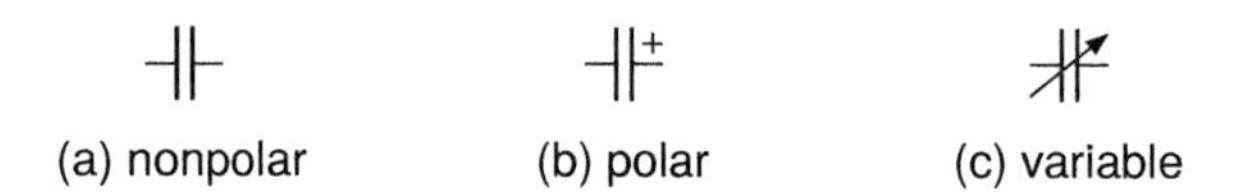

Figure 13.1 Circuit diagram symbols for a capacitor.

The units used to measure the capacitance value of a capacitor are called farads (F). However one farad represents an very large device (they do exist but are unlikely to be encountered in audio electronics). Commonly used capacitors will have values measured in microfarads (μF, often pronounced 'mike'), nanofarads (nF), and picofarads (pF, often pronounced 'puff'). Micro- means divided by a million, nano- divided by a billion, and pico- divided by a trillion. See Appendix A for a full explanation of all the multiplier prefixes which are likely to be encountered working in audio electronics.

When considering the division of capacitors into polar and nonpolar types, as a rule of thumb capacitors smaller than about one microfarad (1μF) will typically be nonpolar types while above this size they will most often be of polar construction. If a polar type is specified in a design it will usually be fine to substitute a nonpolar type of the same size if one is available. Substituting a polar where a nonpolar had been specified is less likely to work satisfactorily.

Capacitor Types

As previously noted the capacitor is probably second to none in the variety of device types and subtypes available. Table 13.1 only scratches the surface of the wide range of component types which can be encountered but it does highlight what are probably both the most common and most important variants utilised in modern audio electronics along with some of the key characteristics which define their operation and use.

For the most part, all these names and designators indicate the nature of the dielectric employed in the construction of the device. The dielectric is the insulating layer which separates the two conductive plates of a capacitor and its precise composition can dramatically affect the performance of a component. As illustrated in Table 13.1 some dielectric types are best used for manufacturing only a limited subset of the wide

Table 13.1 Common capacitor types and typical values for various important characteristics

	Ceramic	Film	Electrolytic	Tantalum
Examples				
Subtypes	NP0/C0G, X7R, Z5U, Y5V	Polyester, Polypropylene, Polystyrene...	Polar, Nonpolar/ Bipolar	Dry slug, Wet slug
	Typical ranges and values for small signal devices –			
Capacitance	10pF to 220nF	1nF to 1μF	1μF to 1000μF	100nF to 63μF
Tolerance	$\pm$10%	$\pm$5% to $\pm$10%	$\pm$20%	$\pm$20%
Voltage	63V/100V	100V	25V/50V	3V to 35V

range of capacitance values encountered. Some types need to be made physically very large in order to achieve larger capacitance values. Higher voltage ratings also result in physically larger devices and some dielectrics will not tolerate high voltages in any case. Size can be an issue as it is often necessary to make circuitry as small as possible when designing and building audio (and other) equipment.

The precise types employed can be of great importance when designing high quality, low noise (and expensive) audio equipment but are generally of lesser importance in the applications under consideration in this book. Amongst the ceramic types the NPO subtype (aka C0G) is an excellent part which compared to other ceramic subtypes introduces minimal distortion into applied audio signals. Similarly in the range of film subtypes polypropylene or PP devices usually demonstrate superior performance to the more commonly encountered polyester or other film subtypes used in audio applications (Self, 2015, pp. 63–74). Needless to say, in both cases this enhanced performance usually comes at a significantly higher cost and the amount of improvement achieved is typically neither justified nor required in nonspecialist applications.

Typical capacitance values commonly encountered range from a few picofarads up to a few hundred microfarads. Different types tend to cover different ranges as shown in Table 13.1. As a rough guide ceramics cover the low end, film types range through the middle of the spectrum, and electrolytics are usually employed for larger values. Tantalum capacitors which tend to be available in mid to high values are usually considered to be somewhat specialist parts. In audio applications their use can be a hot topic for debate on various audio electronics forums. They probably introduce more distortion than other types but the question of whether it is good or bad distortion is a matter of subjective preference. They can provide a large capacitance in a small package.

It is important to observe the voltage ratings of capacitors, which can sometimes be quite low. For a circuit powered by 9V (as most examined here are) it is probably wise to stay above about 16V or 18V ratings for the capacitors used. This should not be an issue except for some tantalum types (it is also worth remembering that tantalum capacitors are very easily damaged by even small reverse voltages, remember they are a polarised type). When looking at a high wattage power amplifier or similar where much higher voltages can be involved then more attention needs to be paid to the voltage ratings of the capacitors used. Again commonly encountered voltage ratings vary from type to type and typical values are indicated in Table 13.1.

Capacitor tolerances tend to be quite wide with ±20% or more not being uncommon especially for electrolytics. The somewhat more expensive polypropylene capacitors often have a somewhat more respectable ±5% rating or better. Generally this is not an issue. In many circuits the precise values of the capacitors used are not critical to their operation, or the variations encountered can be compensated for in other ways.

Capacitor Applications

As previously stated capacitors block low frequencies and pass high frequencies. This makes them of particular interest in audio electronics as the manipulation of an audio signal's frequency spectrum is a fundamental tool in the design of useful audio circuits. In fact capacitors play a number of key roles in the operation of typical audio circuitry. To give a flavour of the wide gamut of uses where capacitors are employed Table 13.2 highlights some of their more important areas of application.

INDUCTORS

As explained at the beginning of this chapter, an inductor is just a coil of wire. Its impedance rises as frequency rises. Inductors are used only quite rarely in audio circuits. They tend to be bulky, expensive, and nonlinear. In most aspects of their behaviour they are the exact opposite of capacitors, so most things which can be achieved with inductors can also be done with capacitors, by reconfiguring the circuit in some fashion. Most of the circuit symbols commonly used to represent inductors reflect their structure well, consisting of either a spiral or a series of loops as shown in Figure 13.2. See Appendix C for some additional symbols also commonly encountered.

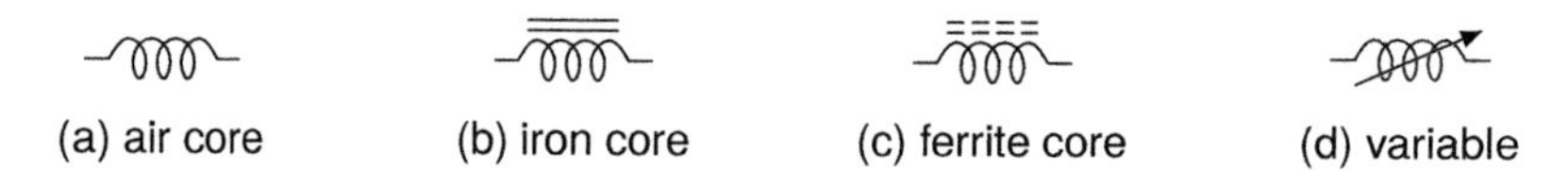

Figure 13.2 Circuit diagram symbols for an inductor.

The variations shown in Figure 13.2 also reflect the fact that an inductor's performance can be significantly altered depending on the nature of the core around which the

Table 13.2 An outline of some common capacitor applications

Application	Description
Blocking and coupling	Capacitors are often used to remove DC and/or subsonic, low frequency components from a signal while coupling the main audio frequencies through to the next part of the circuit.
Filtering	Capacitors are a fundamental part of filter and EQ circuits designed to cut or boost various sections of the audio frequency spectrum in a signal.
Smoothing and storing	Capacitors can be used to adsorb ripple, noise, and interference on a circuit's power lines, and to provide short bursts of high current when signal levels peak.
Suppressing and decoupling	Capacitors can be used in many circuits to attenuate or suppress unwanted signals such as high frequency circuit resonances and RF (radio frequency) interference.
Timing	Capacitors are often used in conjunction with resistors to set the frequency of oscillation in various signal generator and oscillator type circuits.

coil is wound. As indicated, the three commonest methods of construction involve an air core (i.e. no core at all), a solid or laminated metal core, or a ferrite core. Ferrite cores are made from iron dust or filings mixed with a binding agent and moulded into the required shape. An inductor's core material strongly influences its size and performance (much like the dielectric material in a given capacitor).

In many cases inductors are easily recognisable since the coil is often clearly visible, as in the examples in Figure 13.3, although sometimes they are encased inside a plastic casing or a metal can. Some smaller inductors look very similar to standard through hole resistors.

Figure 13.3 Some typical inductors.

The unit of measure for inductors is the henry (H). Most commonly encountered inductors will have values in the millihenry (mH) range with some falling to the level of microhenries (μH). Inductors are on the whole far less commonly employed than capacitors, but there are a few interesting audio circuits which call for an inductor, often being used in some form of a filter or EQ role. Many classic wah pedals (such as the much loved Dunlop Cry Baby) use an inductor in the design of their swept filter. An interesting filter design, combining inductors and capacitors, may be heard generating the characteristic washed out high pass filter sound used to great effect by some early reggae producers. One example of this is what came to be referred to as 'King Tubby's big knob filter' on account of the larger than standard pot knob used to control this particular filter on his mixing desk (Williams, 2012).

RF Chokes and Noise Suppression

One very common source of interference and unwanted noise in electronic circuits is through the pickup of radio frequency (RF) signals from the air. With the stated primary behaviour of an inductor bing that it passes low frequencies and blocks high frequencies, it would seem reasonable to suggest that the inductor might prove an effective tool in the suppression of unwanted RF interference in audio signals, and this is indeed the case. Figure 13.4 illustrates a number of RF choke cores moulded from ferrite material.

Figure 13.4 Ferrite cores for use as RF chokes.

The third in particular (in a hinged plastic case) can commonly be found around the power leads and signal connectors of many electronic devices where it acts as a block to any high frequency signal which tries to pass it. Low frequency signals, and the DC power signals which it is often found working in conjunction with, can pass freely along the wire to which such an RF choke is applied. It may not be immediately obvious where the inductor is in this case. In fact it is formed by the cable itself, running through the ferrite core. Sometimes the cable is looped back to perform a second pass through the core and thus increase the inductance induced, but often simply passing the cable through a core formed tightly around it is sufficient to achieve the high frequency filtering desired.

CHARACTERISTICS OF CAPACITORS AND INDUCTORS

Many of the details of component characteristics examined in the previous chapter on resistors apply equally to capacitors and inductors. These aspects of a component's performance are mentioned only briefly here. For more details on the interpretation and use of these parameters refer back to the appropriate sections of Chapter 12. Table 13.1 shows typical values for the most important capacitor characteristics. Inductors generally display less variation in their available types and variants.

Preferred values – the same E-series preferred values which are used to specify standard resistances also generally apply to capacitors and inductors. Fewer values are typically available, with E6 or even E3 being very commonly offered ranges.

Tolerance – the more limited range of values typically available reflects the fact that capacitors and inductors tend to be manufactured to much looser tolerances then resistors. They are usually specified as $\pm10\%$ or even $\pm20\%$ as compared to typical resistor tolerances of $\pm1\%$ and $\pm5\%$. Tighter tolerance capacitors are available but quickly become prohibitively expensive and are unnecessary for all but the most demanding applications. Tighter tolerances in inductors are rare but not unheard of. In most cases circuits can be designed to provide the desired levels of performance without having to resort to these more expensive parts.

Markings – while a very few capacitors can still be found which use a colour code similar to resistors in order to display their nominal value most capacitors will have their values printed on them. The value is likely to be formatted in one of two ways. On electrolytic capacitors and other physically large devices it is often printed directly (e.g. 470μF). On other types (which tend to be physically smaller) a simple three digit code is usually used. The first two digits indicate how the value starts and the third indicates how many zeros to add to complete the value given in picofarads. So for example 224 means twenty two followed by four zeros, 220,000pF, two hundred and twenty thousand picofarads, or two hundred and twenty nanofarads.

Polarised capacitors use a couple of different marking methods to differentiate the positive and negative terminals. Electrolytic types will usually have a row of minus signs running towards the negative leg. Tantalum capacitors on the other hand tend to mark the positive pin with a plus sign. Some older tantalum types used a colour code with a number of horizontal stripes and a single blob or dot of colour. For these types the rule 'when the dot's in sight the positive's on the right' was used to identify which way round they were to be connected.

Inductors can be commonly encountered using either a colour code or a printed value. Typical values span the micro- and millihenry ranges. The smallest devices can range into the nanohenry region while larger values extend up to several henries.

Voltage rating – a key characteristic of a capacitor is its voltage rating. It holds a similar position to the power rating parameter described for resistors. This is the maximum voltage which the capacitor should ever to exposed to and sometimes these values can be quite low so it is worth keeping an eye on this value when using capacitors. It will often be printed on the body of a capacitor along with the capacitance value.

Current rating – for inductors the corresponding parameter to a capacitor's voltage rating or a resistor's power rating is the current rating most often specified as a DC current rating. Since inductors pass low frequencies and block high frequencies they pass DC (i.e. 0Hz, the lowest possible frequency) best of all. This means that a DC voltage can overload and burn out an inductor very easily by causing a large current to flow through it. An inductor circuit must be designed to ensure that excessive currents are not allowed to flow through the inductor in normal operation.

As with resistors many other parameters and specifications exist covering all aspects of the construction and behaviour of various devices. In the main these details can be ignored in all but the most demanding of applications.

Device Impedance

The impedance of a device measured in ohms (Ω) indicates how easily or with what level of difficulty electricity will flow through that device. Resistors as described in the previous chapter have a resistance (also measured in ohms) which is pretty much constant across the full range of the devices operating conditions. Resistance is just one component of impedance. When capacitors and inductors enter the scene the other side of impedance is encountered. This second component is called reactance. Together resistance and reactance make up impedance. Resistance can be considered as the constant part of impedance while the reactive portion of impedance changes depending on the frequency of the electricity flowing through the component in question.

So resistors have a reactance of zero. Their impedance is purely resistive and so it doesn't change with frequency. On the other hand capacitors and inductors generally have zero (or negligible) resistance but non-zero reactance meaning that their impedance varies as the frequency of the signal passing through them varies. Eq. 13.1 shows how to calculate a capacitor's impedance. Note that just as R has been used previously to indicate resistance in equations, X is now employed to indicate reactance, X_C for capacitive reactance and X_L for inductive reactance. Later when resistance and reactance are combined, Z will appear to indicate total impedance as a function of these others. These are the standard letters used to indicate these quantities in equations. All of these quantities are measured in ohms (Ω).

$$X_C = \frac{1}{2\pi fC} \tag{13.1}$$

As can be seen from Eq. 13.1 the value of a capacitor's reactance depends on the capacitance of the device and the frequency of the signal. So for a given device which has a fixed value (measured in farads) the impedance can be plotted against the frequency of the signal applied. This relationship is graphed in Figure 13.5 for three capacitance values: 22nF, 47nF, and 100nF. As the frequency increases the impedance decreases. This illustrates the capacitor's core behaviour which clearly explains why capacitors block low frequencies and pass high frequencies. At low frequencies the impedance becomes very high and so little signal can get past while at high frequencies the impedance becomes very low and so signals have little difficulty in passing through.

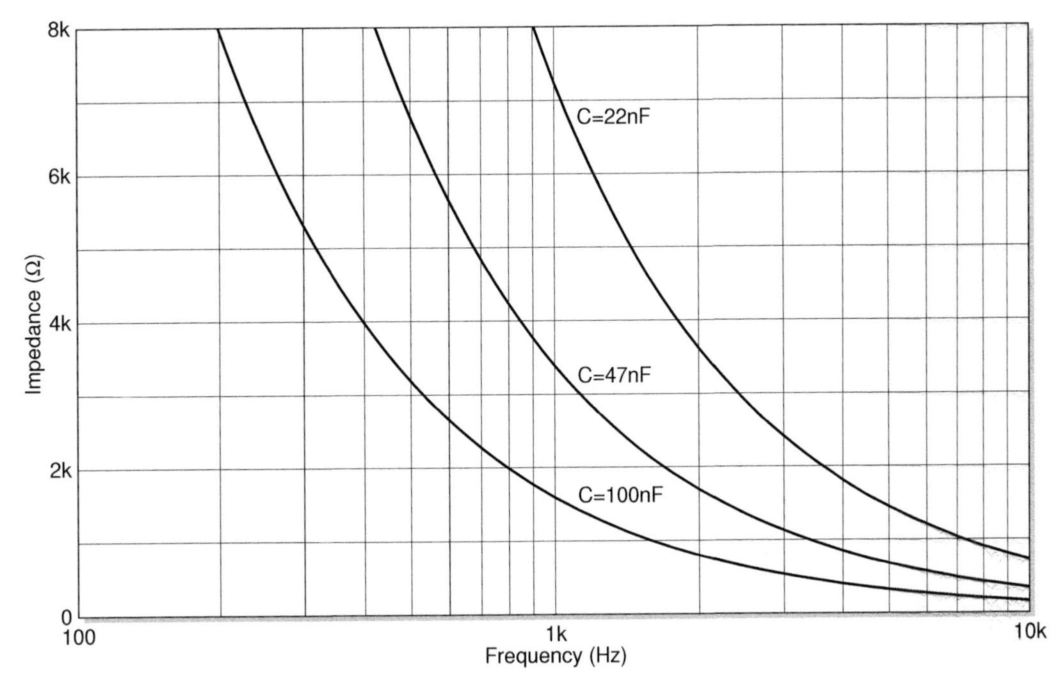

Figure 13.5 Graph of capacitor impedance plotted against frequency.

As has been pointed out previously capacitors and inductors behave in broadly opposite fashions, and so it should be no surprise to see the form of the impedance equation for an inductor as shown in Eq. 13.2. Instead of getting smaller as the frequency grows larger, the inductor's impedance increases along with the frequency. L is the letter conventionally used to indicate inductance in equations just as C is used to represent capacitance.

$$X_L = 2\pi f L \tag{13.2}$$

In Figure 13.6 the behaviours of resistor, capacitor, and inductor are plotted on the same graph in order to highlight their relationships and their differences. The resistor plot consists of a horizontal line indicating that the impedance remains at the same level at all frequencies. As the value of the resistor is changed the horizontal line will appear higher or lower on the graph. So the plot for a 8kΩ resistor would be a horizontal line crossing high up on the graph while that for a 100Ω resistor would appear still horizontal but much lower down.

The curves to illustrate capacitors of differing values would move either towards or away from the origin of the graph depending on the size of the capacitor as is well illustrated by the three plots in Figure 13.5. Larger capacitor values correspond to plots which traverse ever closer to the origin while smaller value capacitors result in plots lying farther out.

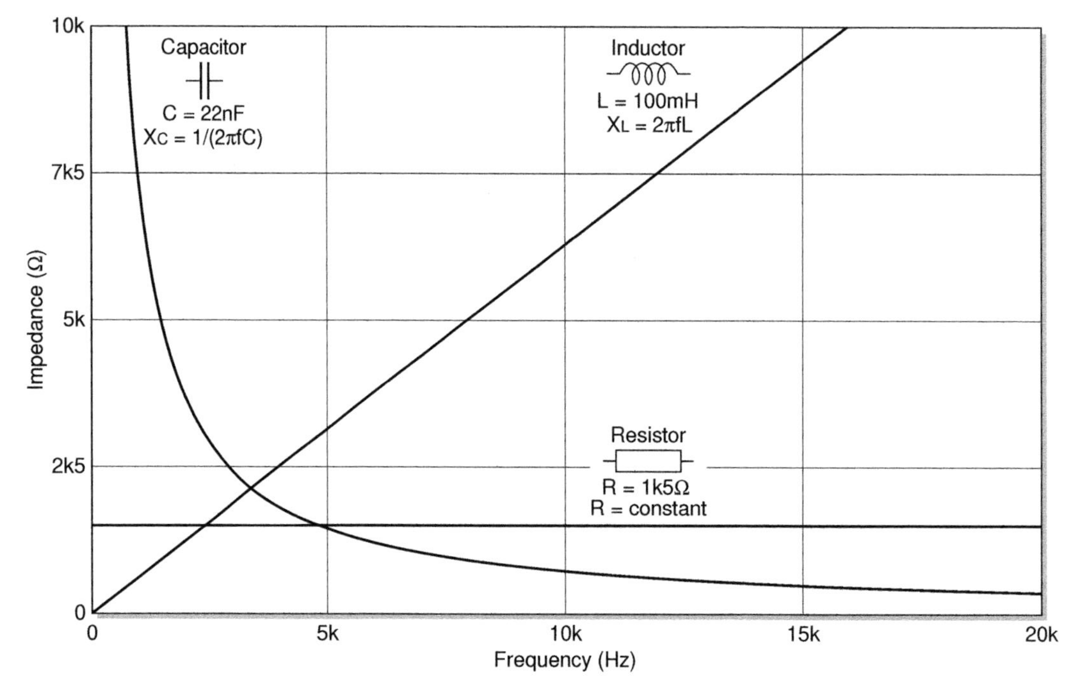

Figure 13.6 Graph of resistor, capacitor, and inductor impedance versus frequency.

It is worth noting that the same information is sometimes plotted using log scales on the Impedance and Frequency axes. This type of rendering results in straight lines instead of curves for the capacitor plots as illustrated in Figure 13.7. It is important to realise that these two representations are entirely equivalent. No new or different information is being displayed by this seemingly different graph. It is just another way of plotting the same impedance functions shown above. The lines for resistance and inductive reactance remain straight lines. Notice that in this second style of plotting the origin of the graph (0,0) will never appear. Moving down and to the left the numbers get smaller and smaller but they never reach zero. This helps to explain how a curved line in the previous plot which gets closer and closer to but never touches zero lines of the X and Y axes can become a straight line in this one.

The third line in Figure 13.6 and Figure 13.7 plotting an inductor's reactance (Eq. 13.2) is a straight line through the origin. This means that when the signal frequency is zero the inductor reactance is zero which can be seen from the inductor's impedance equation. Once again this matches with the key characteristic of inductors that they pass low frequencies and block high frequencies since a low impedance will pass a signal and a high impedance will block it.

Inductors of different values result in straight lines of different slopes all passing through the origin. As the inductance is increased the reactance at any given frequency also increases so larger inductances result in lines of greater slope becoming increasingly close to the vertical and smaller inductances result in shallower slopes becoming increasingly close to horizontal.

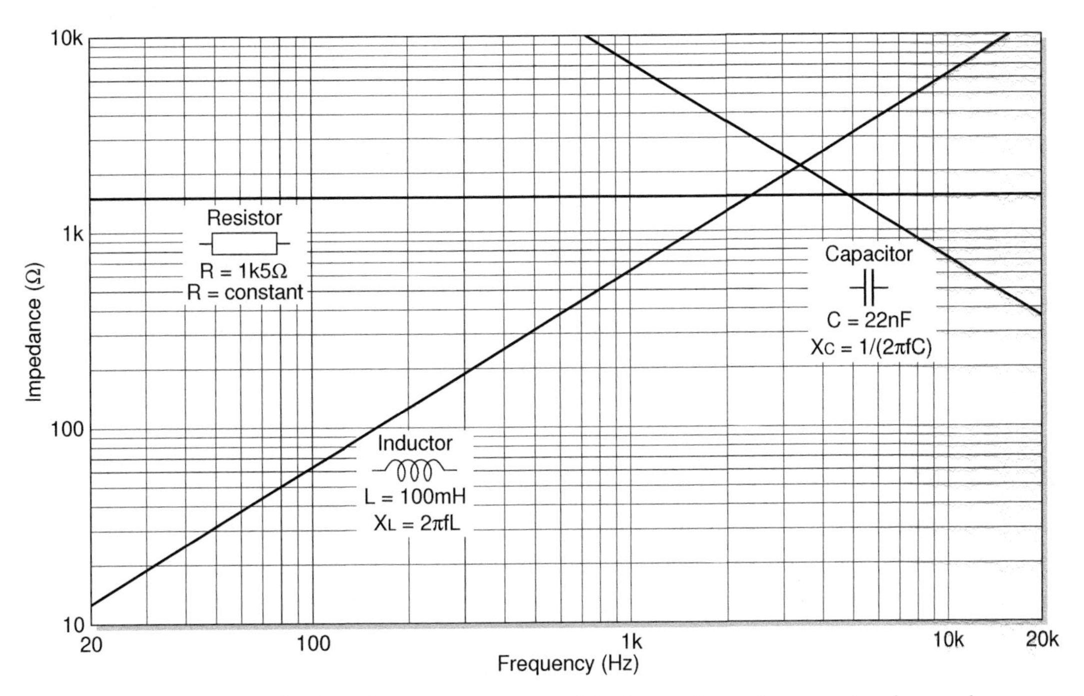

Figure 13.7 Graph of resistor, capacitor, and inductor impedance using log scales.

Notice that the three previous figures all use different combinations of linear and logarithmic scales for their axes even though they are all plotting the same information on each axis, frequency on the X axis and impedance on the Y axis. The use and interpretation of these different graphing scales is more fully examined in Chapter 5 where the four possible combinations of linear and logarithmic scales are examined, along with their relationship to the decibel scale.

Equivalent Series Resistance

The simple ideal models of component behaviour are often sufficient, but they never tell the whole story. For example a real capacitor will exhibit resistance and inductance as well as capacitance. The levels of inductance present are so small that they only become a concern at very high frequencies well beyond the audio. Some capacitors, particularly electrolytics, do however present a small but sometimes significant resistance. This is called ESR or equivalent series resistance.

Generally in film and ceramic types it is small enough to ignore, typically well less than an ohm, but in electrolytics it is often as much as a few tens of ohms. Low ESR is important because the lower the ESR the faster the capacitor can charge and discharge. This is a great benefit because it means that such capacitors can respond faster to large transients. Resistance also means power dissipation.

Sometimes a small ceramic capacitor is used in parallel with a much larger electrolytic capacitor. At first glance it would seem to have little effect on the circuit as it adds just a small bit more capacitance to the large amount already provided by the electrolytic. However at high frequencies the electrolytic's relatively large ERS degrades its performance and the ceramic with its much smaller ERS is able to step in. This arrangement is often seen where digital ICs need decoupling to prevent noise glitches generated by fast switching digital circuitry from contaminating the power supply.

Eq. 13.7 and Eq. 13.8 in the section on RLC circuits which follows are presented in terms of analysing circuits which combine actual capacitors, inductors, and resistors. However they can also be applied here to more accurately calculate a component's true impedance by combining its resistive and reactive elements. This is seldom if ever required in the kind of circuit analysis needed here but it is worth bearing in mind.

Series and Parallel Rules

Rules are introduced in Chapter 12 to calculate the effect of series and parallel combinations of resistors. The same configurations can also be very useful when applied to capacitors and inductors. This can be particularly the case because the values readily available for capacitors and inductors are often far more limited than those available for resistors and so it is often necessary to make up a value required in a particular circuit by combining two or more components of those values which are available.

The four equations given here for series and parallel combinations of capacitors and inductors follow the same pattern as the earlier equations given for resistors. It should be noted however that when compared to the resistor case the two equations for capacitors are reversed while they remain in the original order for inductors. In other words for straight addition of values resistors and inductors are places in series but capacitors are placed in parallel.

It is always the case that one configuration will give a resultant value larger than either of the constituent components while the other configuration gives a resultant value smaller than either on its own. Resistances and inductances get bigger in series and smaller in parallel while capacitances get smaller in series and bigger in parallel.

$$C_{series} = \frac{C_1 \times C_2}{C_1 + C_2} \tag{13.3}$$

$$C_{parallel} = C_1 + C_2 \tag{13.4}$$

$$L_{series} = L_1 + L_2 \tag{13.5}$$

$$L_{parallel} = \frac{L_1 \times L_2}{L_1 + L_2} \tag{13.6}$$

Combining three or more components in series or parallel is simply a matter of multiple applications of the rules provided here. Indeed any network composed solely of one of these three types of components resistors, capacitors, or inductors no matter how large and complex can be simplified down into a single value which can then be used in analysing the behaviour of the original network. Figure 13.8 shows a (somewhat unlikely) network of inductors. The steps below illustrate one way to go about determining the overall inductance of the network. A number of alternative routes through such a process will often exist but they will always yield the same final result. In this case for instance step one might either be to combine L1 and L2 or alternatively L3 and L4 might be combined first and so on. At each stage it is necessary to identify two components which can be combined into one through the application of either the series or parallel rule. Ultimately the entire network is replaced by a single component value.

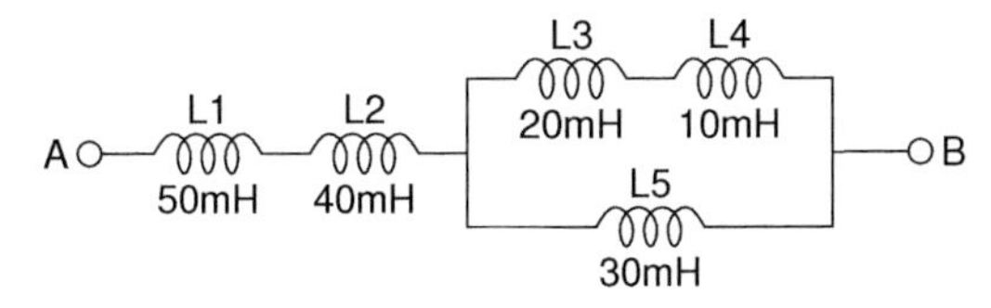

Figure 13.8 Example inductor network.

Step 1: Combine L1 and L2, series rule – $L_x = L1 + L2 = 50m + 40m = 90mH$
Step 2: Combine L3 and L4, series rule – $L_y = L3 + L4 = 20m + 10m = 30mH$
Step 3: Combine L_y and L5, parallel rule – $L_z = \frac{L_y \times L5}{L_y + L5} = \frac{30m \times 30m}{30m + 30m} = 15mH$
Step 4: Combine L_x and L_z, series rule – $L_{total} = L_x + L_z = 90m + 15m = 105mH$

And so the total inductance between points A and B through the inductor network shown in Figure 13.8 is equal to 105mH, one hundred and five millihenries. The same procedure can be applied to resolving any R, L, or C network into a single equivalent value for the purposes of analysis.

RLC Circuits

Up to this point each type of component has been considered in isolation being combined only with other components of its own type. Some very useful circuits (for example simple low pass and high pass filters) can be constructed by combining resistors (R), inductors (L), and capacitors (C). Circuits involving any combination of these three components are traditionally referred to using the corresponding letters, so that for instance an LC circuit is composed solely of inductors and capacitors, and so forth.

In order to understand the operation of such circuits a number of additional equations are needed, describing how the different components combine and interact. A full analysis would include consideration of the effect on the phase of signals of different frequencies passing through such combinations of components. This kind of analysis is more advanced than is required here. In this work only the magnitude component is considered.[a]

Previously Eq. 13.1 and Eq. 13.2 showed how to calculate the impedance of a capacitor or inductor at a given frequency (recall that unlike resistors, capacitor and inductor impedances are frequency dependent). The next step which must be taken is to calculate the frequency dependant impedance of a circuit consisting of different component types in series or parallel. Eq. 13.7 and Eq. 13.8 show how such impedances can be determined.

$$|Z_{\text{series}}| = \sqrt{R^2 + (X_L - X_C)^2} \tag{13.7}$$

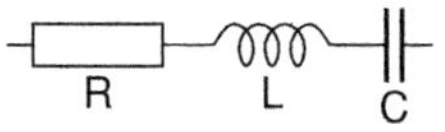

R L C

$$|Z_{\text{parallel}}| = \frac{1}{\sqrt{\left(\frac{1}{R}\right)^2 + \left(\frac{1}{X_L} - \frac{1}{X_C}\right)^2}} \tag{13.8}$$

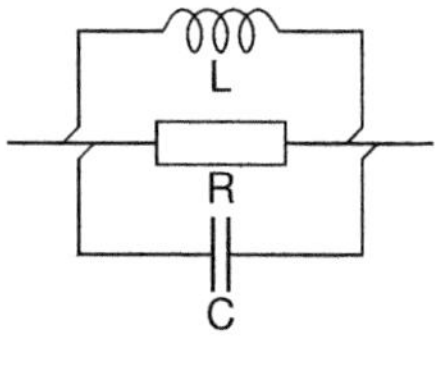

L R C

These equations can be used for RL, RC, and LC circuits by simply dropping the unused term in the appropriate equation. If R is dropped the simplified forms in Eq. 13.9 and Eq. 13.10 can be derived. It should be quite straightforward to see that Eq. 13.9 can be arrived at by setting the value of R in Eq. 13.7 to zero, and similarly Eq. 13.10 can be arrived at by setting the value of R in Eq. 13.8 to infinity (i.e. an open circuit in parallel with the two other components). Eq. 13.9 and Eq. 13.10 are given as simplifications of Eq. 13.7 and Eq. 13.8 because these cases where capacitance and inductance but no (significant) resistance are present are quite common and the associated equations are thus correspondingly useful.

a. The vertical bars |...| bracketing various terms in the following equations are a mathematical notation indicating magnitude or absolute value. They turn any negative number into a positive number.

$$|Z_{\text{series}}| = |X_L - X_C| \tag{13.9}$$

$$|Z_{\text{parallel}}| = \left|\frac{X_L \times X_C}{X_L - X_C}\right| \tag{13.10}$$

Notice that in all four equations the capacitive and inductive reactances are subtracted not added. This might seem counterintuitive at first but is in fact the case and again reflects the equal but opposite natures which have previously been pointed out as existing between capacitors and inductors. In the first two equations the square eliminates any possible negative answer resulting from this subtraction while in the second pair the magnitude operator arrives more simply at the same result as square followed by square root. As a result these equations always yield a strictly nonnegative impedance as is to be expected from a magnitude calculation.

13.1 – The LC Series Rule

In the main text Eq. 13.9 is presented as a simplification of Eq. 13.7, where the resistance term is removed. It is a relatively straightforward matter to see how the one is derived from the other. In this case removing R means replacing it with a wire (i.e. a resistance of zero), as illustrated in the diagrams which accompany the original equations. Replacing R in Eq. 13.7 develops as follows.

$$|Z_{\text{series}}| = \sqrt{R^2 + (X_L - X_C)^2} \qquad \text{RLC series rule from Eq. 13.7}$$

$$= \sqrt{0^2 + (X_L - X_C)^2} \qquad \text{set R equal to zero}$$

$$= \sqrt{(X_L - X_C)^2} \qquad 0^2 = 0\text{, and so disappears}$$

Now the square and the square root almost completely cancel each other out, but not quite. It must be remembered that squaring a negative number yields a positive number, so if $X_L - X_C$ in the equation turns out to be negative, it will be necessary to discard the minus sign. The magnitude operator $|\ldots|$ performs this function. Thus the square and square root can be replaced by a simple magnitude, to arrive at the desired final result.

$$|Z_{\text{series}}| = |X_L - X_C| \qquad \text{LC series rule from Eq. 13.9}$$

13.2 – The LC Parallel Rule

The derivation of the LC parallel rule is a little more involved than the case of the series rule presented above. Again the two equations in question are given in the main text. Eq. 13.8 is the general case, with Eq. 13.10 being the special case where the resistance is removed. In this case the resistance must be replaced by an open circuit (a short circuit as in the previous case would be a wire bypassing the L and C components). An open circuit is represented by an infinite resistance, and so in this case replacing R develops as follows.

$$|Z_{\text{parallel}}| = \frac{1}{\sqrt{\frac{1}{R^2} + \left(\frac{1}{X_L} - \frac{1}{X_C}\right)^2}} \qquad \text{RLC parallel rule from Eq. 13.8}$$

$$= \frac{1}{\sqrt{\frac{1}{\infty^2} + \left(\frac{1}{X_L} - \frac{1}{X_C}\right)^2}} \qquad \text{set R equal to infinity}$$

$$= \frac{1}{\sqrt{\left(\frac{1}{X_L} - \frac{1}{X_C}\right)^2}} \qquad \frac{1}{\infty^2} = 0, \text{ and so disappears}$$

As before, the square and the square root almost completely cancel each other out, but not quite. It must be remembered that squaring a negative number yields a positive number, so if $X_L - X_C$ in the equation turns out to be negative, it will be necessary to discard the minus sign. The magnitude operator $|\ldots|$ performs this function. Thus the square and square root can be replaced by a simple magnitude.

$$= \frac{1}{\left|\frac{1}{X_L} - \frac{1}{X_C}\right|} \qquad \text{square and square root cancel}$$

$$= \frac{1}{\left|\frac{X_C - X_L}{X_L \times X_C}\right|} \qquad \text{normalise and combine fractions}$$

$$= \left|\frac{X_L \times X_C}{X_C - X_L}\right| \qquad \frac{1}{a/b} = \frac{b}{a}$$

$$= \left|\frac{X_L \times X_C}{X_L - X_C}\right| \qquad |a - b| = |b - a|$$

RLC Time Constants

One common use for capacitors is in setting the time constant of a circuit. A time constant might be used to control the frequency of oscillation of an oscillator circuit or the attack and release times in an envelope generator for instance. By extension inductors can be used in a similar fashion but this is much less common as the use of inductors is rare in general. As has been mentioned previously inductors have a few significant drawbacks (size, price, nonlinearity etc.) which generally make their use much less appealing then capacitors in most applications.

The idea of the time constant is that when a capacitor is charged through a resistor the voltage across the terminals of the capacitor rises in a very well defined and predictable way. Consider the circuit shown in Figure 13.9. Initially the switch is in position A and the voltage across the capacitor is zero. As soon as the switch is moved to position B current starts to flow through the resistor and the voltage across the capacitor begins to rise. The manner in which the voltage rises over time is plotted in the graph in Figure 13.9. Initially a large current flows and the voltage rises rapidly but as the voltage across the capacitor rises the voltage remaining across the resistor must fall (KVL) and as such Ohm's law means that the current flowing must fall. Eventually the capacitor is fully charged the entire available voltage is dropped across the capacitor and there is no voltage left across the resistor and hence no more current flows.

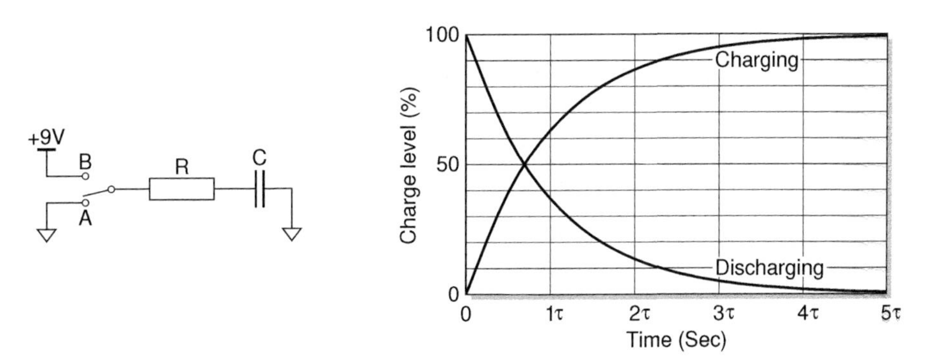

Figure 13.9 Circuit and graph for charging a capacitor through a resistor.

If the switch is at this point moved back from position B to position A the charging process is reversed and the capacitor is discharged through the resistor as per the second plot on the graph in Figure 13.9. The charge and discharge profiles are as illustrated in the graph logarithmic in nature starting off fast and then slowing down as they progress. The rate of charge or discharge can be characterised by the time constant.

$$\tau = \mathrm{RC} \tag{13.11}$$

Here τ (the lowercase Greek letter tau) represents the time constant and as before R is the size of the resistor in ohms and C is the size of the capacitor in farads. The time constant as defined above is the time it takes to charge the capacitor through the resistor to approximately 63% of the final voltage. This also equates to the time taken for it to fall to about 37% of the initial voltage in the case of discharging the capacitor (the charge and discharge curves are identical in shape mirrored around the horizontal).

The actual functions which define the charge and discharge curves are rarely needed. They are simple exponentials and are included below for reference. In them t is time in seconds and the levels vary between zero and one. Multiplying by a hundred yields the percentages plotted in Figure 13.9.

$$v_t = v_{final} \times (1 - e^{-t}) \quad \text{– charging a capacitor or energising an inductor}$$

$$v_t = v_{final} \times e^{-t} \quad \text{– discharging a capacitor or deenergising an inductor}$$

Multiples of τ are often quoted when examining the behaviour of an RC circuit. It is usually taken that 5τ, having allowed the voltage to surpass 99% of the final value, is effectively the stopping point where steady state has been achieved. In other words the capacitor is then taken to be fully charged (or fully discharged as the case may be). See Table 13.3 for commonly used multiples of the time constant and what they translate to in terms of the degree of charging or discharging achieved in a circuit.

Table 13.3 Charge/discharge levels corresponding to various multiples of τ

Time	% of maximum (charging)	% of maximum (discharging)
0.5τ	39.3%	60.7%
0.7τ	50.3%	49.7%
1.0τ	63.2%	36.8%
2.0τ	86.5%	13.5%
3.0τ	95.0%	5.0%
4.0τ	98.2%	1.8%
5.0τ	99.3%	0.7%

Once again it should come as no surprise that a similar analysis is possible for a circuit where the capacitor is swapped for an inductor except that in this case instead of tracking the rise and fall of voltages it is the current flowing through the circuit which first rises to a maximum and then dies away to nothing as the switch is moved from A to B and then back again. In the case of an inductor the time constant τ is calculated as follows.

$$\tau = \frac{L}{R} \tag{13.12}$$

Again τ represents the time constant and as expected L is the inductance in henries and R is again the resistance in ohms. In this case τ is the time taken to energise inductor L through resistor R to 63% of its final current. Initially the current starts off slowly as the growing electric field opposes its flow but as the field is established and slows its rate of change more current is allowed to flow.

Passive Filters

A passive circuit involves only passive components such as resistors, capacitors, and inductors. An active circuit includes active elements like transistors and ICs, and requires a source of power in order to operate. Generally speaking amplification of a signal requires active circuitry. Passive circuits can cut the power level of a signal, but they can not boost it. Some passive circuits (most notably those involving transformers, as explored in the next chapter) can boost voltage at the expense of current or current at the expense of voltage, but the power in the signal is at best maintained at a constant level, or more likely drops. Only an active circuit can boost the power of a signal (by drawing power from a power supply). Recall from Watt's law that power in watts is the product of voltage and current ($P = I \times V$), and so it can be seen how voltage might be traded against current without any increase in power.

Filters are circuits designed to alter a signal's frequency makeup by passing some frequencies and rejecting others. The most commonly encountered filter types are the familiar high pass filter (HPF) and low pass filter (LPF). Standard passive RC high pass and low pass filter circuits along with their respective frequency response characteristics, shown as graphs of signal attenuation measured in decibels plotted against frequency in hertz, are given in Figure 13.10 and Figure 13.11 below. Filters such as these are characterised in terms of their cutoff frequency. The formula for calculating the cutoff frequency in a standard RC filter is shown by Eq. 13.13. This relationship holds for both high pass and low pass RC filters. The cutoff frequency of any of the filter types described here is defined as the frequency at which the output level falls 3dB below the corresponding input level. This is often also referred to as the −3dB point or half power point.

$$f_c = \frac{1}{2\pi RC} \tag{13.13}$$

By substituting the expression for the RC time constant from Eq. 13.11 this equation can be rewritten as $f_c = 1/2\pi\tau$. It is instructive to note that this version of the equation also works in the case of the corresponding RL filters examined later as can be seen by combining Eq. 13.12 and Eq. 13.14. Notice from the frequency response graphs that these simple filter types do not cut off very quickly. The signal level, although falling all the time remains significant for quite some distance beyond the nominal cutoff frequency. More elaborate filter designs can achieve a faster rate of fall but for these basic types the rate of fall (referred to as the slope of the filter) is constant at a modest −6dB per octave.

The Passive High Pass RC Filter

The operation of the passive RC high pass filter as shown in Figure 13.10 is quite straightforward to decern from a basic examination of the circuit. Consider what happens to a low frequency signal component and a high frequency signal component arriving at IN. Recall that the fundamental action of a capacitor is that it blocks low frequencies and passes high frequencies. Thus the series RC arrangement present in this circuit acts as a frequency dependent voltage divider. For low frequencies the top impedance associated with the capacitor is very high and the voltage at OUT is pulled closer to ground by the relatively small resistance of R. As the frequency climbs higher the situation is reversed with the resistance of R now dominating thus sending OUT up towards the level present at IN.

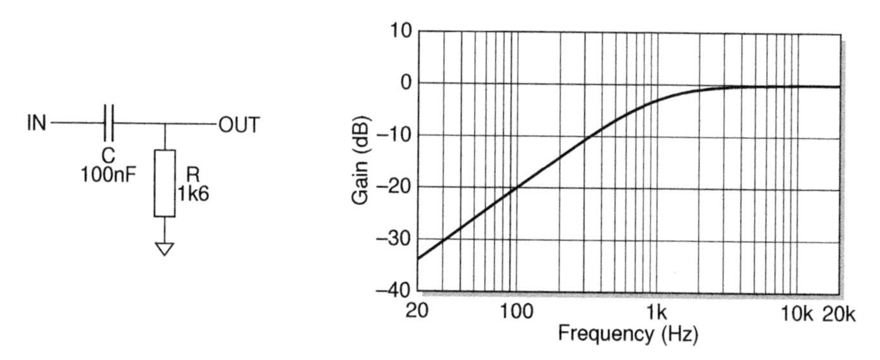

Figure 13.10 Simple RC high pass filter circuit and frequency response.

A quick check can be performed using Eq. 13.13. Inserting the known values for R and C the expected cutoff frequency for the circuit shown can be calculated.

$$f_c = \frac{1}{2\pi RC} = \frac{1}{2\pi 1k6 \times 100n} \approx 995\text{Hz}$$

Looking to the accompanying graph, the cutoff frequency at −3dB can be seen to be very close to 1kHz confirming that roll off is commencing at the correct point. It is also noted that these simple RC filters should (once sufficiently beyond the knee) have a slope of −6dB per octave (or equivalently −20dB per decade, these two amount to the same thing). This slope can be confirmed using either of these metrics. Looking at the line from 20Hz to 200Hz (one decade) it can be seen to rise from about −34dB to about −14dB, a change of 20dB in one decade. Similarly at 100Hz the line crosses −20dB, and at 200Hz is again at about −14dB, thus a change of 6dB over one octave. These observations help to confirm the accuracy of the graph which is plotted in Figure 13.10.

The Passive Low Pass RC Filter

Since the specified component values in Figure 13.10 and Figure 13.11 are the same, the high pass and low pass cases shown are completely symmetrical, and so all the analysis performed up to this point also applies to the RC low pass filter in Figure 13.11 with the appropriate reversal of resistor and capacitor terms as needed. The cutoff frequency and slope are both unaltered, but in this case the graph falls with increasing rather than decreasing frequency.

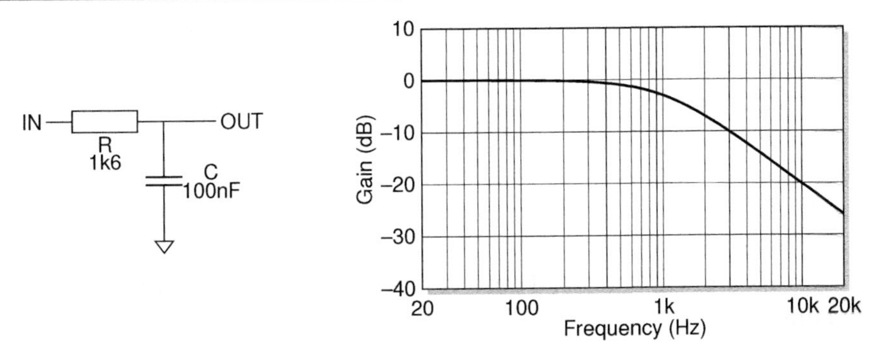

Figure 13.11 Simple RC low pass filter circuit and frequency response.

RL High Pass and Low Pass Filters

As noted before, any circuit involving capacitors is likely to have a closely corresponding inductor based circuit. Figure 13.12 illustrates the equivalent RL based high pass and low pass filter circuits, corresponding to the RC circuits just presented, while Eq. 13.14 shows the RL form of the cutoff frequency equation.

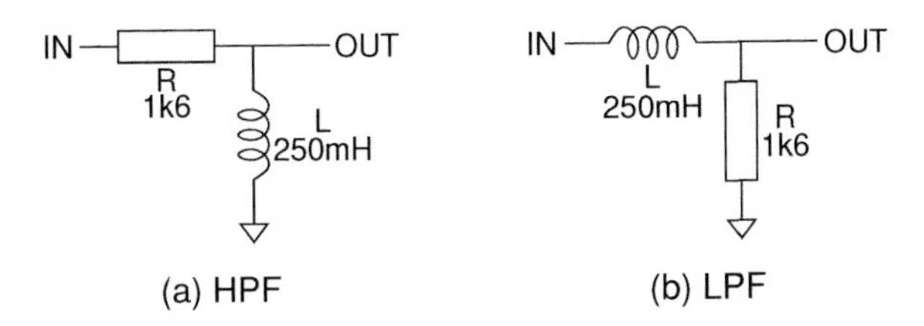

Figure 13.12 RL high pass and low pass filter circuits.

$$f_c = \frac{R}{2\pi L} \tag{13.14}$$

The component values in these four high pass and low pass filter circuits are chosen to be commonly available values to give corner frequencies close to 1kHz. The two RC filters actually have cutoff frequencies at about 1019Hz, while the corresponding RL configurations lead to frequencies of approximately 995Hz. Given component tolerances and expected deviations from ideal performance these are quite close enough to be considered equal for most purposes. These values can be confirmed using Eq. 13.13 and Eq. 13.14.

Standard Filter Section Configurations

As alluded to above, the basic filter configurations presented so far are just the simplest of a large family of associated passive filter types. Analysis of the performance of these more elaborate variants gets very mathematical very quickly, and is not pursued further here. For the reader who wishes to get deeper into filter design many good references exist (e.g. Niewiadomski, 1989; Williams and Taylor, 2006). Some of the more commonly encountered configurations are illustrated in the figures below.

Figure 13.13 shows the basic configuration already examined (referred to as an 'L' section), alongside two variations on the theme referred to as 'T' section and 'Π' section configurations respectively. All three filter sections shown here implement high pass filters. Unsurprisingly, any of the three can be converted to a low pass filter simply by swapping resistors and capacitors. These various configurations are often encountered in audio circuits.

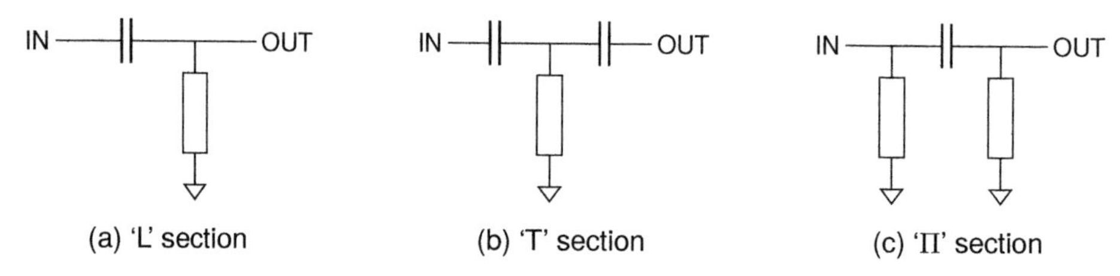

Figure 13.13 Standard filter section types.

Figure 13.14a shows the twin-T configuration. It should be clear that this is an HPF and an LPF wired in parallel. The result is a band stop filter or notch filter. The LPF half allows frequencies below the notch to pass, while the HPF half allows frequencies above the notch through. The labelling given here represents the commonest, and simplest, arrangement, where the capacitor to ground (labelled 2C) is twice the size of the other two capacitors, and similarly the resistor to ground (labelled R/2) is half the size of the other two resistors.

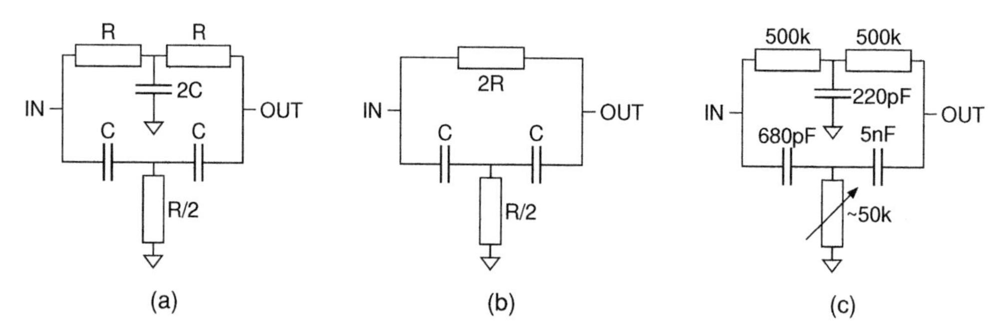

Figure 13.14 The twin-T and bridged-T filter configurations.

Figure 13.14b shows a simplification of the twin-T which is occasionally encountered, called the bridged-T configuration. The capacitor to ground is dropped, and as a result the two resistors labelled R, now being simply in series, can be replaced by a single resistor of twice the size (2R).

Finally, Figure 13.14c shows a variation on the standard twin-T with non-matching component values selected. This example is another taken from the 'Circuit Snippets' collection; the circuit in question is called the 'Idiot Wah', and the twin-T implements the standard wah swept notch filter. The variable resistor on the lower 'T' represents the effects familiar foot pedal control.

As noted in the comments accompanying the 'Idiot Wah' circuit from which the shown values are taken, 'the unequal capacitors in the T network give a more useful, more musical range than the typical matched values'. These observations will of course be a matter of individual taste, but as always the general rule is, experiment. The original circuit as found in the 'Circuit Snippets' collection also includes a switch on the 220pF capacitor to ground, allowing the twin-T to be converted into a bridged-T configuration as in Figure 13.14b, for added sonic possibilities.

LC Filters

So far all the filters presented have involved a combination of resistor and either capacitor or inductor. Another common variation is to drop the resistor and just combine capacitors and inductors. Figure 13.15 illustrates basic high pass, low pass, and band pass LC filters. LC filters have advantages and drawbacks when compared to their RC and RL cousins. They avoid the inevitable insertion loss of a resistance based filter; a resistor must drop voltage and dissipate power at all frequencies, whereas capacitors and inductors become virtually invisible in their pass bands. LC filters also exhibit twice the slope of corresponding RC and RL filters; −12dB per octave (−40dB per decade), as opposed to only −6dB per octave (−20dB per decade).

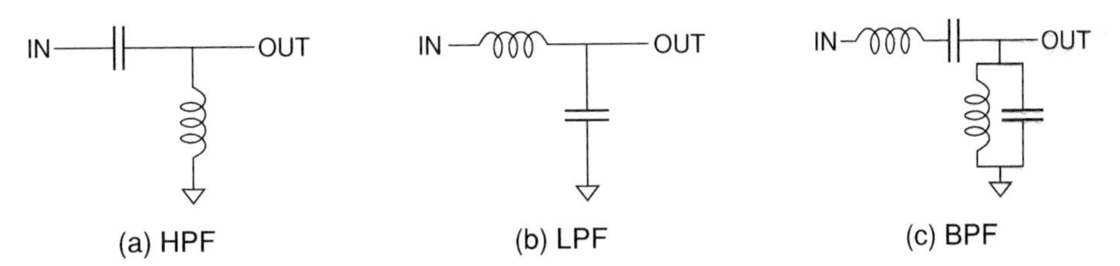

Figure 13.15 Simple LC high pass, low pass, and band pass filter circuits.

However, as illustrated in Figure 13.16, care must be taken in order to avoid excessive resonances and other response irregularities. The response variations shown are a result of different interfacing impedances of the circuitry to which the filter is connected. The under damped case, with a bit of resonance coming in at the cutoff frequency is a characteristic which is often actively encouraged in audio circuits, where 'resonant filters' can provide an excellent character to the tone of the circuit. Once again, the importance of well chosen interfacing impedances becomes clear, as the surrounding circuit plays a crucial role in the performance of the filter section.

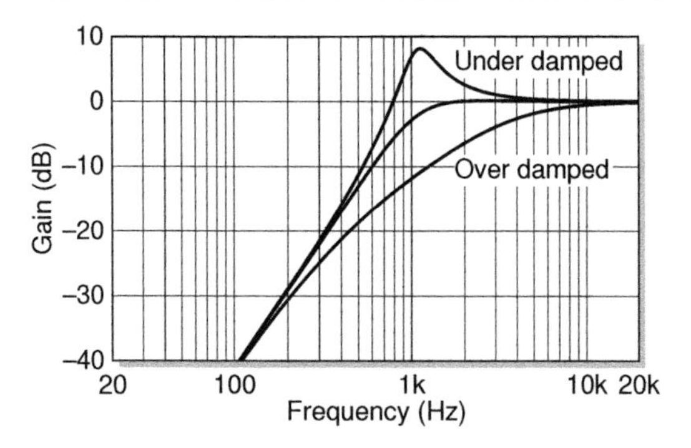

Figure 13.16 Simple LC high pass filter frequency response with resonance.

LC Resonance Frequency and Characteristic Impedance

As was previously seen in Figure 13.6, capacitor impedance falls with rising frequency and inductor impedance rises with rising frequency. As such there is always a single frequency at which the two components in any circuit consisting of a capacitor and an inductor will have the same impedance. This is represented in Figure 13.6 as the point at which the two lines on the graph cross. It thus follows naturally from Eq. 13.9 that for a capacitor and inductor in series there is a frequency at which the combined impedance of the circuit equals zero (when X_C equals X_L).

Similarly if the capacitor and inductor are instead placed in parallel then the impedance at this same frequency will instead rise towards infinity.[a] Figure 13.17 illustrates this behaviour, plotting impedance against frequency for circuits consisting of a capacitor and an inductor in series and in parallel respectively.

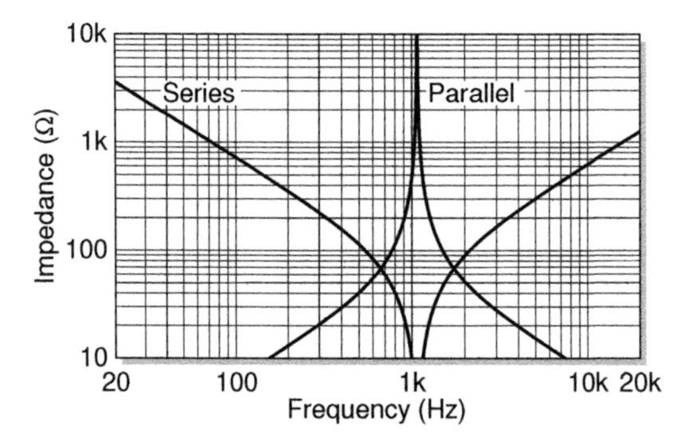

Figure 13.17 LC series and parallel impedance against frequency.

This frequency where the impedances become equal is called the resonant or natural frequency of the circuit, commonly indicated f_0, and is readily calculated for any given capacitor-inductor pair by the equation shown in Eq. 13.15.

a. Remember that in mathematics dividing by zero yields infinity, $1/0 = \infty$.

$$f_0 = \frac{1}{2\pi\sqrt{LC}} \tag{13.15}$$

It is a simple matter to derive this equation directly from the two impedance equations previously given in Eq. 13.1 and Eq. 13.2. Simply set the two expressions equal to each other and rearrange in order to make frequency the subject of the equation. The reader is encouraged to attempt this derivation as an exercise. This resonance property can be very useful and is widely used in, for instance, tuned circuits in radio receivers. It also represents the cutoff frequency which is achieved in an LC filter circuit.

A simple LC circuit also has a characteristic impedance, Z_0, as defined by Eq. 13.16. This value can be important when designing LC circuits. As alluded to above, in order to control the resonance peak which can be exhibited by an LC filter and maintain a reasonably flat frequency response in the pass band the interfacing circuitry before and after the filter should have impedances close to this characteristic impedance. Alternatively a desired degree of resonance can be achieved by moving away from this optimum interfacing impedance in order to derive some interesting sonic effects when using these filters.

$$Z_0 = \sqrt{\frac{L}{C}} \tag{13.16}$$

Without getting into a detailed derivation or analysis, it is instructive to note that multiplying the expressions for capacitive and inductive reactances (Eq. 13.1 and Eq. 13.2) yields L/C. This quantity has units of ohms squared and so applying a square root operation gives a quantity with units of ohms. While not a proof of the validity or meaning of the expression given for the characteristic impedance these observations do illustrate the relationship which the given equation bears to the reactances of the component parts of the circuits in question.

More often the design question to be answered is not 'What f_0 and Z_0 correspond to given values of L and C?' but rather the other way round. If a filter with a specific cutoff frequency and characteristic impedance is required, what values of L and C should be chosen? The mathematical rearrangement of Eq. 13.15 and Eq. 13.16 to find expressions for L and C is not complex. The interested reader is encouraged to work through the derivation for themselves. The resultant equations are shown below.

$$L = \frac{Z_0}{2\pi f_0} \tag{13.17}$$

$$C = \frac{1}{2\pi f_0 Z_0} \tag{13.18}$$

The shape of these equations should come as no surprise. This is the third time the same basic form has been encountered in this chapter, with terms rearranged as appropriate. The first instance is the capacitor and inductor reactance equations (Eq. 13.1

and Eq. 13.2) where the reactance (X_C or X_L) at frequency f is established. The second stems from the RC and RL filter cutoff frequency equations where in combination with resistor R they yield cutoff frequency f_c, and the third instance, encountered here, has the characteristic impedance Z_0 and resonant frequency f_0 provide the ohms and hertz terms respectively.

Eq. 13.17 and Eq. 13.18 can be useful in the design of an LC filter, but possibly even easier to use and of even more utility is the kind of reactance chart shown in Figure 13.18. This kind of graphical lookup table, which equates to a pictorial representation of the two equations, is common in engineering references (e.g. Ballou, 2008a, p. 271), and proves a useful and easy to use design tool. Consider for example a design goal to reproduce the stepped LC high pass filter described in Williams (2012). Ten cutoff frequencies are listed for the ten steps of the high pass filter ranging from 70Hz to 7.5kHz.

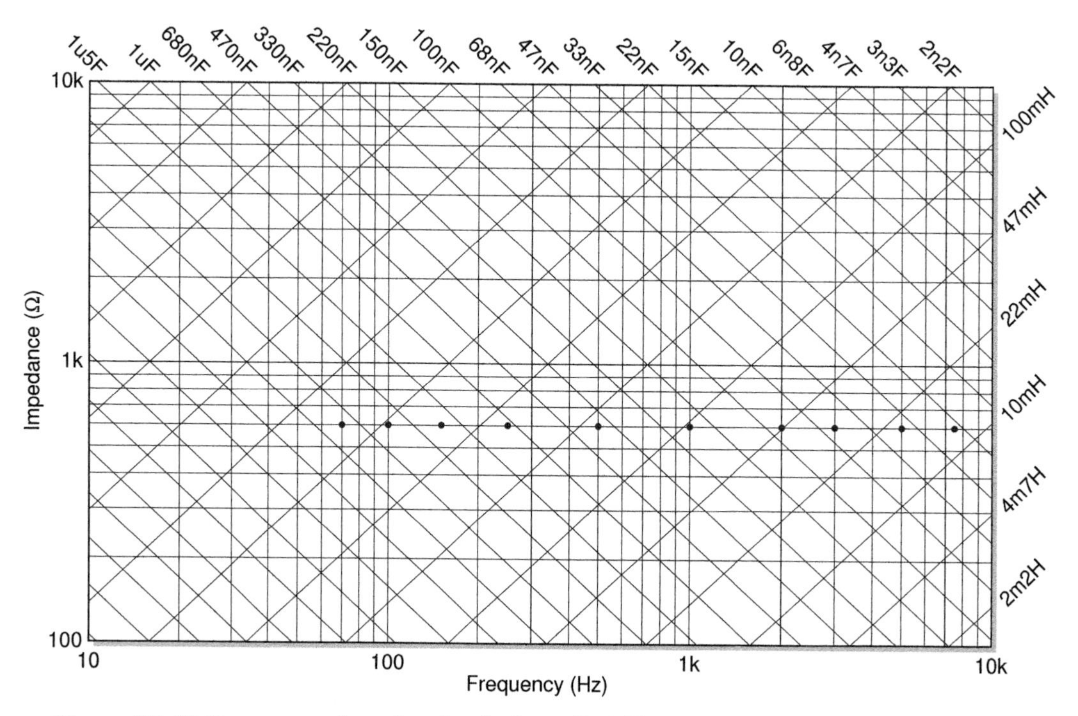

Figure 13.18 Reactance chart for the design of LC filters. The dots along the 600Ω line indicate the ten target frequencies in the example filter design task discussed in the text, simplifying the task of selecting suitable L and C values.

Clearly the switched filter sections will each interface with the same drive and load circuitry and as such all filters should share the same characteristic impedance in order to achieve consistent interfacing. Further research indicates that 600Ω is the most likely original characteristic impedance. Using the reactance chart it is a simple matter to take a horizontal line at the 600Ω level and identify points along this line for each of the desired cutoff frequencies, and thus appropriate LC pairs can be selected for each of the ten filters to be implemented.

It is important to also recognise that in the real world capacitors and especially inductors are only available in a limited set of values. By marking only the available values on the chart it is a simple matter to choose the best fit values for L and C. Furthermore if, for instance, the accuracy of the cutoff frequency is more important, while closely matched characteristic impedances are deemed to be less critical it is a simple matter to scan the chart visually to make the best possible design choices. Similarly, if extra resonance is to be preferred over excessive damping, again the chart allows a quick scan to yield the optimum values. Attempting the same procedure using Eq. 13.17 and Eq. 13.18 is much more cumbersome.

References

G. Ballou, editor. *Handbook for Sound Engineers*. Focal Press, 4th edition, 2008a.

G. Ballou. Resistors, capacitors, and inductors. In G. Ballou, editor, *Handbook for Sound Engineers*, ch. 10, pp. 241–272. Focal Press, 4th edition, 2008b.

P. Baxandall. Negative-feedback tone control. *Wireless World*, 58(10):402–405, 1952.

S. Niewiadomski. *Filter Handbook – A Practical Design Guide*. Newnes, 1989.

S. Niewiadomski. Sharper by design. *Practical Wireless*, 80(9):34–35, 2004.

D. Self. *Small Signal Audio Design*. Focal Press, 2nd edition, 2015.

B. Vogel. *The Sound of Silence – Lowest-Noise RIAA Phono-Amps: Designer's Guide*. Springer, 2008.

A. Williams and F. Taylor. *Electronic Filter Design Handbook*. McGraw-Hill, 4th edition, 2006.

S. Williams. Tubby's dub style: The live art of record production. In S. Firth and S. Zagorski-Thomas, editors, *The Art of Record Production: An Introductory Reader for a New Academic Field*, ch. 15, pp. 235–246. Routledge, 2012.

13.3 – RC Filter Frequency Response Equation

The form of the frequency response plot shown in Figure 13.10 can be derived using only relationships previously examined. This derivation is not essential to the work which follows but it is worthwhile to appreciate the way in which these various aspects come together to describe the situation. The four equations needed are –

$$\mathrm{dB} = 20\log\left(\frac{V_1}{V_2}\right)$$ Eq. 5.6 the decibel equation

$$V_{out} = V_{in} \times \frac{R_2}{R_1 + R_2}$$ Eq. 12.3 the voltage divider equation

$$X_C = \frac{1}{2\pi f C}$$ Eq. 13.1 the capacitor impedance equation

$$|Z_{series}| = \sqrt{R^2 + (X_L - X_C)^2}$$ Eq. 13.7 the RLC series impedance equation

The graph in Figure 13.10 plots decibels against frequency and so the basic form of the function needed is clearly that of the decibel equation. Thus what remains is to determine the values for V_1 and V_2 in Eq. 5.6. Clearly these are the output and input voltages respectively marked as OUT and IN in the circuit fragment, since the ratio of output to input is exactly the relationship desired.

As previously stated the circuit can be considered a frequency dependent voltage divider. From a rearrangement of the voltage divider equation it is clear that $V_{out}/V_{in} = R_2/(R_1+R_2)$ and so the required voltage ratio can be expressed in terms of the ratio of the impedances developing it. In this case R2 the lower resistor in the voltage divider is simply the 1k6 resistor from the circuit and R1 becomes X_C the impedance of the capacitor C as given by Eq. 13.1.

Finally, remembering that complex impedances in series can not simply be added like pure resistances the RLC series equation (Eq. 13.7) is employed to find the required expression to be substituted for $R_1 + R_2$. There is no inductance here hence X_L is zero, and the minus sign disappears when the square is applied so the bottom half of the ratio becomes $\sqrt{R^2 + X_C^2}$. Substituting back into the decibel equation, all this leads to a final form –

$$\mathrm{dB} = 20\log\left(\frac{R}{\sqrt{R^2 + X_C^2}}\right) = 20\log\left(\frac{R}{\sqrt{R^2 + \left(\frac{1}{2\pi f C}\right)^2}}\right)$$

This is the actual function plotted in the graph in Figure 13.10. X_C is the frequency dependent term, and since the circuit shows R as being 1k6 and C as 100nF everything needed to plot the graph is known.

14 | Transformers

Transformers transform voltages up and down

On one level the transformer is a very straightforward device, and yet transformers can also represent a complex topic requiring much know how and a significant amount of analysis in order to specify and deploy them effectively. Fortunately this degree of detail is not required for the kind and level of application considered here. In simplest terms a transformer consists of nothing more than two coils of wire (referred to as windings) mounted close together but not electrically connected. A signal in one winding induces a corresponding signal into the second. To aid this signal transfer the coils of wire are wound around a core made from iron, ferrite, or similar ferromagnetic material. While many other details are needed to fully characterise a typical transformer, this is as much of a description as is required here. Figure 14.1 illustrates the basic structure of a transformer.

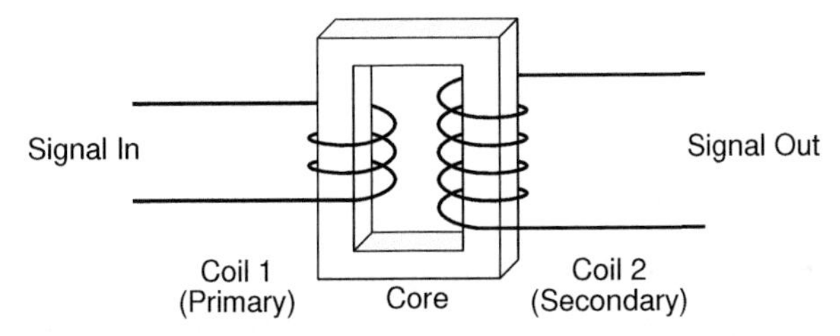

Figure 14.1 The basic structure of a transformer, consisting of two windings wrapped around a common core.

The most familiar description of the function of a transformer is that it changes the voltage level of a signal either up or down, but this is not the only thing that transformers do, and often it is more of a side effect of their intended function in a circuit, and not their primary role. In addition to voltage and current transformation transformers can also provide electrical isolation, signal balancing, polarity inversion, and impedance adjustment. All these aspects of a transformer's operation are examined in the sections which follow.

Transformers come in a wide variety of shapes and sizes. Figure 14.2 illustrates two of the commonest forms of transformer which are encountered in audio electronics. Transformers are also seen in numerous variations of these formats but the ones shown here are typical. Size, shape, weight, and electrical connections can all vary widely depending on the particular device involved and the application for which it is intended.

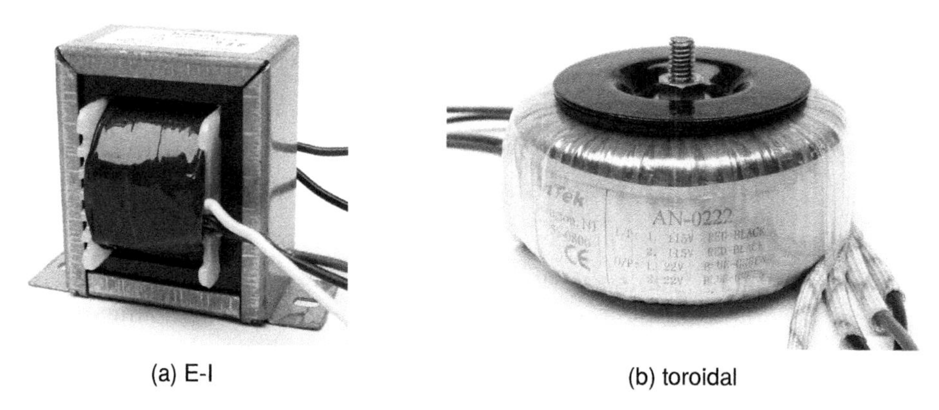

(a) E-I (b) toroidal

Figure 14.2 Examples of the two basic transformer form factors.

Given that two coils is about all that a transformer consists of, it should come as no great surprise that the key characteristic which allows its performance to be analysed is the relationship between these two windings. Specifically what is of interest is the number of turns of wire in each winding. As such, the most fundamental metric defining the action of a transformer is its turns ratio, which is defined as the number of turns of wire on the secondary winding divided by the number of turns on the primary. This relationship is expressed mathematically as Eq. 14.1 in the Transformation Equations section (p. 244), where transformer behaviour is presented in a more formal context.

How Transformers Work

Chapter 2 introduced the idea of electromagnetic induction in the section on electricity and magnetism. There the process is described in terms of a conductor moving through a magnetic field in order to generate electricity. Transformers operate on the same basic principle, except that here the conductor remains stationary while the field itself moves (by growing and diminishing as the signal in the primary winding varies).

When current flows in a conductor, an electric field forms around that conductor. If the current changes so does the field. An alternating current (like a sine wave or an audio signal) is constantly changing and hence generates an ever varying field. If a second conductor is placed close to the one carrying the signal, the electric field being generated by the first surrounds the second, and can thus induce a current to flow in this second conductor. This describes the fundamental action which underlies the operation of a transformer. A signal in one wire induces a signal into another close by.

To achieve better signal transfer the two coils are wrapped around a core which helps to transfer the energy from the first coil into the second. The core can take a number of forms but generally transformers will come in one of two basic formats which might broadly be designated E-I and toroidal (see Figure 14.3), based on the physical shape and structure of the core. The core material will often be laminated, that is to say instead of being a solid piece of material it is built up of layers each one insulated from its neighbours. This helps improve the efficiency of operation of the transformer. Alternatively the core may be moulded from a ferrite material (basically a form of metallic dust in an insulating paste). This serves the same function as the laminations, again optimising the efficiency of the device by minimising the amount of energy lost in transit.

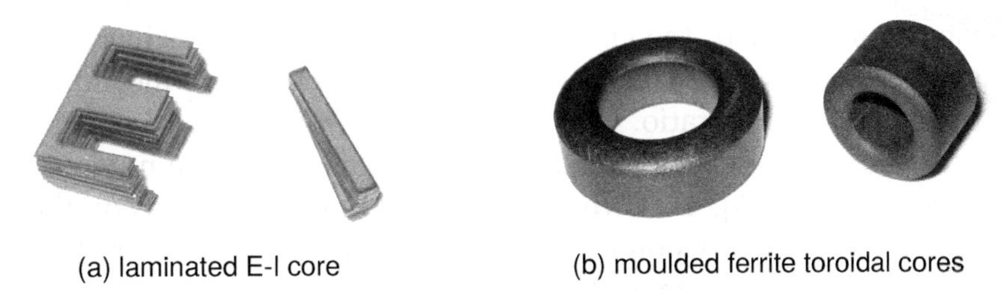

(a) laminated E-I core (b) moulded ferrite toroidal cores

Figure 14.3 Transformer cores: (a) illustrates a laminated E-I core removed from its transformer windings and separated into its collection of E and I shaped parts. (b) shows two toroidal cores moulded from ferrite material.

The two coils or windings of a transformer are designated the primary and the secondary. The incoming signal is sent through the primary coil, and the resulting signal emerges from the secondary. In general, transformers can be used in either direction but will usually be optimised for operation in a particular orientation. In other words either of the two windings might be used as the primary, with the other acting as the secondary. The action of a transformer (in terms of stepping a voltage either up or down) is reversed if the primary and secondary are swapped. One of the most challenging aspects of transformer design and manufacture addresses how these windings are configured. While for some applications it is enough to simply wind one coil and then the other on top of it, often a more elaborate approach is needed in order to achieve the performance required. Bi-filar winding for example involves spinning the two coils simultaneously so that the turns of each are interwoven. When wound in parts, the primary and secondary may be shielded from each other to avoid unwanted interactions. These and other details explain the wide range of types and varying qualities of transformers which may be encountered.

The practicalities of designing and building a high quality transformer are much more complex than the details which are examined here. Without such attention to detail a transformer will quickly start to introduce distortion and bandwidth limitations and variations, and to pick up noise and interference. In pro-audio circuit design the selection of the best transformer for a particular job can involve a lengthy and detailed analysis. High end transformers can be very expensive. Lo-fi audio transformers suitable for experimentation and use in DIY projects can on the other hand be bought quite cheaply from the major online electronics component distributors.

Transformation Equations

Turns Ratio

One of the commonest ways to characterise a transformer is as either step up or step down. In a step up transformer the voltage level of the signal emerging from the secondary winding is larger than the original signal level which is presented at the primary. By contrast, in a step down device the output level is lower than the input level.

As previously mentioned, the characteristic of a transformer which defines how it will manage this task of either stepping up or stepping down the incoming signal voltage level is called the turns ratio. A transformer's turns ratio (k) is calculated as the number of turns of wire on the secondary winding divided by the number of turns on the primary, as shown in Eq. 14.1 where N_{sec} is the number of turns on the secondary and N_{pri} is the number of turns on the primary. Notice that if there are fewer turns on the secondary than on the primary then k is a number less than one. This identifies the transformer as a step down transformer. Conversely if N_{sec} is larger than N_{pri} then k is greater than one and it is a step up transformer. An equal number of turns on the primary and secondary means $k = 1$ and the transformer is what is sometimes referred to as an isolation transformer, which passes a signal unchanged in level.

$$k = \frac{N_{sec}}{N_{pri}} \tag{14.1}$$

It is worth noting that there is a commonly encountered alternative format in which the turns ratio can also be written. In mathematics a ratio is often expressed as two numbers separated by a colon, so another way of presenting a transformer's turns ratio would be $N_{pri} : N_{sec}$. However, consider a transformer with 1,000 turns on its primary winding and 200 on its secondary. By Eq. 14.1 $k = {}^{N_{sec}}/_{N_{pri}} = {}^{200}/_{1000} = 0.2$. This could also be written using the ratio notation as 1000:200, but in fact it will almost always be simplified down to use smaller numbers in the appropriate ratio, in this case 5:1. Note that in the ratio notation N_{pri} comes first but in the calculation of Eq. 14.1 it is N_{sec} which comes first.

Power Transfer

The idea that an electronic component without a power source can make a small signal into a bigger signal (say turning 9V into 1000V) may at first seem a bit too much like getting something for nothing. This is of course not the case. It is important to remember that what transformers really do is trade voltage against current, they can step up or step down the incoming voltage, or leave it unaltered, but if the voltage goes up the current must go down by the same factor, and vice versa. Ultimately the amount of power coming out of the secondary equals the amount of power going into the primary (minus some small losses incurred in the process), as expressed by the rather trivial expression in Eq. 14.2. The efficiency of a transformer indicates how great these losses will be. A good transformer used in a correctly designed circuit will typically exhibit a very high efficiency, usually well above 90%. This means that only a small fraction of the power coming into a transformer will be lost as heat. Most will be passed on to whatever circuit is connected to the secondary.

$$P_{sec} = P_{pri} \tag{14.2}$$

Voltage Transformation

While power remains approximately constant, the other attributes of an incoming signal will usually be altered when it emerges. The turns ratio is key to understanding how a given device will affect any signal. This quantity can be used to see what effect a transformer will have on the signal voltage and current transferred from the primary to the secondary. It also tells how impedances will be mirrored through the transformer. Input and output impedances are important quantities in audio systems. It is often necessary to ensure that the appropriate impedances appear at various points in a system in order for it to operate efficiently, with minimum distortion and maximum signal transfer. Eqs. 14.3–14.5 specify how the transformer turns ratio defines the action of any given transformer. Eq. 14.3 states that the voltage which will appear on the secondary will be k times the voltage which is placed on the primary.

$$V_{sec} = k \times V_{pri} \tag{14.3}$$

Current Transformation

As stated above increasing voltage means decreasing current and vice versa. As such it is entirely predictable that the form of Eq. 14.4 for current transformation entails division rather than multiplication by k. And so it is that the current on the secondary winding is the current on the primary divided by the turns ratio.

$$I_{sec} = \frac{1}{k} \times I_{pri} \tag{14.4}$$

Impedance Transformation

The interpretation of Eq. 14.5 can at first be a little difficult to grasp. Put into words this equation might be expressed as, the apparent impedance when looking into the secondary is k squared times the actual impedance attached to the primary. This is a very important concept in audio electronics because input and output impedances play a key role in successful audio electronic circuit design.

$$Z_{sec} = k^2 \times Z_{pri} \tag{14.5}$$

14.1 – Transformer Current Transformation Relationship

It is a simple matter to derive the current transformation equation given the power and voltage transformation relationships previously specified.

We have stipulated in Eq. 14.2 that the power remains constant, and from Watt's Law (Eq. 9.2, p. 141) we know that power equals amps times volts ($P = I \times V$). This is all the information needed to derive the current transformation equation.

$$
\begin{aligned}
P_{sec} &= P_{pri} && \text{Eq. 14.2}\\
\Rightarrow\ I_{sec} \times V_{sec} &= I_{pri} \times V_{pri} && \text{Watt's law}\\
\Rightarrow\ I_{sec} &= I_{pri} \times \frac{V_{pri}}{V_{sec}} && \text{Rearrange}\\
V_{sec} &= k \times V_{pri} && \text{Eq. 14.3}\\
\Rightarrow\ \frac{V_{pri}}{V_{sec}} &= \frac{1}{k} && \text{Rearrange}\\
\Rightarrow\ I_{sec} &= \frac{1}{k} \times I_{pri} && \text{Substitute to yield Eq. 14.4 as required}
\end{aligned}
$$

Schematic Symbols

The three most commonly encountered versions of the circuit symbol used to indicate a transformer in a schematic diagram are given in Figure 14.4. There is no difference between the meaning of these three symbol styles and they are entirely interchangeable. The close relationship between inductors and transformers can be seen in the fact that inductor circuit diagram symbols also come in the same three common variants, as illustrated in the symbol table in Appendix C.

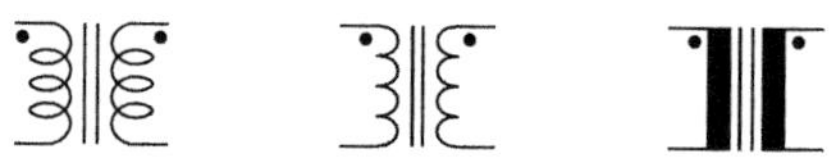

Figure 14.4 Three common variants on the transformer circuit diagram symbol.

Many variations on these basic forms may be seen; Figure 14.5 illustrates a number of possible variants, each one indicating some additional information about the device being represented. The meaning and significance of all these different symbols are explained below. Transformers can have different numbers of input and output connections. Both input and output can consist of one or more coils or windings, and each winding, as well as its end points may have one or more taps (midpoint connections) at various positions along its length. Transformers can also be made with a few different types of core structure and material, and may or may not also include a screen

14.2 – Transformer Impedance Transformation Relationship

The impedance transformation equation can be derived from the voltage and current transformation relationships already established. It becomes a simple matter if we recognise the fact that one equation can be divided by another. In other words, if $a = b$ and $c = d$ then $a/c = b/d$.

$$\frac{V_{sec}}{I_{sec}} = \frac{k \times V_{pri}}{\frac{1}{k} \times I_{pri}} \quad \text{Eq. 14.3 divided by Eq. 14.4}$$

$$= k^2 \times \frac{V_{pri}}{I_{pri}} \quad \text{Simplify}$$

$$\Rightarrow Z_{sec} = k^2 \times Z_{pri} \quad \text{Apply Ohm's law to yield Eq. 14.5 as required}$$

or shield connection which can be used to minimise noise pickup by the device. They might also be enclosed in a metal can to further shield them from their surroundings. Almost all these options and alternatives can be indicated directly on the circuit symbol and as such there are lots of variations on the basic symbols shown here which may be encountered.

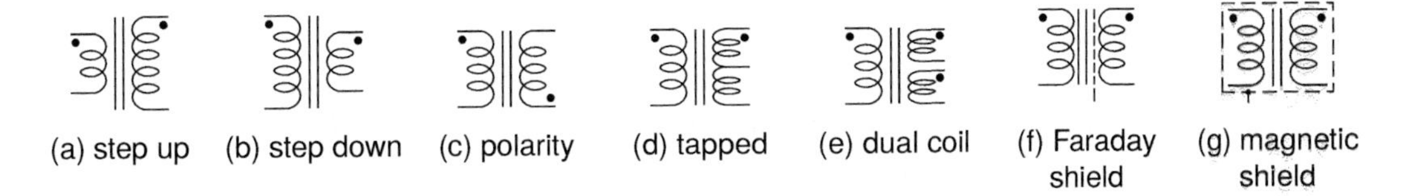

Figure 14.5 Transformer symbols may include additional information.

The symbols in Figure 14.5 have the following interpretations:

Step up – with more turns indicated on the secondary winding than on the primary, the turns ratio (k) is greater than one, and thus the output voltage level is greater than the input. This is referred to as a step up transformer.

Step down – this time the primary has more turns, indicating a k less than one, and a reduction in voltage on the output relative to the input. This is referred to as a step down transformer.

Polarity – the dot inserted beside one of the connection points on every winding indicates which connection represents the start of the winding. A signal passing through a transformer can always be extracted with its polarity either unaltered or reversed. Attention to the position of the dot ensures the desired polarity is maintained.

Tapped – one or more taps may be present on either or both windings. These provide alternative connection points to accommodate different input or output signal levels. A tap at the midpoint of a winding is called a centre tap and is often labelled CT. Taps can however be placed anywhere along the length of a winding.

Dual coil – either the primary or secondary side (or both) may consist of more than just a single winding. Multiple coils may be wired in different ways to provide different turns ratios, as well as increased signal handling capabilities. Wiring two coils in series yields a winding with an increased number of turns, while parallel wiring can increase the maximum signal level which can be handled.

Faraday shield – a connection point to a line between the windings indicates the presence of a Faraday shield or electrostatic shield. Grounding this point can minimise the effect of inter-winding capacitance, which can be important in some applications.

Magnetic shield – this takes the form of what is usually called a mu-metal can. This is a special metal can which a transformer is placed inside in order to prevent external magnetic fields from inducing hum or other interference into the signals passing through a transformer.

Transformer Types and Applications

The basic principles underlying the operation of a transformer are extremely straightforward (a signal in one coil induces a signal in another, with relative properties governed by the turns ratio k). However real world transformers involve a range of compromises and competing priorities which leads to a wide array of different transformers designed for different applications. A detailed treatment of how to design or select the best device for a particular application involves considerable technical analysis and is beyond the scope of this book. A good introduction to the topic can be found in Whitlock (2008).

When designing high end audio equipment the transformers deployed can represent a significant proportion of the overall system cost. At the other end of the spectrum however, small inexpensive devices, often generically referred to as audio matching transformers or coupling transformers can be acquired at very modest expense. These devices will not deliver pristine audiophile sound, but they are ideal for experimentation and learning. The example transformer circuits throughout this chapter and elsewhere (with the exception of the PSU circuit) can be built utilising just such transformers, and more details of suitable devices and where to get them are provided below.

Table 14.1 includes details on a small range of transformers from various manufacturers. The devices from Xicon probably represent the best options for general experimentation. They are relatively inexpensive, and the soft metal legs allow them to be inserted easily into a breadboard, making them ideal for building and testing various circuits with the minimum of effort. The Monacor and the first Jensen are transformers specifically designed to be used in DI boxes. While the offerings from Xicon do not fit the bill too closely, something like the 42TL013 or 42TL019 would be perfectly adequate to build a circuit for educational purposes. The topic of transformers and DI boxes is examined later. Looking further ahead to Chapter 16, the 42TL016 would be a prime candidate for the ring modulator circuit described there.

Table 14.1 Specifications for a range of transformers

Manufacturer	Part number	Impedance pri.	Impedance sec.	Turns ratio	$k = \frac{N_{sec}}{N_{pri}}$	Price guide
Jensen	JT-11-DMCF	680	80	1:1	1	High
Jensen	JT-DB-E	141k	150	11.52:1	0.087	⇓
Monacor	DIB-110	47k	600	10:1	0.1	
OEP	A262A2E	150	600	1+1:2+2[a]	1, 2, 4[b]	
Xicon	42TL013	1k	8	12.6:1	0.08	
Xicon	42TL016	600	600	1:1	1	
Xicon	42TL019	10k	600	5.128:1	0.195	Low

a. Indicates dual windings on both primary and secondary sides.
b. Different turns ratios are possible depending on how the dual windings are wired.

The three Xicon devices all come in the same format, with centre taps on both primary and secondary. The OEP has dual windings on both sides, hence the odd specification of the turns ratio. By wiring these windings in various combinations of series and parallel, k values of 1, 2, and 4 can be achieved. The OEP also sports an electrostatic shield. The Monacor is centre tapped on its secondary, as illustrated in the schematic in Figure 14.7. In the case of the Jensen devices, the JT-11-DMCF provides the bare essentials with just a primary and a secondary winding, while the JT-DB-E is a more elaborate affair with a total of seven connections. Four provide access to the primary and secondary as usual, while the final three connect to dual Faraday shields (one for each winding) and a magnetic shielding can. These examples are by no means exhaustive of the types and applications encountered. They are a snapshot of a range of potentially useful devices.

More broadly in audio electronics transformers are encountered in two main contexts. Power transformers take mains voltage and transform it to the various voltage levels needed in any given system. As many different output voltages as are needed can be tapped off the secondary side of the transformer. Power transformer are designed for the amount of power they can transfer (unsurprisingly), and good designs also aim to minimise and contain the potentially large magnetic fields which they can generate. Such fields can represent a major source of interference if care is not taken to contain their effects. On the other hand power transformers (unlike audio or other signal transformers) care little about the faithful reproduction of the signal shape as it is passed through the device. Linearity is not such a concern here.

The second class of transformer encountered in the area of audio electronics along with power transformers are of course audio transformers themselves, which are the primary focus here. These can be further subdivided into a number of different categories depending on their characteristics and intended function. Such categories include output transformers, input transformers, matching transformers, and coupling transformers. Audio transformers are used in the interfacing of audio circuits, and as such, unlike power transformers, they aim to pass signals as undistorted and as noise free as possible.

In DIY audio circuits a transformer will occasionally be found employed as an inductor, where the second winding is usually left unconnected. On the whole transformer windings make for rather poor inductors, but experimentation can yield some interesting results nonetheless.

In considering the particular tasks which a transformer might perform in a circuit, there are a number of key elements to be addressed. The functions which a transformer may serve include voltage and current transformation, polarity Inversion, electrical (aka galvanic) isolation, signal balancing, and impedance adjustment.

Voltage and current transformation are the functions already addressed most explicitly in the chapter so far. The transformer in a power supply for example takes mains voltage and converts it to a different (typically much lower) level as the first step in generating a DC power output signal. The full AC to DC power supply process, entailing transformation, rectification, and smoothing is considered in Chapter 16 on didoes (with the final optional regulation step addressed in Chapter 18). Similarly, a transformer may be used to drive an audio transducer. As mentioned previously, transformers can trade voltage and current. A circuit suited to driving a moving coil loudspeaker is unlikely to be able to successfully drive a piezo sounder (the piezo needs more voltage than the speaker, but not nearly as much current). Inserting an appropriate transformer can take a signal suitable for a speaker and convert it into something better able to drive the piezo.

Polarity inversion comes for free with a transformer. Simply reversing the connections to either the primary or secondary winding of the transformer will provide an inverted signal to the following circuitry. However, transformers are big and they are expensive, so it is unlikely that one would be included in a circuit solely for this purpose. If one is to be utilised for other purposes, it does provide a simple means to providing this functionality in addition.

Electrical (galvanic) isolation refers to a situation where there is no direct electrical connection between two points in a circuit. As the two sides of a transformer are not directly connected, transformers provide just such an isolation between their primary and secondary sides. This can be desirable on two grounds, firstly as a safety feature limiting the potential for electric shock, and secondly it can provide a means for eliminating noise from a system by interrupting unwanted ground loops.

Signal balancing describes a method for transferring signals which has the potential to avoid picking up noise during transmission. Transformers provide a very straightforward method for converting an unbalanced signal into a balanced one. This is a good thing to do before sending a signal over any distance, especially in a noisy environment.

Impedance adjustment is another function which, like voltage and current transformation, has been addressed already in the main discussion above. The effect is described in Eq. 14.5, and represents a major aspect of the way in which transformers are used to improve the interfacing of signals across circuit with different output and input characteristics.

Transformers and DI Boxes

Many if not most DI boxes (both passive and active) incorporate a transformer. Many passive DI boxes contain little else. Transformers provide for most of the important tasks which the typical DI box is designed to accomplish. In particular signal balancing, impedance adjustment, and galvanic isolation are key aspects of what a DI box is designed to be used for. Figure 14.6 shows a very simple passive DI circuit which accomplishes all three tasks, with a TS jack input, an XLR output, a ground lift switch, and at its heart a 10:1 step down transformer. Passive DIs typically employ a transformer with a turns ratio in the vicinity of 10:1 to 14:1.

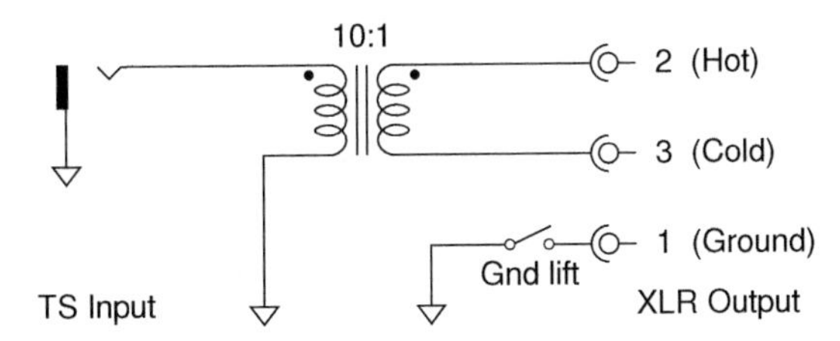

Figure 14.6 Circuit for a very simple passive DI box using a transformer with a 10:1 turns ratio ($k = 1/10 = 0.1$).

Typically a relatively short unbalanced TS lead from a guitar, synthesiser, laptop computer, or other unbalanced sound source is plugged into the input of a DI box, and a potentially much longer balanced XLR cable connects to the DI's output and takes the signal on to its destination. As alluded to above, a number of benefits can accrue from using such an arrangement rather than simply making the connection with a long TS to XLR cable.

Using just a cable means the full length of the connection is unbalanced. The main benefit of a balanced connection is its ability to reject noise which is picked up along the length of the cable. The balanced output of the DI can form one end of a balanced connection and thus reap the benefits in reduced noise and interference.

The importance of appropriate output and input impedances at either end of an interface has been highlighted previously. In most cases an impedance bridging arrangement is desirable. This is when the input impedance is significantly higher than the corresponding output impedance. Usually a factor of about ten is cited as appropriate. (Impedance matching, where the output and input impedances are equal, is also common in some situations, especially for digital or other high frequency connections. For the kind of analog, audio frequency connections considered here, bridging is preferred.)

When a relatively high output impedance source (like an electric guitar) is to be connected to a standard audio input, it is usually beneficial to insert a DI box to achieve a more appropriate ratio of output to input impedances. The 10:1 transformer in the example above corresponds to a turns ratio of $k = 1/10$. From Eq. 14.5 it can be seen that the resulting transformation factor is $k^2 = 1/100$. This means that the high output impedance of the electric guitar, when viewed from the output of the DI box will be divided by a factor of 100, bring it down to a level much more suitable for interfacing with a typical microphone or line level input.

The third benefit attributed to the transformer was galvanic isolation. Sometimes this is desirable, and sometimes it is not. The ground lift switch allows or defeats this isolation depending on its position. When the switch is closed, the ground connections on the primary and secondary sides are joined and isolation is defeated. This most likely to be the desired situation for something like an electric guitar. Linking the grounds holds the ground potentials on either side together, and thus maximises signal headroom.

However if the equipment on both sides of the DI box has a connection to mains earth ground (e.g. a synthesiser connecting to a sound desk, both mains powered), then the connection through the closed ground lift switch completes a ground loop, and ground loop noise is likely to ensue as a result. Opening the switch breaks the loop and prevents the noise currents from flowing.

Figure 14.7 provides the schematic diagram for a fully fledged passive DI box including parallel output, 0dB/−20dB/−40dB switchable pad, and ground lift switch. It is a worthwhile exercise to confirm that the resistors shown in the pad circuit do indeed provide the specified levels of attenuation. The pad setting is selected using a dp3t switch (see Chapter 15). In position one (as shown in the schematic) the two halves of the switch combine to bypass the resistors completely. This is the 0dB, unpadded setting.

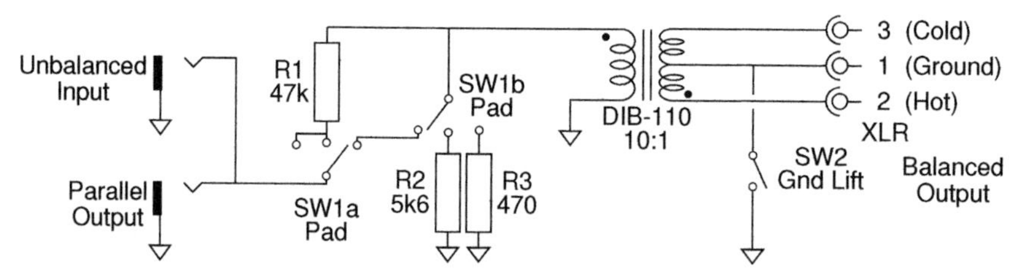

Figure 14.7 Full circuit diagram for the DIB-100 DI box from img Stageline. This DI box uses the Monacor DIB-110 10:1 transformer – they are in fact the same company.

In position two R1 = 47k and R2 = 5k6 come into play, forming a voltage divider between the input and the connection feeding the transformer. The voltage divider rule can be used to see what fraction of the signal makes it into the transformer, and then the decibel equation can be used to turn that fraction into an attenuation in decibels.

$$\begin{aligned}
V_{out} &= V_{in} \times \frac{R_2}{R_1 + R_2} && \text{Voltage divider rule} \\
\Rightarrow \frac{V_{out}}{V_{in}} &= \frac{R_2}{R_1 + R_2} && \text{Rearrange} \\
dB &= 20 \log \frac{V_1}{V_2} && \text{Decibel equation} \\
&= 20 \log \frac{R_2}{R_1 + R_2} && \text{Substitute} \\
&= 20 \log \frac{5k6}{52k6} && \text{Insert values} \\
&= -19.5dB && \text{Calculate}
\end{aligned}$$

With an aim point of −20dB for the first pad, this looks perfect. Anything less than a decibel out is considered to be negligible. Exactly the same calculation can be used for the second pad position. In this final position the switch selects R1 and R3, so the 5k6 is replaced with 470. This leads to a pad level of −40.1dB as shown below, again well within the required range.

$$\begin{aligned}
dB &= 20 \log \frac{470}{47{,}470} && \text{Insert values} \\
&= -40.1dB && \text{Calculate}
\end{aligned}$$

LEARNING BY DOING 14.1

BUILD A PASSIVE DI BOX

Materials

The main requirement is the transformer of your choice. The 42TL013 or the 42TL019 from Xicon will suit. Neither is perfect but both will work fine for test and experimentation purposes. The Monacor or even the Jensen DI transformers would be good if available, although these will not connect so easily to a breadboard, and might be a bit expensive for just experimenting. If you plan to build a DI box to keep and use, these are perfect.

Also needed are resistors, switches, connectors, signal source, oscilloscope, and soldering equipment, if you want to build a circuit to keep (using the stripboard design in Appendix D or otherwise).

Practice

Figure 14.6 and Figure 14.7 provide two possible approaches to a simple passive DI. Depending on the supplies you have available, finalise a design for the circuit you are going to build and test. If your transformer has a centre tap on its secondary, the ground lift scheme from Figure 14.7 is suitable, otherwise Figure 14.6 shows an alternative approach.

Test your circuit by sending a test signal through it. A 1kHz sine wave at about 0dBu ($\approx 0.775V_{rms}$) is usually a good starting point. Using the oscilloscope, measure the level of the signal which emerges and observe any visible distortion or noise which has been introduced.

Apply the pads and see how this affects the signal level. Don't expect to see exactly 20dB or 40dB of attenuation. Input and output impedances will play a role in determining the final signal level.

Play some music through the circuit. Does it come through clearly, does the pad affect it as expected. Does the ground lift have any effect on the sound? Try different sources and amplifiers, both mains powered and battery operated.

Transformers as Output Drivers

A typical small signal transistor does not have the current handling capability to drive a loudspeaker directly. Feeding its output through a suitable transformer can significantly improve this situation. It won't go very loud, and it won't be very clean, but it can prove an interesting effect. The extravagantly titled 'Ultra Class A Superdrive Power Amp' from Escobedo's 'Circuit Snippets' collection provides a very simple circuit in this mode. If the suggested Radio Shack transformer can not be found, it could reasonably be replaced with a Xicon 42TL013 or something similar. Notice that the secondary impedance of this transformer is listed in Table 14.1 as being 8Ω. This is something of a giveaway that this transformer is most likely specifically intended for interfacing directly to an eight ohm loudspeaker.

In the example above, the device is used in its intended orientation, as a step down transformer, trading voltage for current, to better drive the low impedance load of the loudspeaker. Another common kind of audio transducer besides the loudspeaker is the piezo driver. However, a piezo element presents a very high impedance and, as such, an amplifier designed to drive a standard loudspeaker is unlikely to be able to get much life out of a piezo. What is needed is more voltage, and since the high impedance piezo also needs very little current, the transforming properties of a transformer should be able to convert an output suited to driving a loudspeaker into one more able to handle a piezo.

Figure 14.8 illustrates this situation – Figure 14.8a shows a 386 audio power amplifier connected directly to a loudspeaker as it is intended to be used. In this configuration the circuit will operate as expected and sound will emerge from the loudspeaker in the normal fashion. In Figure 14.8b however the loudspeaker has been replaced with a piezo driver. There is nothing wrong with the circuit per se, but the piezo will produce

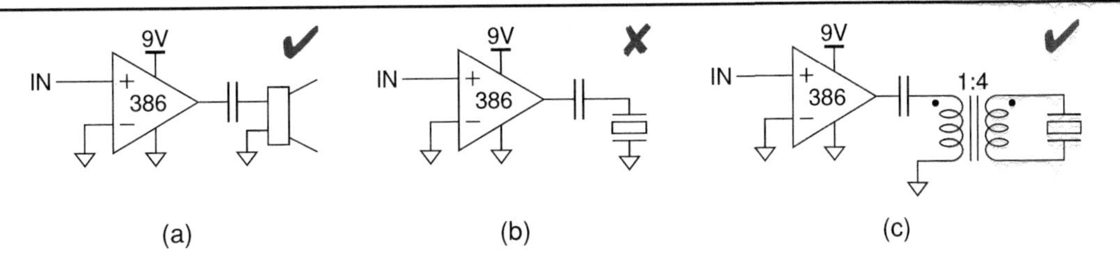

Figure 14.8 Using a step up transformer in order to trade current for voltage so as to better drive a piezo sounder from an amplifier output designed to drive a much lower impedance load. (a) The 386 as it was intended to be used. (b) Trying and failing to drive a high impedance load. (c) Using a transformer to better match source to load.

a very weak signal if any at all. There simply isn't enough voltage available to get the piezo to move very much.

A step up transformer placed between the amplifier and the piezo driver will increase the voltage available, and get the piezo moving sufficiently to generate some sound. The transformers described so far are mostly designed as step down devices, but by reversing their connections between primary and secondary they can be used in the opposite direction in order to step a signal up. Therefore, the configuration shown in Figure 14.8c can be implemented using, for example, a 42TL013 or 42TL019 transformer with primary and secondary connections swapped.

LEARNING BY DOING 14.2

TRANSFORMER DRIVEN PIEZO

Materials

- Amplifier (either build a simple 386 amp or use a pre-built unit)
- Loudspeaker
- Piezo element
- Transformer (turns ratio 1:4 ($k = 4/1 = 4$) or higher)

Theory

The sequence of circuits in Figure 14.8 makes for a good progression to follow in a practical examination of these principles. The transformer is the crucial part here. Make sure the device you choose has a reasonably high turns ratio; a 1:1 isolation transformer is no use, it will not boost the voltage at all. Identify the primary and secondary windings, and determine whether you will need to use the device in reverse or not. A step up transformer is needed (k greater than one), so to use a step down device it must be connected the 'wrong' way round.

Practice

Gather the necessary materials, and build the amp if necessary.

Step 1: Connect a sound source to the input

Step 2: Connect the output to ta loudspeaker; sound should emerge as usual

Step 3: Replace the speaker with a piezo; little or no sound should be heard

Step 4: Interpose the transformer; the sound level should increase significantly

Step 5: Piezos work best attached to a resonator; press the piezo against various surfaces – small tins and tubes can work well

Step 6: Try reversing the transformer; no sound should be heard

References

G. Sowter. Soft magnetic materials for audio transformers: History, production, and applications. *Journal of the Audio Engineering Society*, 35(10):760–777, 1987.

B. Whitlock. *AN-002: Answers to Common Questions About Audio Transformers*. Jensen Transformers, 1995.

B. Whitlock. Audio transformer basics. In G. Ballou, editor, *Handbook for Sound Engineers*, ch. 11, pp. 273–307. Focal Press, 4th edition, 2008.

15 | Switches

Switches route, connect, and select electrical signals

The humble switch seems a trivially simple topic at first glance but there is much more to this versatile component than might be immediately apparent. In its most basic form it is indeed a very straightforward device; the function of a light switch or a doorbell buzzer is no more than to make and break a single electrical connection. This is achieved using what would be called a single pole single throw or spst switch. Poles and throws are two key concepts for switches, and a switch might have (in theory) any number of each – more on this later.

Figure 15.1 illustrates the three basic roles which a switch can play in the operation of an electrical circuit, as given in the one line description at the top of the chapter. Before getting into the details of these and other operational aspects of switches, an overview of common switch styles is presented. The term 'style' is used here to refer to the mechanical design as opposed to the electrical characteristics of a switch.

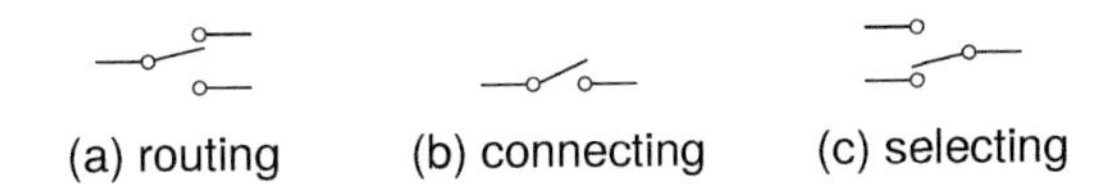

Figure 15.1 The three basic functions which a switch can perform are illustrated here. Consider signals flowing from left to right in the three examples above. In routing, the incoming signal can be routed to either one of two different destinations. In connecting, the incoming signal can be switched in or out of the circuit. In selecting, just one of the two incoming signals is selected to continue on.

Unlike most electronic components, the switch is also a prime candidate for home-made variants, from tilt switches to finger contacts and much more beside. Collins (2009) includes some interesting suggestions for experiments in this area.

Switch Styles

Switches come in many shapes and sizes, such that the same electrical functionality may be achieved using a wide variety of physically different devices. The brief catalogue which is presented in Table 15.1 is by no means exhaustive. It does however cover most of the major categories which are commonly encountered in audio technology and DIY audio electronics building, and one or two other less common types besides. Most of the switch styles illustrated here will encompass several variations and subtypes. These vary dramatically in physical size. Some include locking devices to prevent accidental operation. Others may have indicator lights to show their current state. They might have legends, adjustable actions, or other special features. All these options and more might need to be considered when selecting what switch to use in a given application.

Switches are one of the main hardware interface components encountered, along with such things as pots and faders, connectors, indicator lights, and meters. As such not only their electrical characteristics but also their mechanical action and physical design are important. For instance the push switch style might be the most appropriate for playing notes on an instrument, while a toggle switch might be better suited to selecting modes of operation (perhaps engaging filters or selecting waveforms in a synth). Similarly, less often accessed functions (perhaps the pad select on a DI box) might be most conveniently accomplished using a slide switch, and critical controls like the main power on a large console might benefit from a key switch to avoid accidental power down.

Determining the electrical characteristics required in a particular application is only the first step in selecting the right switch for the job. Many factors can come into play in the selection of the most appropriate switch style for a particular application including cost, availability, ease of use, aesthetic look and feel, robustness, footprint and profile, in addition to electrical characteristics such as voltage and current ratings. When experimenting or when building one off circuits the deciding factor is often availability – what suitable switch can be most readily found amongst the components currently to hand. So long as it provides the required number of poles and throws, the exact switch style is often less critical.

Poles and Throws

As alluded to above, the number of poles and the number of throws are two key descriptors specifying the functionality of any given switch. An individual switch might control any number of separate things simultaneously. Perhaps it is desired to be able to turn the signal on and off to a pair of loudspeakers with a single switch. This will involve breaking two independent connections at the same time, and so the switch used will need two poles; effectively two totally separate switches are actuated by a single control (be it a toggle, push button, slider, or something else).

This leads on to the related property described in the number of throws. Throws indicate the number of different places one connection can be switched between. Returning to the example above, if instead of merely muting the loudspeakers in question, the goal is perhaps to switch between three different sets of speakers to allow easy comparison of a sound source on a number of different systems. For three sets of speakers a

Table 15.1 Examples of common switch styles encountered in audio technology and DIY audio electronics (and a few less common styles as well)

Style	Example	Notes
Toggle		Toggle switches are probably the most common switch type used in DIY audio electronics projects.
Rocker		Rocker switches are very commonly used for ON-OFF type applications. They often include a status light within the rocker.
Slide		The mechanism in slide switches is often cheaper and hence less robust than other switch styles.
Push		Push switches can be very compact. They are commonly found in keypads and keyboards. They can be momentary or latching types.
Pull		Pull switches are usually latching but can also be momentary in operation. Almost always single or dual throw.
Rotary	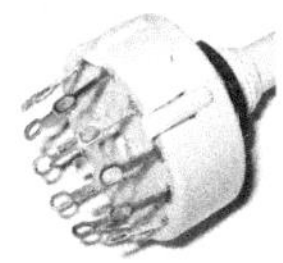	Rotary switches are particularly good when a large number of switch positions are needed. The illustration shows an 'sp12t' type – a single pole, and twelve throws.
Key		The power on high end gear might be controlled using a key switch. Special programmer or configuration modes may also be accessed this way.
DIP		DIP switches are used to set seldom changed options. DIP stands for dual inline package, and refers to the layout of legs in two rows.
Reed		Reed switches are actuated by a magnet. Good for detecting when two objects come close to each other, or when they move apart.

three throw switch will be needed – in this case a dual pole three throw (dp3t) switch, as a stereo signal is being switched between three different pairs of loudspeakers.

The number of poles indicates how many separate signals can be switched.
The number of throws indicates how many different paths can be selected.

When naming the type of any given switch, for one pole or throw the term used is 'single', for two the term is 'double' or 'dual', and after that numbers are used. So for instance a very common combination would be the 3pdt (three pole dual throw) switch, allowing three separate things to be switched between two different paths each. The foot switch on a guitar effects pedal (e.g. Figure 15.2) is very often of this type; one pole switches the input signal path between bypassed and affected, the second pole switches the output signal similarly, and the third pole activates a status LED to let the user know the current state of the pedal.

Figure 15.2 A dpdt and a 3pdt foot switch. A standard foot switch for something like a guitar effects pedal will usually be implemented using a heavy duty latching push switch, most often of the three pole dual throw (3pdt) variety.

The number of poles in a switch does not commonly exceed about three, but switches with higher numbers of throws are quite common. For instance some audiophile hi-fi volume controls use a dual pole switch with a large number of throws (perhaps 24 or more) instead of the more conventional volume pot. This means that the volume changes in discrete steps rather than in a continuous fashion, which may seem like a backwards step. However it makes it possible to keep the left-right matching much tighter than is typically achievable with a pot, avoiding shifts in the stereo image as the volume is altered.

Naturally a switch pole can be used in either a one-to-many or a many-to-one configuration (routing vs. selecting). In other words either a single incoming signal can be routed to a number of different destinations or a number of different incoming signals can be selected between, sending only one of them on to the next part of the circuit. A multi-pole switch can have different poles working in either fashion. In the guitar effects pedal example above, the first pole is being used one-to-many (routing the incoming signal either into or around the pedal's circuitry) while the second pole in the same switch is being used many-to-one (selecting the pedal output either from the main or the bypassed signal path).

Though it may be obvious, it is perhaps worthwhile mentioning that when using a single throw switch it can be wired up either way around without affecting the functionality, as all it is doing is making and breaking a single connection; no alternate routing in or out needs to be considered.

One final purely practical aspect to the selection and use of switches is that it is always possible to use a switch with more than the required number of either poles or throws. The unused poles/throws are simply left unconnected. By and large this would be considered wasteful as switches with higher pole or throw counts will usually be more expensive, but sometimes it is more convenient to make do with what components are readily to hand. In particular dual throw switches are often used in place of their single throw counterparts with one of the contacts left unconnected. One unused throw may be conveniently left unconnected as an 'off' position, but leaving more than one throw unused is likely to result in a confusing experience for the user, as multiple switch positions do the same thing.

Momentary vs. Latching

A light switch and a doorbell would both most commonly use an spst switch (in both cases the required action is simply to make or break a single connection). There is however an important difference between the actions of these two devices. When a light is switched on it should stay on until switched off again, while with a doorbell it should ring when pressed and stop when released. The former requires a latching action (sometimes called 'successive operation' for a push style switch), while with the latter the action is referred to as 'momentary' or 'biased'. A momentary switch only stays switched as long as it is held (like a door bell). A latching switch remains in its current position until it is switched to a new position by the user (like a light switch). The keys on a keyboard are going to be momentary switches while the routing buttons on a console will be latching. In circuit diagrams the labelling 'MOM' is sometimes found beside a switch in order to indicate that the specified component is intended to be a switch with a momentary action.

Radio Buttons

The term radio buttons refers to a set of separate but interconnected latching push switches. The idea is that only one of the buttons can be latched into its pushed down setting at any given time. When one of the buttons is pressed, whichever other button was previously latched down automatically pops up. The name comes from this type of buttons most common original use. On old radio sets the user could often select the band using a set of radio buttons. Car radios used to provide a selection of pre-tuned stations in the same way. Similar functionality could be achieved using a rotary switch for instance, although unlike the radio button solution, it would not be possible to move between two non-adjacent settings without passing through the intervening positions.

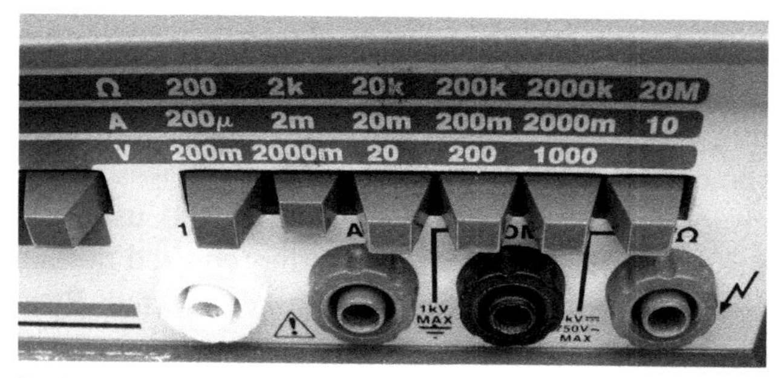

Figure 15.3 Radio buttons used to select between measurement ranges on a benchtop multimeter. Here the second of six is currently selected.

Normally Open (NO) vs. Normally Closed (NC)

A single throw momentary style switch (think door bell buzzer or keyboard key) can act either as an open switch which is closed when pressed, or a closed switch which is opened when pressed. Some fairly self explanatory terms are used to describe switches of these two types. A switch which is closed when pressed and is otherwise open will be referred to as 'Push-to-Make' or 'Normally Open' (often abbreviated NO). Conversely a switch which is open when pressed and is otherwise closed will be referred to as 'Push-to-Break' or 'Normally Closed' (often abbreviated NC).

The former action is by far the more common (the doorbell and keyboard examples cited above are both of this type) but the latter exists also and can sometimes be more appropriate for a given application. In particular the dead man's switch style will usually be a momentary, normally open switch type, the idea being that the switch must be kept actively engaged to maintain a connection. Locomotives, power boats, and some sports vehicles employ dead man's switches so that the engine will stop automatically if the driver is away from the controls for whatever reason. This might be in the form of a foot switch which must be kept depressed, a cutoff tether clipped onto the drivers clothing, or an under seat switch needing a persons weight to keep it actuated.

Two standard circuit diagram symbols exist for push-to-make and push-to-break style switches, as illustrated in the 'spst NO' and 'spst NC' entries in Table 15.2. Imagine pushing down on the vertical bar to see what the switch does.

Break-Before-Make vs. Make-Before-Break

Switches involving two or more throws can be manufactured in two slightly different variants called break-before-make (aka BBM or nonshorting) and make-before-break (aka MBB or shorting). As with the terms of the previous section, the names here are fairly self explanatory. In the first instance (break-before-make) when the switch is moved from one position to another the initial connection is disconnected before any connection is made to the next throw contact. By contrast, with a make-before-break switch the new connection is made before the original one is severed. By far the more common of the two is break-before-make, and unless otherwise specified this is the type which should be used. Sometimes it is important to use one type and sometimes the other, and sometimes it does not actually matter too much which is used.

In some circumstances a break-before-make changeover used in an audio circuit may result in a loud click or thump in the audio signal passing through the circuit, which can in some circumstances be much diminished by the use of a make-before-break style switch. On the other hand it will often be the case that make-before-break operation can connect together two things which should never be connected together. Even an extremely brief undesirable connection of this kind can result in serious and permanent circuit damage. As such it is important to insure that the type of switch used is suited to the job at hand. Most often this will be a break-before-make type.

Positions vs. Throws – On/Off Switch Position Designators

Sometimes a switch descriptor will carry additional labelling, with one of the following (or other similar) designators being appended to the name (see for instance the third switch symbol in Table 15.2):

1. on-(on)
2. on-off-on
3. on-off-(on)
4. (on)-off-(on)
5. on-on-on

There are a few extra characteristics being shown in these unusual looking codes, indicated by the brackets and the 'on' and 'off' labels. Each 'on' or 'off' represents a switch position. Brackets indicate a momentary rather than a latching action and the 'off' positions indicate places where there is no electrical connection available. These 'off' positions are not counted as 'throws' when describing the switch. This explains one way in which (as shall become clear) a switch with three positions might nonetheless be described as only a dual throw switch. With these guidelines in mind the five designators above can be interpreted as follows:

1. A switch labelled on-(on) is straightforward. It is a dual throw switch with one of its two throws being of the momentary persuasion. It normally sits in one position and will only stay in the other position for as long as it is held there. Once released its spring loaded action returns it to its resting position. This might be used to implement some kind of a doorbell circuit, but a doorbell is much more likely to be an spst off-(on) style, as nothing needs to be connected when the buzzer is not pressed.

2. An on-off-on switch is a three position, dual throw switch where the middle position makes no electrical connection (and so is not counted as a throw). All three positions are latching (the moving contact stays there when put there). See Table 15.2 for a circuit diagram representation of this kind of configuration. Off-on-on would also be perfectly possible but this is not a common or standard part.

3. The on-off-(on) style is a simple variation of the previous example where the switch latches when moved off centre in one direction but springs back when moved off centre in the other direction.

4. The (on)-off-(on) variant is another dual throw type following the same pattern as the previous two, but this time both contact positions are momentary, springing back to the central off position when the switch lever is released. This type is sometimes found implementing a kind of a 'dead man's switch' in a motion control system, with forward and reverse available in the two off centre positions, both of which are exited as soon as the switch lever is released.

5. On-on-on could of course be used to describe a totally standard three throw switch, with three latching throws. However using this additional labelling would be redundant. The tags examined here are used to highlight non-standard switch types. It might however seem a suspicious designation to be found describing a dual throw switch. How can a dual throw switch have three 'on' positions? In fact dpdt on-on-on and 4pdt on-on-on are both quite common variants. In the dpdt case each movement of the switch only changes the connection of one of the two poles (see Figure 15.4): in position one both poles are at throw one, in position two one pole is at throw one and one throw is at pole two, and in position three both poles are at throw two. The 4pdt variant is effectively just two dpdt on-on-on switches side by side in one package, actuated by a single movement.

Position 1 Position 2 Position 3

Figure 15.4 The switching action of a dpdt on-on-on switch. In position one both moving contacts connect with the top throw. In position two the left contact only has moved to connect with the bottom throw. In position three both moving contacts have shifted to connect with the bottom throw.

Strangely, quite often when the dpdt on-on-on switch type is encountered it is wired to provide the switching action of an sp3t switch. Actual sp3t switches are relatively uncommon. Figure 15.5 illustrates how a dpdt on-on-on switch is wired so as to act as an sp3t switch. A connection is made from the second throw of the first pole to the moving contact of the second pole. The result is that the moving contact of the first pole is connected successively to: pole one throw one, pole two throw one, and finally pole two throw two. This gives the effect of a simple sp3t switch. Dpdt on-on-on switches are also commonly found in electric guitars, wired for pickup selection. This application is described at the end of the chapter.

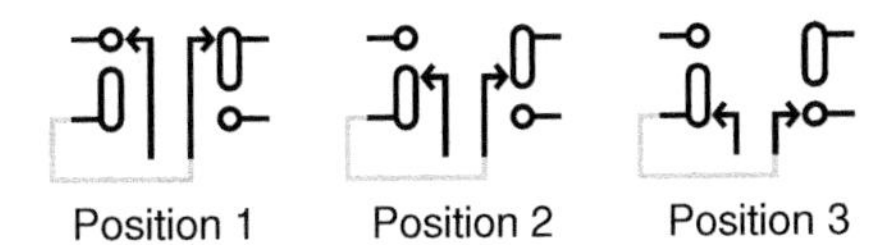

Figure 15.5 A jumper wired as shown linking throw two of pole one to the moving contact of pole two turns a dpdt on-on-on switch into an sp3t switch.

Circuit Symbols

Most of the variations in style and operation described above are not indicated directly on the standard circuit diagram symbols used to show switches. Where particular non-standard options are preferred or required the circuit diagram or accompanying documentation will include notes to make any particular requirements explicit. Table 15.2 illustrates the most commonly encountered circuit diagram symbols used to indicate switches in a circuit.

As with all circuit diagram symbols, it is common to also encounter variations and non-standard symbols representing switches. The intended meaning of such symbols is usually quite easy to decern but occasionally a little research may be required to confirm a particular interpretation.

Table 15.2 Switch symbols used in circuit diagrams

Circuit diagram symbols for various types of switch				
spst	spdt	spdt on-off-on	spst NO	spst NC
		NC		
dpst	dpdt	dpdt (alt.)	sp5t	spst relay
		SW1a SW1b		

A few of the switch variants included in Table 15.2 merit comment, or may need a bit of explanation. The third entry in the table shows an example of a switch with three positions but only two throws. As explained previously, in this on-off-on variant no connection is available when the switch is in its middle position. The 'NC' at this position stands for 'No Connection', not to be confused withe 'NC' for 'Normally Closed'. The context should always make clear which of these two interpretations of this particular abbreviation is intended.

Following the logic of the designators described in the previous section it can be seen that the fourth and fifth switch types in the table (spst NO and spst NC) could equally well be labelled spst off-(on) and spst on-(off) respectively. The normally open (NO) and normally closed (NC) labelling convention is the standard in this case.

The two dpdt symbols in the table are just two possible alternatives for showing the same dual pole dual throw switch in a circuit. If it is convenient to draw the two switch sections close to one another then the first style might suit, with a dashed line linking the two (or more) switch sections. However if the various sections of a multi-pole switch end up in disparate parts of the circuit diagram it will probably be more convenient to label each one separately, as shown in the second dpdt symbol style illustrated, where SW1a and SW1b indicate two poles of the same physical switch, designated SW1.

Electro-Mechanical and Electronic Switches

Extending the scope of the topic beyond simple mechanical switches, four more involved approaches to switching electrical signals may be considered: electrically actuated mechanical switches (relays), mechanically actuated electronic switches (touch switches), switching transducers (sensor switches), and purely electronic switches. These four categories are examined briefly in the sections which follow.

Relays

A relay is like a normal switch packaged up with a little electromagnet. When a voltage energises the electromagnet it pulls on the switch actuator and moves it from one position to the other. Relays will be either single throw or dual throw styles. The relay circuit symbol in Table 15.2 illustrates an spst off-(on) style. Figure 15.6 shows a 3pdt mains voltage relay capable of switching three independent signals, each of up to 10A, between two separate contacts. Relays may also incorporate various other switch types, such as spdt on-(on) or spst on-(off) variants.

Figure 15.6 A 3pdt mains voltage relay. The large coil visible below forms the electromagnet, while the switching terminals can be seen above.

In each of the three designators presented above, the label for the second position is enclosed in brackets, marking it as being momentary. This is because when the signal energising the relay's electromagnet is removed the switch will snap back to its original position. Typically relays are actuated using a relatively small signal level energising the electromagnet (via the connections to the rectangle with a line through it in the relay circuit symbol). The signal which is being switched (connected to the switch terminals as shown in the circuit symbol) can be much larger. In the relay illustrated in Figure 15.6 up to ten amps may flow through the switch contacts. This is a very large current and one useful application of a relay is to keep such large currents well away from the control circuitry which switches them. Similar relays are commonly used in car wiring to send a large current from the battery to the starter motor when the key is turned in the ignition. It would be unwise to allow such a large current to flow anywhere near the ignition key (and hence the person turning it). A relay can be used to perform the switching remotely.

Touch Switches

These switches have no moving parts. They can be particularly useful in outdoors applications and in harsh environments, as they are generally robust and can be sealed effectively against water and other contaminants. Resistive touch switches are often used in outdoors locations on elevator buttons, ticket dispensers, and vending machines. Capacitive touch switch technology is the basis of many touch screens on mobile phones, tablets, and computers. Piezo touch switches can be actuated while wearing gloves as they do not require skin contact in order to work. This is in contrast to resistive, and to some extent capacitive styles, which generally do require contact or very close proximity to actuate. Other variations on these basic approaches also exist involving various methods for detecting touch or proximity. The present section is not intended to be in any way a detailed review of this class of switch. The intention is simply to flag the existence and primary features of such switches. These devices tend to be more expensive than traditional mechanical switches and will rarely be encountered in audio electronics applications, although homemade variants can be useful and interesting.

Resistive Touch Switches need two electrodes which need to be brought into physical contact with something electrically conductive (for example a finger) to operate. They work by completing a circuit, providing a path for a small signal current to flow thus allowing an electronic switch to operate. Resistive Touch Switches are much simpler in design and construction than capacitance switches. Placing a finger across the two contact points results in a closed state. Removing the finger turns the switch to its open state. This kind of simple contact switching can be useful in developing unconventional user interface designs. Touching a simple pair of contact points might set an oscillator running or modify a filter for example.

Capacitive Touch Switches need only one electrode to function. The electrode can be placed behind a non-conductive panel like perspex or glass providing great flexibility in the construction and installation of this kind of switch. The switch works using body capacitance – it is able to detect changes in capacitance. When a person touches it, this alters the capacitance and triggers the switch. These devices can also be used as short-range proximity sensors since physical contact is not required in order for the effect to be used. DIY versions are possible and can be interesting to experiment with.

Piezo Touch Switches are based on the mechanical bending of a piezo element (see Chapter 20). This approach enables touch interfaces with any kind of material so long as some degree on movement or vibration can be transmitted through the surface to the piezo sensor behind. Piezo touch switches can be constructed so that a fairly light touch is enough to actuate them even when they are mounted behind stiff materials like stainless steel. Piezo touch switches can also be actuated through contact by any material, unlike the resistive and capacitive styles already described which require direct skin contact or limited other alternatives. Again, piezo transducers are fertile ground for interesting DIY experimentation.

Sensor Switches

Many of the devices in this category are not so much switches in their own right, rather they can be used to detect various kinds of events and provide the actuating signal to an electronic switching circuit. Sensors for all kinds of situations and events are possible.

Light sensors often use simple photoresistors (see Chapter 12) but can also be based on photodiodes, phototransistors, and other light sensitive devices. photoresistors are cheap and extremely easy to use, and as such are found in a lot of simple DIY audio projects. They can be used to control the cutoff frequency of filters or the pitch of audio oscillators amongst many other possibilities. Figure 15.7 shows an optical sensor switch. When an object is placed in the slot, it breaks the light beam passing from one side to the other, and the switch actuates. These devices are commonly found in ticket dispensers and similar machines. Spinning a punched disc of card through the slot could generate an interesting pattern of switching events.

Figure 15.7 A slotted optical switch; when an object is placed in the slot between the LED and the phototransistor, the switch actuates.

Sound sensors are deployed in a wide variety of applications from baby monitors to disco light controllers to clap switches and beyond. Sound sensor boards are available as expansion modules for the arduino and similar microcontrollers. The actual sensor in these is often an electret condenser microphone element, as described in Chapter 20.

Pressure sensors might be considered in two subtypes: mechanical pressure sensors and fluid pressure sensors. The former would include the piezo touch switches described earlier, while the latter would detect the pressure in air, water, or other fluids. Along with piezo elements, pressure sensitive resistive strips can make very versatile tactile interfaces in audio electronics projects. Fluid pressure sensors are of less interest in this area, and are more often found in industrial settings.

Thermostats and temperature sensors are good for monitoring the heat dissipation of circuits and systems but present few obvious opportunities for integration into the kinds of circuits most often of interest here. The thermo-switch shown in Figure 15.8 uses a bimetallic strip to actuate a mechanical switch at a set temperature (this one switches at 80°C).

Figure 15.8 A bimetallic temperature sensing switch.

Proximity switches are used to detect when two objects are coming close together. The commonest and simplest type is probably the magnetic reed switch, as illustrated in Figure 15.9. When the magnet is moved close to the glass tube, the fine metal arm within is bent away from its contact point, opening the switch.

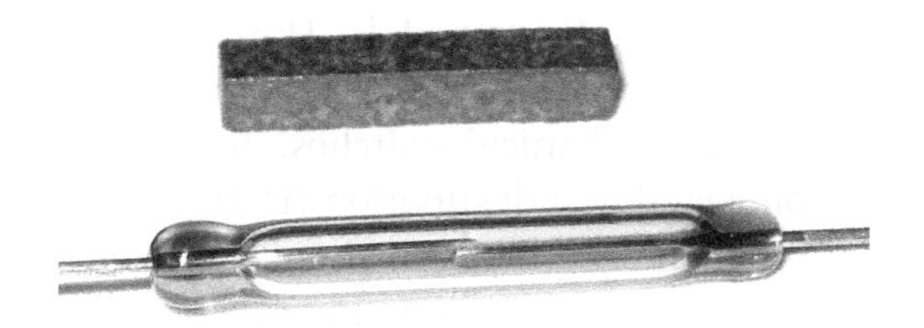

Figure 15.9 A reed switch with the bar magnet used to actuate it shown above.

Many other varieties of sensor switching are also possible. The simple mercury tilt switch used to be quite common; a blob of mercury in a glass bulb with two electrodes. When the bulb is tilted so that the mercury touches the electrodes, an electrical contact is made. A homemade tilt switch is an easy thing to make with a metal ball bearing and some nails. More elaborate tilt sensor modules can be used to monitor the motion of machinery, providing information on how much and in what direction it has been tilted. Again, the potential applications in DIY audio projects are probably more limited.

Electronic Switching

There are a number of ways of switching a signal electronically. Computers and other digital devices are at their core primarily extremely large and complex arrays of switches. Probably the commonest and most straightforward way to achieve electronic switching is with the use of transistors. The routing and muting of audio signals in a sound desk is often achieved using simple circuits based on field effect transistors (see Self, 2015, p. 590), as is the bypass switching in many guitar effects pedals. The broad topic of switching electrical signals using transistors is considered more fully in Chapter 17 which introduces and explains the various types of transistor and their typical applications.

Another common approach to electronic switching is using analog switch ICs such as the CD4016BE and CD4066BE. These devices are part of the 4000 series of CMOS ICs discussed in Chapter 18. The CD4051BE and CD4053BE analog multiplexers are other members of the same family, which can also be very useful in simple switching applications. A number of popular arpeggiator and sequencer type noise making circuits uses the 4051 in particular, to great effect. Driven by a simple square wave LFO, it can rapidly switch between a sequence of circuit settings, producing repeating runs of tones.

INSTRUCTIVE EXAMPLES

This section details a number of particularly common switching circuits encountered in audio electronics.

Effect Bypass Switching

As mentioned above, various electronic switching methods are often used to implement bypass switching in electric guitar effects pedals. However, many still favour a simple mechanical switching approach. Figure 15.10 shows four possible ways to implement bypass switching using simple mechanical switches. In all cases the basic idea is that a signal source (Src) is connected to a destination (Dst), with the signal either directly connected, or passing through an intermediate circuit (here labelled FX) depending on the switch position. Each scheme represents an improvement on the ones before it.

A simple spdt switch routing the source either to the bypass wire or into the FX circuitry might be considered. Unfortunately this leaves the output of FX connected to the destination at all times. Such circuitry can be noisy, especially when its input is left floating, so this is clearly not a great idea.

If an spdt switch is all that is available, then moving it to the destination side provides a better solution. It means that the FX circuit is still being driven by the source – not ideal – but at least its output has been isolated when the switch is in the bypass position.

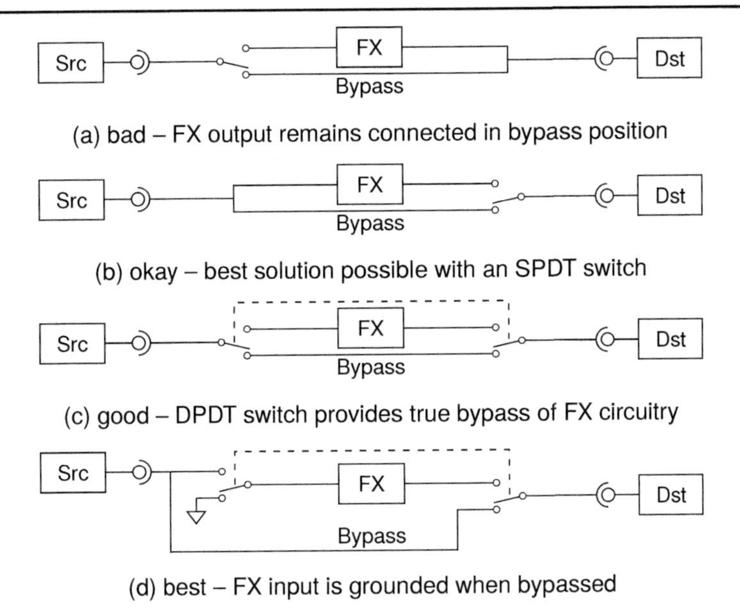

Figure 15.10 Bypass switching options.

A much better and much more commonly encountered solution is to use a dpdt switch, so that FX can be isolated both at its input and its output. This a good solution, however the floating FX circuitry can still produce some noise which might find its way into the signal passing close by. In particular, leaving the input of FX floating can result in oscillations and other undesirable mechanisms taking hold.

The forth configuration again uses a dpdt switch, but wires it in a more elaborate fashion. Now when FX is bypassed, its input is also grounded, minimising the chances of undesirable activity in the circuit while it is not in use.

In fact, the foot switch used to implement bypassing on many guitar effects pedals is often actually a 3pdt type, like the one shown in Figure 15.2. The third pole is usually used in order to switch a status LED to show when the effect is in circuit. The full wiring for such a switch is shown in Figure 15.11. This simple LED circuit is discussed in Chapter 16.

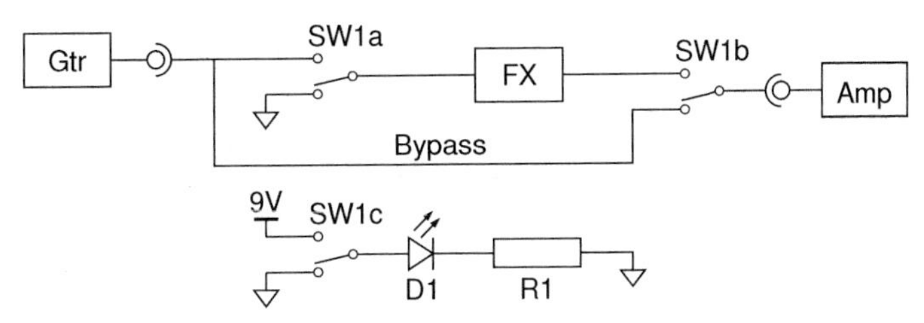

Figure 15.11 Bypass switching with indicator LED.

Guitar Pickup Selection

One of the switch types described above was the dpdt on-on-on switch. One common use for a switch like this might be in the pickup selection circuitry for a two pickup electric guitar. Three pickup configurations usually involve a selection switch with more throws, but with just two pickups the usual options are just: one, the other, or both. The wiring shown in Figure 15.12 achieves this set of options. In the top position only the neck pickup is selected, in the middle position the two pickups are paralleled together, and in the bottom position (illustrated) only the bridge pickup gets through. NC in the diagram stands for 'no connection'.

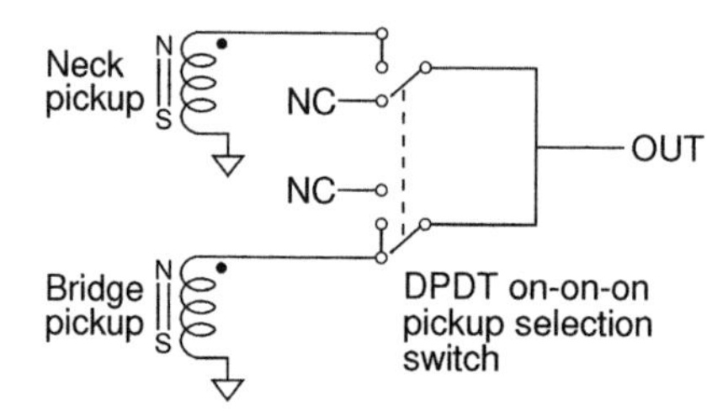

Figure 15.12 Typical electric guitar pickup selection switch wiring for a two pickup, three way selector configuration.

Headphone Jack Switching

Switching functionality can sometimes be encountered built into other components. The commonest example of this is the switching jack. TS and TRS jacks can come in switching and non-switching varieties. These switching jacks are encountered regularly in some very common audio circuits, from headphone jacks that cut off the loudspeakers, to combined mono-stereo inputs, to patch bay semi-normalled connection points. The idea is simple enough; for each contact inside the jack, there are two terminals connected together. When a jack plug is inserted it breaks these connection as it connects itself to one of the two terminals.

The arrangement is illustrated in Figure 15.13. Before the headphones are plugged in, two pairs of contacts send the audio through to the left and right loudspeakers.

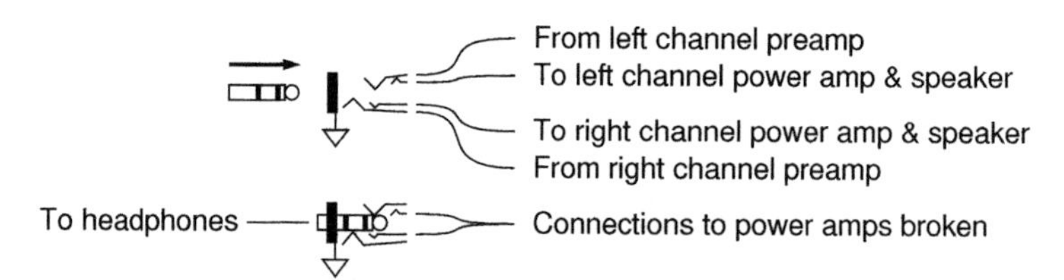

Figure 15.13 Using a switching jack to disconnect the loudspeakers when a headphone jack plug is inserted.

Plugging in the headphones pushes the terminals apart and breaks the connections to the loudspeakers. The signal instead comes out through the jack plug, and goes to the headphone drivers.

The same switching mechanism can be used to achieve a number of useful signal routing functions in audio equipment. Sometimes one signal is being rerouted to a different destination, as in the example here. In other cases it is a question of a new signal being switched into the signal path, replacing a previous input connection.

Power Supply Switching

Figure 15.14 illustrates a commonly utilised configuration which achieves two useful functions through the combined wiring of the input and power supply connector jacks. Firstly, on/off switching is implemented without the need for a separate switch, and second, automatic selection between battery and mains powering is achieved.

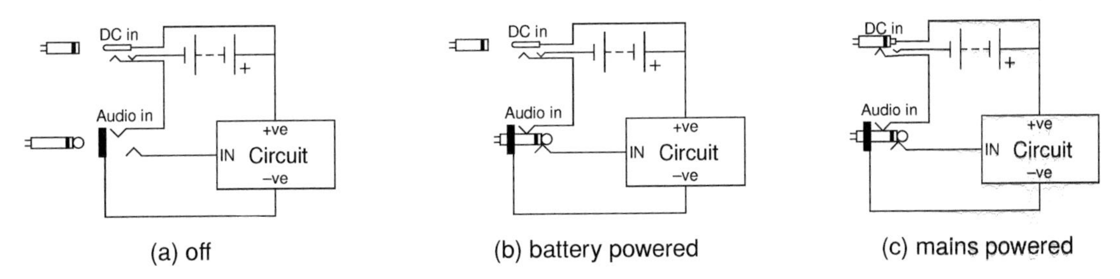

Figure 15.14 Power supply ON/OFF and source switching scheme commonly used in electric guitar effects pedals.

In situation (a) the circuit is turned off (even if the power jack were inserted). This is because the negative connection to the power supply is wired to the ring terminal of a TRS 'Audio in' jack. The circuit is intended to be used with a TS jack plug (i.e. a guitar lead), so the use of a TRS jack may seem odd. Recall however that in a typical circuit, the jack sleeve connection goes to ground, and this is where the power supply negative should get connected in order to complete the loop and power up the circuit.

When a standard TS jack plug is inserted into the TRS input jack the long barrel to the jack plug sleeve shorts together the TRS jack's ring and sleeve, and thus powers up the circuit. Situation (b) shows this arrangement, with the circuit running under battery power. If a power supply is plugged into the DC input jack, this should take precedence over the battery, so a switching power jack is used for the DC input. In situation (c) a power jack plug has been inserted, actuating the switch and disconnecting the battery, allowing the circuit to get its power from the external power supply instead.

References

N. Collins. *Handmade Electronic Music*. Routledge, 2nd edition, 2009.

D. Self. *Small Signal Audio Design*. Focal Press, 2nd edition, 2015.

16 | Diodes

Diodes let current flow in one direction but not the other

In Chapter 3 diodes were introduced at a theoretical level, described in the context of the semiconductor materials from which they are fabricated. Here that introduction is expanded into a more practical examination of what diodes do and how to use them. The diode is the simplest of all semiconductor devices, made (usually) by sticking two pieces of semiconductor material together, one p-type and one n-type, in order to form a p-n junction.

A diode's two terminals are referred to as the anode and the cathode, as shown in Figure 16.1. The fundamental action which characterises the behaviour of a diode is that it allows current to flow in one direction but not the other. In the situation where a diode's anode is at a positive voltage relative to its cathode, the diode is said to be forward biased, and current can flow. When the anode is negative relative to the cathode, the diode is referred to as being reverse biased, and no current flows.

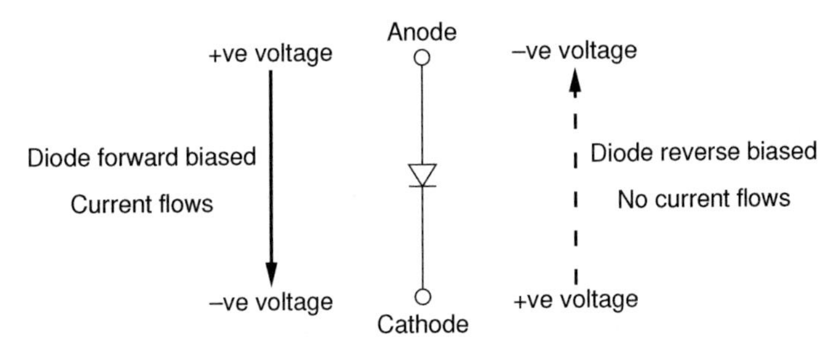

Figure 16.1 Electric current always flows from positive to negative. Diodes allow current to flow only when the anode is positive relative to the cathode. When the cathode becomes more positive, no current flows.

An ideal diode would behave as a short circuit (very low resistance) in the forward biased direction and as an open circuit (very high resistance) in the reverse biased direction. The behaviour of real diodes is a little more complex but this simplified description is often sufficient to provide an understanding of what a diode will do in a particular situation. It is however the deviations from this ideal behaviour which is of most interest in such applications as distortion effects, where diodes often find application in audio circuits. These deviations from the ideal are addressed in the following section on the electrical characteristics of diodes.

In some contexts diodes are also referred to as rectifiers. The two terms are more or less interchangeable but rectifier is most often reserved to refer to a device which can handle a comparatively large current, while diode refers to a device designed to work with lower level signals. Such diodes will sometimes be specifically referred to as small signal diodes in order to emphasise the distinction. As a rule of thumb, if it has a glass body it's probably a small signal diode, if it has a black plastic type body, then it's more likely to be referred to as a rectifier. In many circumstances a broad range of diodes of both types will work equally happily in a given circuit. Figure 16.2 shows some typical examples of the kinds of diodes which are commonly encountered in audio electronics. Notice that in all cases (except for LEDs) there is a stripe on one end of the body. This indicates the cathode connection of the diode.

Figure 16.2 Some typical diodes: a 1N4148 small signal diode, a 1N4001 rectifier diode, and three LEDs – rectangular, 5mm round, and 3mm round.

Electrical Characteristics

As mentioned above, the simple 'one way valve' description of a diode, while often adequate, does not fully characterise the real world behaviour of these devices. This ideal model is indicated on the graph in Figure 16.3. In the reverse biassed condition the ideal graph runs horizontally along the negative X axis, indicating that zero current will flow no matter what the reverse bias voltage is. In the forward biassed condition the graph runs straight up the positive Y axis, indicating a zero voltage drop and unconstrained current flow.

The most important deviation from this ideal model is that a real diode does not start allowing current to flow as soon as the anode voltage exceeds that of the cathode. As explained in Chapter 3, a depletion layer exists at the p-n junction which forms a barrier to current flow. It requires a certain amount of voltage in order to overcome this depletion layer. Only then can current start to flow. Even at voltages just above this

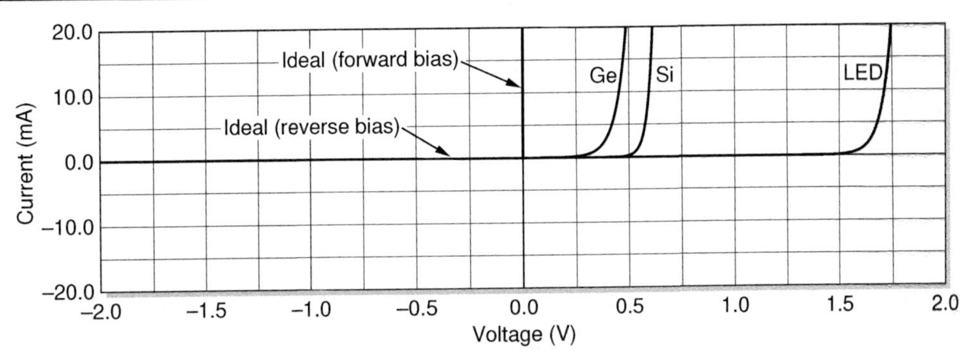

Figure 16.3 Diode i-v characteristics, showing the idealised behaviour alongside plots for germanium, standard silicon, and LED devices.

switch on level current initially encounters significant resistance, and the diodes need a little more voltage in order for the graph of the current level to start climbing steeply. The three plots in Figure 16.3 illustrate this behaviour for three different types of diode.

As seen on the graph, the forward bias voltage for conduction varies between different kinds of diode. Germanium based devices start to turn on at a very low forward bias voltage, typically round about 0.3V, but they also advance more gradually as the voltage increases. Thus the Ge plot has a slightly gentler slope than the Si plot next to it (notice that the lines are slowly converging as they rise). Standard small signal Si diodes turn on more abruptly at round about 0.6V. LEDs typically need more than a volt to turn on, and the voltage varies quite a bit with different LEDs, most notably with LEDs of different colours.

Real world diode behaviour is much closer to the ideal in the reverse bias direction. All diodes do however have some small reverse leakage current. If the graph scale were expanded in the reverse bias region, the lines could be seen dropping just a little bit below the axis, indicating some slight reverse bias current flowing. Table 16.1 puts some numbers on these quantities for various commonly encountered diodes. The columns are as follows:

Part number – the serial number used to order the part, or when searching for data sheets or general information on a particular device.

Type – the class or type of diode in question. See the table footnote for details.

Maximum reverse blocking voltage (V_R max) – diodes can only take so much reverse voltage before they breakdown (and usually burn out). Note in particular that LEDs can not withstand much reverse bias voltage before breaking down (just 5V in the two examples shown here).

Maximum reverse leakage current (I_R max) – as mentioned above, diode reverse blocking is not perfect. Some minimal leakage current always gets through. Mostly it is negligibly small. The 1N5817 is a notable exception at 1mA; in general Schottky diodes exhibit poor reverse leakage characteristics.

Table 16.1 Typical specifications for some standard diodes of various types

Part number	Type[a]	V_R max	I_R max	V_F max[b]	I_F max	P_{tot} max	C_D max	Package
1N4148	Si	100V	25nA	1.00V	0.2A	0.50W	4pF	DO-35
1N914	Si	100V	25nA	1.00V	0.3A	0.50W	4pF	DO-35
1N4001	Si	50V	30μA	1.00V	1.0A	3.00W	15pF	DO-41
1N4007	Si	1000V	30μA	1.00V	1.0A	3.00W	8pF	DO-41
1N5408	Si	1000V	5μA	1.00V	3.0A	6.25W	30pF	DO-16
6A6	Si	600V	10μA	0.90V	6.0A	5.40W	150pF	R-6
1N5817	Sch	20V	1mA	0.45V	1.0A	1.25W	125pF	DO-41
1N34A	Ge	60V	30μA	1.00V	50mA	80mW	<1pF	DO-7
1N270	Ge	100V	100μA	1.00V	40mA	80mW	<1pF	DO-7
OA90	Ge	30V	20μA	1.00V	50mA	80mW	–	DO-7
Red	LED	5V	50μA	1.80V	7mA	14mW	11pF	⌀5mm
Blue	LED	5V	10μA	4.50V	30mA	105mW	100pF	⌀3mm

a. Si = silicon, Sch = Schottky, Ge = germanium, LED = light emitting diode.

b. The V_F max values quoted may not be as expected. V_F is highly dependant on the current at which it is measured. The values given here are taken from data sheets which quote test conditions ranging from 5mA (for the 1N34A) up to 6A (for the 6A6).

Maximum forward voltage (V_F max) – the standard spec. is for when the diode is considered to be 'fully on', i.e. conducting a significant current. This explains the unexpectedly high values quoted for the three Ge devices. Germanium diodes are usually described as turning on at a lower voltage than their silicon cousins, but recall that they also exhibit a gentler slope on their i-v graph, and so the two types converge moving further along the graph.

Maximum forward current (I_F max) – the maximum forward current the diode can handle without damage. The six silicon types are arranged in ascending I_F max. The first two would generally be referred to as small signal diodes, while the latter four would be classified as rectifiers.

Maximum total power dissipation (P_{tot} max) – closely tracks with I_F max. LEDs and germanium devices in particular can not usually dissipate very much power.

Maximum diode junction capacitance (C_D max) – low junction capacitance can be important in some applications. This is one place where germanium diodes outperform their silicon counterparts. At the other end of the scale, Schottky diode tend to have fairly poor capacitance metrics.

Package – the size and shape of diodes can vary significantly. Common package outlines are illustrated in Figure 16.4.

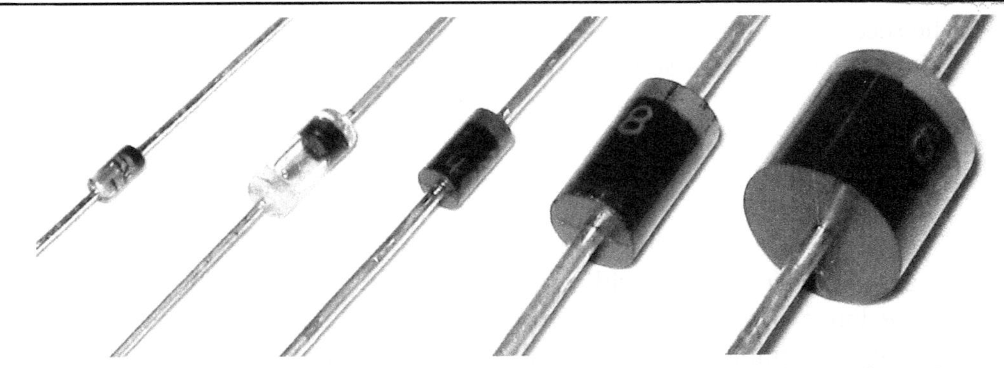

Figure 16.4 Standard diode package outlines: DO-35, DO-7, DO-41, DO-16, R-6.

LEARNING BY DOING 16.1

DIODE CURVE TRACING – MANUAL

Theory

This exercise demonstrates two ways of manually plotting a diode's i-v characteristic curve. One uses a variable resistance while the second uses a variable voltage. The circuits used for these two approaches are shown below.

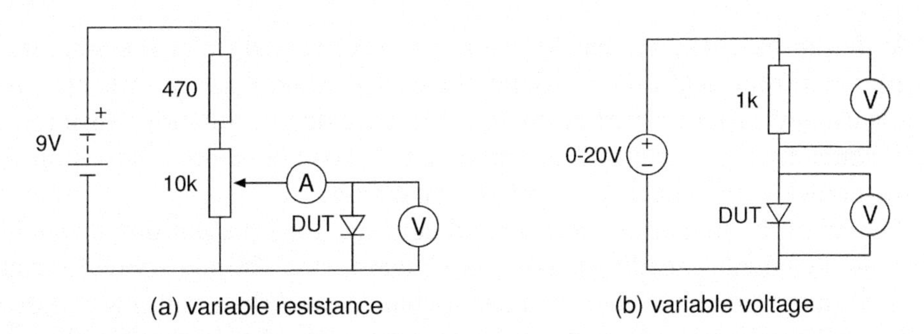

Because of the steeply rising current once conduction begins, it is always important to ensure that you strictly limit the maximum current which can flow in circuits like these. A maximum of 20mA should keep pretty much any diode you wish to test in safe territory; always check the data sheet for each device you work with to find its limits.

In the case of the first circuit we can estimate an absolute maximum current by assuming a diode drop of 0V. Then with the pot all the way up Ohm's law tells us that $I = V/R = 9/470 = 19mA$. The real world diode voltage drop builds in a bit of a safety margin on top of this, so the configuration looks good. If a different supply voltage is to be used, then a new value should be calculated for the 470Ω resistor.

For the second circuit the calculation is even simpler. The resistor is being used as what is often called a current sensing resistor. The voltage across it can be translated directly into a value for the current flowing through it (and hence through the diode). The choice of 1kΩ for its value makes the transformation trivial; again Ohm's law is used. In this case it tells us that the voltage read in volts can be converted directly into the current in milliamps. So again, assuming a diode drop of 0V, limiting the variable power supply to 20V limits the possible current flowing to 20mA, with the actual diode drop once again providing a safety factor.

Once the limits not to be exceeded are established and understood, the data gathering phase is simple enough in both cases, and then a graph can be plotted. Repeat the procedure for a few different diode types and compare the results.

Method 1

Build the circuit shown in (a), leaving out the diode, and leaving the power disconnected for now. Set the ammeter and voltmeter to appropriate ranges (we are expecting a maximum current of 20mA, and a maximum voltage of a couple of volts, depending on the diode chosen).

Step 1: Turn the pot all the way down and insert the diode to be tested (DUT stands for device under test, in case you were wondering). Confirm that the diode has been inserted the correct way round. Connect the power. Both current and voltage should read zero.

Step 2: It can be tricky to decide where to make readings; at the start its all voltage with no current, and then suddenly the current shoots up for little change in voltage. It is usually easier to start at the top and work down. Smoothly turn the pot up to max, watching the current all the time. If it heads above our 20mA limit turn the pot down quickly, and check for errors in the circuit.

Step 3: Record the maximum voltage and current. Turn the pot down watching for a substantial change in readings, and record again. Don't bother trying for round numbers, just take a reading when you see a change, and then continue on down. Keep going until you reach zero volts and zero amps. You should aim for about ten or so readings.

Step 4: Repeat Steps 1–3 for all the diodes you want to test.

Step 5: Disconnect power from your circuit. Get some graph paper and plot your results (or use a plotting tool, even Excel does a pretty good job).

Method 2

Build the circuit shown in (b), leaving out the diode, and leaving the power supply turned down to zero for now. Set the two voltmeters to appropriate ranges (we are expecting a maximum of 20V on the top meter, and a maximum of a couple of volts on the bottom meter, depending on the diode chosen).

Step 1: Insert the diode to be tested. Confirm that the diode has been inserted the correct way round. Confirm the supply voltage is zero. Both voltmeters should read zero.

Step 2: It can be tricky to decide where to make readings; at the start its all the lower voltmeter, and then suddenly the upper meter starts to move with little happening on the lower one. It is usually easier to start at the top and work down. Smoothly turn the power supply up to a maximum of 20V, watching the upper meter across the current sensing resistor all the time. It should never exceed 20V.

Step 3: Record the maximum voltage and current (remember volts on the top meter equal milliamps). Turn the pot down watching for a substantial change in readings, and record again. Don't bother trying for round numbers, just take a reading when you see a change, and then continue on down. Keep going until you reach zero volts and zero amps. You should aim for about ten or so readings.

Step 4: Repeat Steps 1–3 for all the diodes you want to test.

Step 5: Disconnect power from your circuit. Get some graph paper and plot your results (or use a plotting tool, even Excel does a pretty good job).

LEARNING BY DOING 16.2

DIODE CURVE TRACING – AUTOMATIC

Theory

Manually plotting diode curves can be instructive, but there is an easier way using a function generator and an oscilloscope in X-Y mode. There is one major precaution which must be observed. The function generator and the oscilloscope MUST NOT share a common signal ground. Most mains powered lab equipment has its signal ground hardwired to the mains earth ground connection. If you do not have access to either a floating signal output from a function generator or floating oscilloscope inputs then this exercise is not for you. A battery powered or double insulated function generator is one option. Notice in the diagram that the scope's ground clips need to go between the resistor and the diode, which will leave them one diode drop above the function generator's ground. Connecting these two points through the earth ground would allow much current to flow, and bad things would happen.

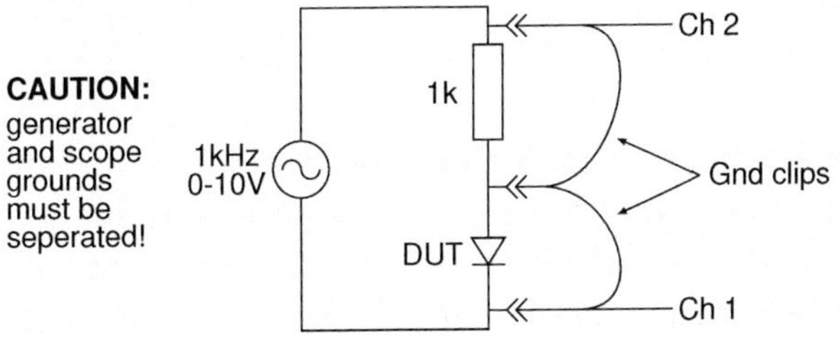

The function generator wants to be able to output a signal amplitude of at least about ten volts peak. We are using the same 1k current sensing resistor as before, so ten volts translates into a maximum current of about 10mA. Something like the ubiquitous 1kHz sine wave signal works well for the output wave shape.

When an oscilloscope is operated in X-Y mode, instead of plotting the input to each channel against time, the two channels are plotted one against the other. This allows characteristic graphs such as the diode i-v curve we are looking for to be generated. The graph we want displays the current flowing through the diode against the voltage across it. The channel one probe, connected directly across the diode, will clearly give the voltage we are looking for (at least minus it, as the ground clip is at the top of the diode, so we will want to invert this channel). The second probe is connected across the resistor, and as we have observed already the voltage across this resistor is directly proportional to the current we seek, so we have the two parameters needed.

If you want to plot an LED's graph watch out for its maximum reverse voltage. This can be quite low for LEDs. If your function generator has a DC offset capability, this can be used to avoid reverse biasing the diodes under test by keeping the drive voltage strictly positive.

Practice

To display a diode's characteristic i-v curve on the oscilloscope screen follow these steps.

Step 1: Configure the function generator for a 1kHz sine wave. Set the amplitude to zero for now.

Step 2: Configure the oscilloscope as follows (accounting for ×10 probes if necessary).

Channel 1: 0.5 volts/div, DC coupling, inverted.

Channel 2: 2.0 volts/div, DC coupling, X-Y mode.

Step 3: Construct the circuit as shown below. Keep the function generator output turned down to zero for now. You should get a dot in the centre of the oscilloscope screen.

Step 4: Slowly turn up the amplitude on the function generator. The diode graph should grow out on the screen as the amplitude is increased, first sticking to the X axis, and then rising steeply as the diode voltage is reached and current starts to flow.

Diode Types

In addition to standard diodes (small signal or rectifier) there are a number of other common types (along with quite a few more esoteric variants which are not covered here). The three variants most likely to be encountered in audio electronics are light emitting diodes (LEDs), Zener diodes, and Schottky diodes. Also of interest in the context of audio electronics are the germanium variant of standard small signal diodes. These four categories are discussed below.

Light Emitting Diodes

LEDs are very familiar as the glowing indicator lights encountered on the front panels of almost every piece of audio technology. As with all diodes, it is important to differentiate between the anode and the cathode when connecting an LED into a circuit. In the case of LEDs there are two common signifiers to look out for, indicating the cathode or negative pin of the device, a shortened leg and a flat region on the side of the LEDs body. LEDs are just diodes that emit light, and can occasionally found used in a circuit where a normal diode might be expected, rather than exclusively as a front panel indicator light. So long as the lower than usual V_R max is taken into account when using an LED in a circuit, there is no reason not to try one and see how it sounds.

Zener Diodes

Of the very many special purpose diode types which exist, the most common is probably the Zener diode. Zeners are wired into a circuit in the reverse orientation to normal, and are used to provide very well defined reference voltages. While many diodes can be easily damaged or destroyed by reverse biasing them all the way to their breakdown voltage, Zener diodes are designed to operate at this point on their characteristic curve, and this is how they provide their stable voltage reference. Figure 16.5 shows the i-v characteristic curve for a three point one volt Zener diode. Its forward characteristic is more or less identical to a standard silicon diode, but in reverse bias it conducts at a specified voltage much lower than a normal diode's V_R max.

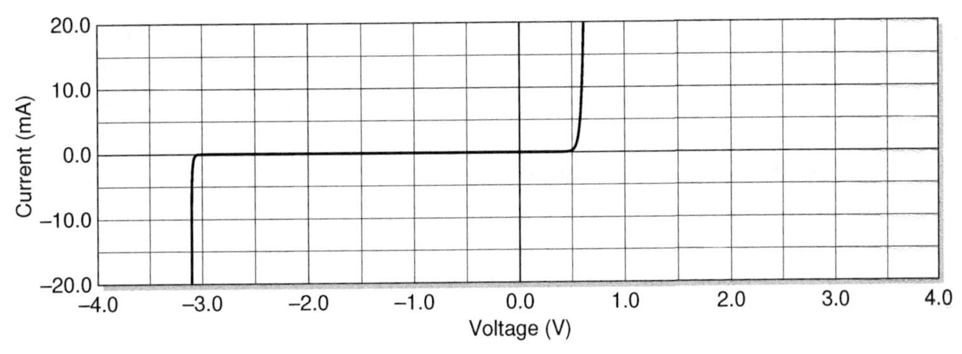

Figure 16.5 Typical i-v characteristic for a 3V1 Zener diode. Zener diodes are designed to have very specific reverse breakdown voltages. This one breaks down at −3.1V.

Zener diodes can be seen in two of the circuits included in the circuit catalogue in Appendix D. In both cases they are performing their usual job of providing a fixed reference voltage. In the phantom powered electret microphone circuit the Zener utilised is a 12V device, while a 5V1 Zener can be seen deployed in the noise gate circuit. Notice in both cases that the diode is inserted into the circuit counter to the normal diode orientation. The anode connects to the low voltage side, with the cathode wired at a point of higher potential, in order to reverse bias them and bring them into conduction at their designed breakdown voltage.

At first glance the device in the electret mic circuit may appear to be inserted the wrong way round, but notice that its top (the anode) is connected to ground, while its bottom connects through to the +48V coming into the circuit from the phantom power supply. Sometimes the layout conventions outlined in Chapter 7 on circuit diagrams are not strictly adhered to. In this instance it simply leads to a much neater schematic diagram to arrange things with the ground point in the middle of the diagram.

Schottky Diodes

Another commonly encountered type of diode is the Schottky diode. These have a low switching voltage and operate very quickly, making them ideally suited to high frequency circuits. These high switching speeds are generally of no great benefit in audio circuits which operate in a very limited bandwidth when compared to the likes of radio communications circuitry. Their low switching voltage is however sometimes used to advantage in audio circuits. The main disadvantages of Schottky diodes are their relatively large reverse leakage current and junction capacitance. They also tend to have a lower value for V_R max, and so can not withstand as large reverse voltages as most standard diodes can. The specifications for the 1N5817 Schottky diode included in Table 16.1 illustrate these points well.

Schottky diodes do not actually consist of a standard p-n junction as standard diodes do, but rather are formed by the junction of a semiconductor with a metal. This is what gives them their low forward voltage drop and fast switching action. A silicon diode has a typical forward voltage of 600–700mV, while the Schottky's forward voltage is usually in the range 150–450mV. This lower forward voltage requirement allows higher switching speeds and better system efficiency. A Schottky diode can be seen employed in the low power audio amplifier circuit in Appendix D. Here advantage is being taken of its low switching voltage. It is being utilised as a reverse voltage protection measure (the 386 IC in this circuit is easily damaged by reverse polarity power connection). The low forward voltage drop means that as little voltage as possible is lost to the rest of the circuit in normal operation. There are certainly better reverse polarity protection measures which could be employed, but this has the advantage of being trivially simple to implement.

Germanium Diodes

These days diodes are almost exclusively made from silicon (Si) semiconductor material, but germanium (Ge) based diodes used to be much more common. While germanium semiconductors have fallen out of general usage in most electronics applications due to their poor performance in most respects when compared to their silicon equivalents, some audio circuits call specifically for germanium diodes for their particular sonic characteristics. In particular, many distortion circuits specify (or at least suggest) Ge diodes. The gentler slope of their i-v characteristic curve means that they can produce a softer distortion effect than their silicon counterparts. Silicon diodes switch from fully off to fully on over a very small voltage range and as such the clipping they introduce can be very harsh. Germanium diodes provide a slightly gentler alternative.

Their very low initial conduction voltage relative to silicon devices also makes them better suited when dealing with very low level signals. A small signal might be all but used up by the time it has managed to switch on a silicon diode, but the same signal will be able to drive a reasonable amount of current through a germanium device. Sometimes a Schottky diode would also work well in such applications, however the much larger junction capacitance of a Schottky can often rule them out, making germanium devices the best option.

Schematic Symbols

The circuit diagram symbol used to signify a standard diode is an arrow with a line at its end, as shown in Figure 16.6a. The arrow indicates the direction in which current is allowed to flow, while the line represents the fact that current is prevented from flowing back in the other direction. A number of variations on this basic symbol also exist, used for showing various special types of diode. Those corresponding to the diode types described earlier are included here, others exist, used to represent some of the additional, less common types which might occasionally be encountered.

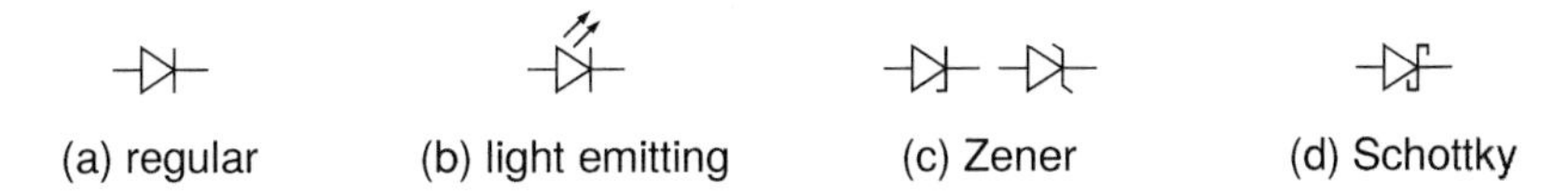

Figure 16.6 Diode circuit symbols for the diode types most commonly encountered.

The symbol in Figure 16.6b indicates an LED (the two arrows represent light coming out of the device), while Figure 16.6c shows two different variants which are both used to represent Zener diodes. Figure 16.6d shows the symbol usually used to signify a Schottky diode. In all cases the connection on the left hand side of the symbol is the anode, and connection on the right hand side is the cathode. Remember however that Zener diodes will usually be connected into a circuit with their anode at the lower voltage side rather than the usual way around for the other diode types.

Circuits

Diode Current Limiting Resistor

When wiring up an LED as a standard indicator light as shown in Figure 16.7, a current limiting resistor is always needed. The value of the resistor depends on three things, the voltage drop and working current of the LED, and the DC voltage level which will power the circuit. Once these are known, the optimum resistor value can be calculated. This value is then usually rounded down to the nearest E12 standard resistor value.

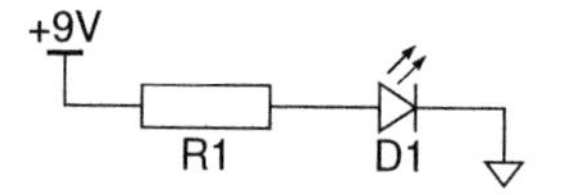

+9V
R1
D1
D1: Red LED, $V_f = 1.85V$, $I_f = 20mA$

Figure 16.7 How to choose a suitable current limiting resistor for an LED.

Typical values for a standard red LED might be something like 2.1V and 20mA for normal operation. Details will be found in the manufacturer's data sheet for the specific device to be used. If a 9V power supply is to be used then the calculation proceeds as follows.

Step 1: Voltage across resistor: $V_{R1} = V_{total} - V_{diode} = 9 - 2.1 = 6.9V$

Step 2: Value of resistor: $R_1 = \frac{V_{R1}}{I_{R1}} = \frac{6.9}{0.02} = 345$

Step 3: Next largest E12 value: $R_1 = 390\Omega$

Reverse Polarity Protection

Some components can be damaged if power is inadvertently connected to a circuit the wrong way round. Various approaches can be taken to mitigate or prevent any potential problems. One of the simplest solutions is to connect a diode in series with the power supply. If the connections are reversed, the diode will be reverse biased and no current will flow, thus saving components from potential damage. The down side to this simple solution is that the voltage dropped across the forward biased diode in normal operation is lost to the circuit it is protecting.

In order to minimise the impact, the circuit shown in Figure 16.8 indicates a Schottky diode as opposed to a standard silicon diode. As has been explained, the Schottky diode has a lower forward voltage drop than the more common silicon diode, round about 0.3V, versus about 0.7V. This may not seem like much, but when a circuit is powered from a small power supply of maybe 5V or even 3.3V, as some are, it can make all the difference. For even lower insertion losses the MOSFET solution presented in Chapter 17 can be used.

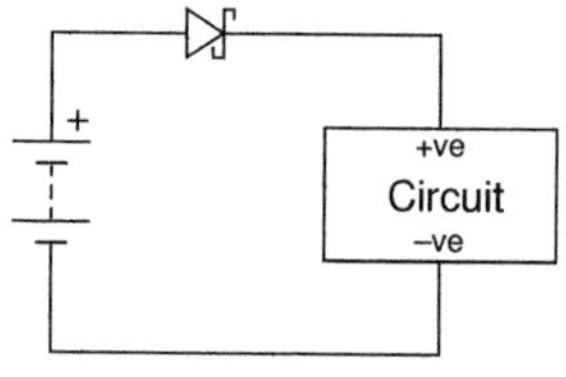

Figure 16.8 Rudimentary reverse polarity protection using a Schottky diode.

Diode Ring Modulator

One of the most widely recognised diode circuits (by name if nothing else) is the classic diode ring modulator circuit (Figure 16.9). The circuit takes its name from the ring of diodes at the heart of the circuit. The ring modulator combines two input signals, effectively multiplying them together. This can produce a rich mixture of harmonic and anharmonic frequency components in the output signal, which can result in interesting metallic and bell like sounds. The right combination of inputs are needed in order to obtain really good results, and the signals levels need to be well balanced for best effect.

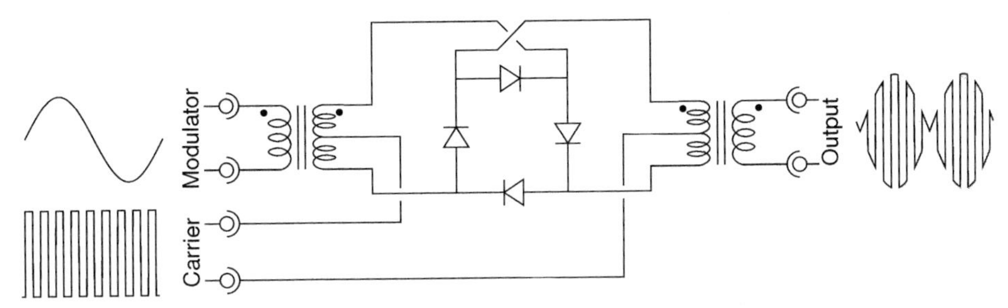

Figure 16.9 Classic diode ring modulation circuit.

AC to DC Power Supply Circuit

This circuit has some visual similarities to the ring modulator of the previous section, but it is completely different. The most important thing to notice is that, although there are again four diodes in a loop, here they do not form a ring, but rather a bridge. Notice that in the ring modulator ring the four diodes chase each other's tails around the circle. Here, on the other hand, the four diodes all flow left to right. This means that the loop here operates in a very different way. The configuration is called a bridge rectifier, and as Figure 16.10 illustrates, it takes the bottom half of an AC signal and folds it back up on itself. This process is called full wave rectification, and is described in Chapter 5: Signal Characteristics.

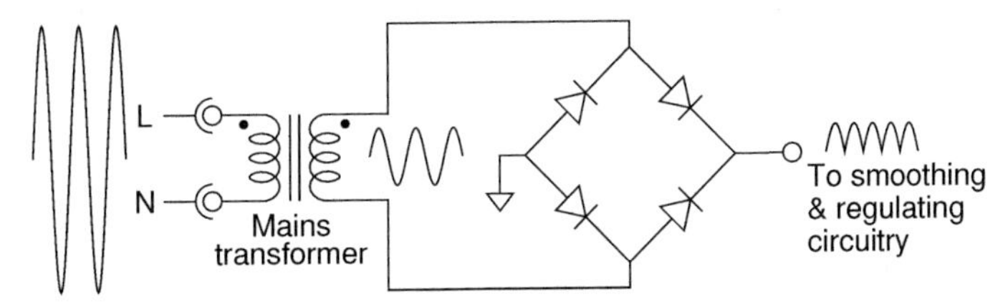

Figure 16.10 A diode bridge rectifier in a standard power supply application.

The circuit shown in Figure 16.10 is the business end of a typical linear power supply circuit. The transformer takes in mains voltage and converts it down to a more user friendly level. The bridge rectifier then full wave rectifies it. This is usually followed by a large smoothing capacitor and possibly a regulator circuit, to turn it into a nice steady DC voltage ideal for powering DC circuits.

An variation on this circuit, using a transformer with a centre tap output, allows the four diode bridge in Figure 16.10 to be replaced with just two diodes to achieve the same full wave rectified result. Penfold (1994) closes with a simple mains power supply project which uses this approach. It adds a large 1000μF smoothing capacitor, followed by an LM317 based regulator circuit similar to Figure 18.9, to produce a well regulated DC power supply suitable for powering most of the circuits in this book.

LEARNING BY DOING 16.3

DIODE CLIPPING AND CROSSOVER DISTORTION

Theory

Clipping distortion is an effect very often used in audio circuits, and diode clipping is probably the commonest way of introducing distortion into a signal. Crossover distortion is encountered less often, but can make for an interesting change. The two simple configurations explored here illustrate some of the possibilities.

In configuration (a) a pair of back to back diodes shunt the signal to ground when it exceeds the diode's switch on voltage.

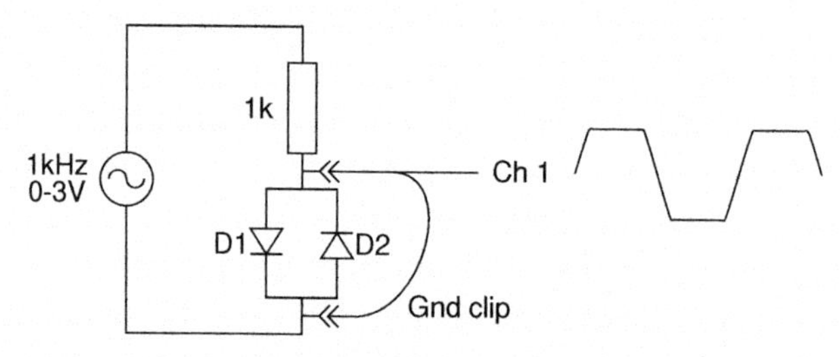

(a) peak clipping distortion

In configuration (b) the back to back diodes block signal until the switch on voltage is exceeded, after which they conduct freely.

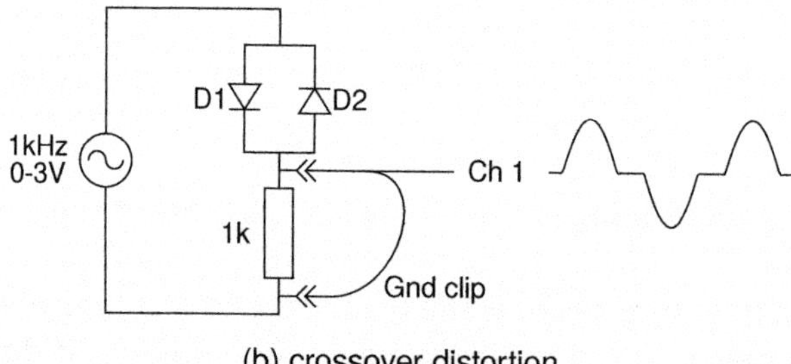

(b) crossover distortion

Practice

Build each of the circuits shown above. Start with the function generator turned down. Watch the evolving wave form emerge as the amplitude of the input signal is slowly brought up.

In both cases the back to back diode pair can be modified and augmented in various ways. Try different types of diode: germanium, LED, Schottky. A matching pair is not a requirement, try different types together, try two diodes in series in one arm. There are also alternative components which can be inserted: tube diodes, MOSFETs or tube triodes wired as diodes (see Chapter 17 and Chapter 19). All these variations can alter the tone and character of the distortion produced.

As well as viewing the effects these various configurations have on a simple sine wave, it is of course going to be interesting to actually listen to them. Distorted sine waves don't make for much of a listening experience. For best effect an electric guitar signal will need to go through a booster before hitting these circuits.

References

J. Falin. *Reverse Current/Battery Protection Circuits*. Texas Instruments, 2003.

Maxim. *AN 636: Reverse Current Circuitry Protection*. Maxim Integrated Products, 2001.

R. Penfold. *Practical Electronic Musical Effects Units*. Bernard Babani, 1994.

A. Sedra and K. Smith. *Microelectronic Circuits*. Oxford University Press, 7th edition, 2014.

17 | Transistors

Transistors amplify, buffer, and switch electrical signals

Of all the topics covered in this book the transistor is the one with the greatest potential to quickly become overwhelmingly complex. The aim here is to introduce the most relevant and practically useful information without straying too far into the technical and mathematical background which necessarily underpins a more academic approach to the subject. While not attempting to enable the design of anything but relatively straightforward circuits from scratch, the intention is to give enough of an understanding to allow circuit designs to be followed, understood, and to some extent modified and built upon.

There is no getting away from the fact that transistors and transistor circuits form a complicated area of study. The knowledge and skills necessary to analyse or design an arbitrary transistor based circuit from the ground up are considerable and indeed many books have been dedicated to addressing the subject (see, for example Evans et al., 1981; Evans, 1972; Severns, 1984; Zumbahlen, 2008). Fortunately it is not necessary to be able to fully characterise a transistor circuit in order to understand its basic operation. As such, the material presented in this chapter is designed to provide just that level of detail and technical analysis which is required to successfully utilise transistors in a broad range of simple but useful audio circuits. It also provides a solid grounding and foundation on which to build a more detailed and in depth understanding for those who may wish to proceed and delve deeper into this particularly rich area of analog audio electronics.

Transistor Types

In Chapter 3 on semiconductors an outline was provided describing the basic structure and operation of the three important classes of transistors: BJTs, JFETs, and MOSFETs. This is as much background theory as is required to appreciate the general workings of these devices within most circuits.

The broad behaviour that all transistors hold in common is that a signal at one pin controls the electrical characteristics between the other two. The exact behaviour can be quite different from one type to another, but this basic description is a good starting point to understanding what is going on.

Device Classes

Figure 17.1 presents the transistor family tree for the three classes of device mentioned above. The first split differentiates between bipolar junction transistors (BJTs) and field effect transistors (FETs). FETs are further subdivided into junction FETs and metal oxide semiconductor FETs. These names are all about the structure and functioning of the devices, but the details are unimportant. Each type of transistor comes in two complimentary polarities (NPN and PNP for BJTs, and n-channel and p-channel for FETs).

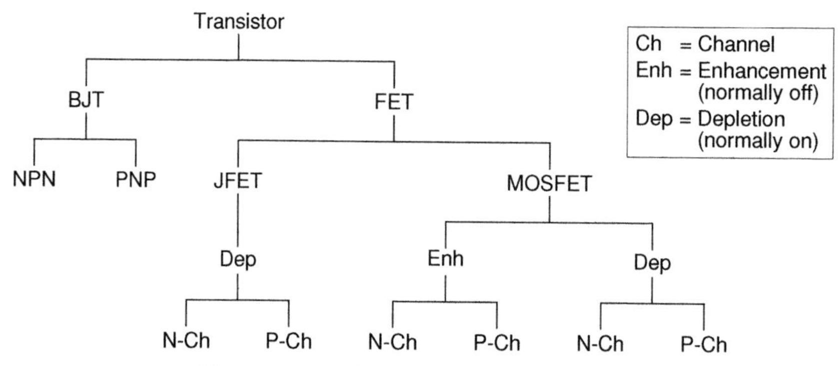

Figure 17.1 The major transistor types.

Thousands of different devices exist across these various classes, all designed to have particular characteristics to address specific application areas. For instance transistors may be designed for small signal, high power, low noise, high speed, and many other possible considerations. It is necessary to refer to an individual device's data sheet in order to discover its particular characteristics. Some of these characteristics are introduced for a few common devices in the sections dealing with individual device types which follow.

Darlington Transistors

One particular subtype of the BJT class which merits special mention is the Darlington transistor. These devices are bipolar junction transistors with very high gain. In other words they amplify a signals by a very large factor. A typical BJT may have a gain of 100 or so, meaning that the output can be as much as a hundred times the size of the input. Darlingtons are actually made by linking two standard BJTs within a single package, with one following the other. If the first can amplify by 100, and its output is fed into the second, also with a gain of 100, then the final signal emerging from the device might be amplified by as much as a factor of $100 \times 100 = 10{,}000$. The circuit symbols introduced above include ones specially for Darlington NPN and PNP transistors.

Package Outlines

A rough impression of the power handling capability of any given device may be gained by looking at the physical size of the package in which it comes. Figure 17.2 illustrate a number of different packages in which transistors can be found. Clearly the little TO-92 (used for very many of the commonest small signal devices) will not be able to handle nearly as much power as some of the larger outlines such as the TO-3 at the other end of the spectrum. The holes present in the larger package formats are provided in order that a heat sink might be attached to the device to further improve its power handling capabilities by increasing its ability to dissipate the heat generated when large currents flow through such devices.

Figure 17.2 Transistors come in a wide variety of standard shapes and sizes. The devices pictured here illustrate just a selection of the wide range of package styles which exist. The TO in the names stands for 'Transistor Outline'. Most of the devices encountered in this chapter come in a TO-92 package.

Note also that the two final package outlines only have two legs, which probably seems odd since transistors have three terminals. The third connection in the case of these particular types of package is made to the metal body of the device itself. Circuits involving such large, high powered devices are beyond the scope of this book. Transistors such as these will most often be seen in the output circuitry of big power amplifiers.

Circuit Symbols

Figure 17.3 illustrates the various circuit diagram symbols used to represent transistors of different types in schematic diagrams. The details of their interpretation are addressed separately as each class of device is examined. Notice for now that each symbol has a connection point entering from the left. This represents the connection which can be thought of as the 'control' pin mentioned above. Two more connections exist, one entering from the top and one from the bottom. These can usually be envisioned as carrying the output signal which is being controlled by the action at the first. While other working arrangements can apply, this is the most useful general scheme to remember.

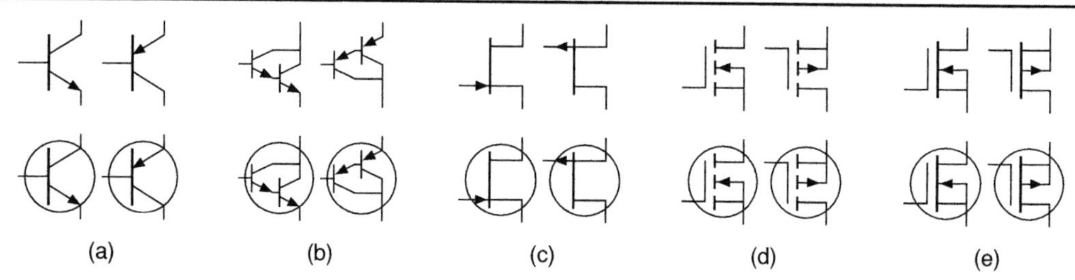

Figure 17.3 Transistor symbols for all the major transistor types are shown. Symbols may be drawn with or without an enclosing circle. Each pair shows the symbol for an n type (NPN or n-channel) device followed by the corresponding p type (PNP or p-channel) device: (a) standard BJTs, (b) Darlington BJTs, (c) JFETs, (d) enhancement MOSFETs, and (e) depletion MOSFETs.

Notice also that all the symbols include an arrowhead (or in the case of the Darlington devices, two). These helpful to differentiate between the various transistor types, and are also used to indicate the correct orientation of the device when it is connected into a circuit. It is important to observe the correct connections when using transistors. The identification of which physical leg corresponds to which connection on the circuit diagram is considered below in the section on transistor basics.

General Use and Operation

An important first step in successfully utilising any transistor is to have a clear understanding of what the three terminals are referred to as, and how to identify which is which, both in a circuit diagram and on the physical device. The task is made slightly more complicated by the fact that the terminals have different names for BJTs and for FETs (they do at least remain the same through all the different variations which compose the FET family). Figure 17.4 provides a visual guide to the identification of transistor terminals on the standard symbols for both classes of device. The base (in BJTs) and the gate (in FETs) are easy. The other two can be easily mixed up. The thing to remember when working out which is which is that the arrow identifies the emitter and source respectively.

Once it is clearly understood how to correctly designate the terminals on the circuit diagram symbols, the next step is to transfer this information successfully onto the three pins of an actual device. Since the transistors used in the circuits in this book all come in the common TO-92 package outline, this is the example which is used here. The identification of the pins in the case of other package types follows the same pattern.

The key thing to remember is that the pin layout varies from one device to another, so it is important to look up the pinout information each time a new device type is encountered. Figure 17.5 illustrates the kind of format in which the information is usually presented in a device's data sheet. The particular device shown here is a 2N3904,

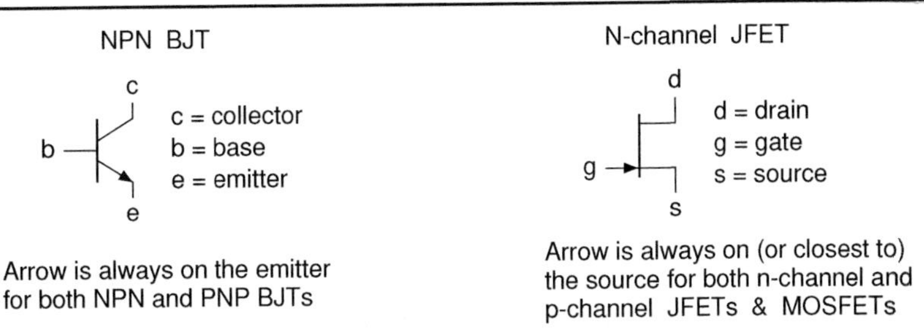

Figure 17.4 Transistor terminal names. With so many different variants it is important not to get the pins mixed up. It can be particularly easy to confuse the collector and emitter on a PNP BJT, and likewise the drain and source on a p-channel FET.

which is a very common and popular small signal NPN BJT. This transistor is used in a number of the circuits in this book.

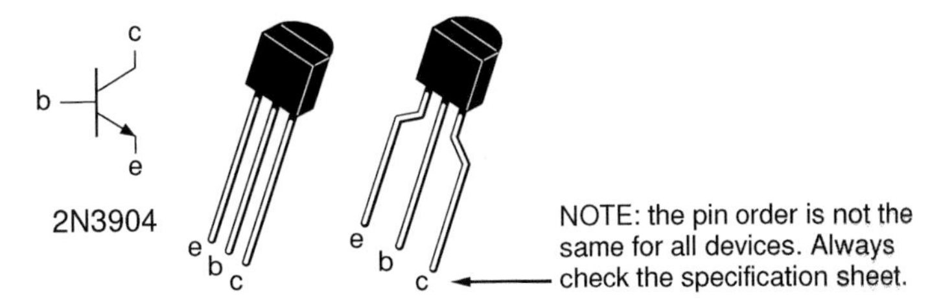

Figure 17.5 Correctly identifying which pins on a physical transistor correspond to each terminal on the circuit diagram symbol is crucial to arriving at a working circuit. A diagram similar to this will usually be found on a device's data sheet.

In this case the base, which is the middle pin on the standard symbol, is also the middle pin on the physical device. It is particularly important to note that although this is common, it should not be assumed that it will always be the case. Also be sure to check which way round the device is being shown when the pins are labelled. While some data sheets will show a TO-92 package with the flat face forward as is illustrated here, others will picture it the other way round, with the curved face forward. It is attention to these small details which spells the difference between success and failure for a circuit builder just getting familiar with the way things work.

Bipolar Junction Transistors (BJTs)

BJTs are made from three alternating layers of semiconductor material, either NPN or PNP, hence the names for the two basic types of BJT which have been introduced earlier. The three pins on a BJT are called the base, collector, and emitter. BJTs are extremely common in all sorts of circuits, and it is necessary to be able to determine what such circuits are doing in order to modify and experiment with them successfully.

BJT Characteristics

The commonest way in which the workings of a BJT tend to be introduced is by way of a pair of graphs such as the ones shown in Figure 17.6. What these two graphs are showing is how much current will flow through the collector (I_c) depending on the voltages applied between the various terminals. The two important voltages shown here are V_{be} and V_{ce}. The names state clearly what these voltages are. V_{be} is the voltage from base to emitter, while V_{ce} is the voltage from collector to emitter. The final quantity labelled on graph (b), I_b, is the current flowing in the base.

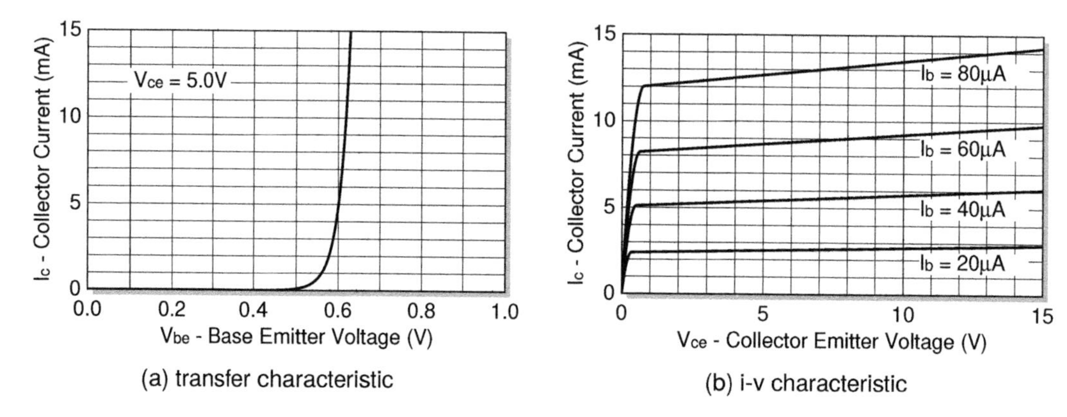

Figure 17.6 Typical NPN BJT characteristic curves.

Armed with all this information the interpretation of these graphs becomes relatively straightforward. The transfer characteristic in (a) shows how much current will flow in the collector for any given voltage applied between base and emitter. Notice in the corner of the graph it is specified that $V_{ce} = 5.0V$. The whole graph will shift around as V_{ce} is changed, but its overall form will stay basically unaltered.

Similarly the i-v characteristic in (b) shows collector current against collector-emitter voltage. The current first rises steeply and then levels off. A family of four plots are shown, illustrating how the collector current is controlled by what is happening at the base. Recall that the base can usually be seen as the control pin on a BJT.

What graph (a) is saying is that if V_{be} is at or close to zero then no collector current flows, but as V_{be} increases above about 0.5V, I_c starts to climb dramatically. To a good approximation, graph (a) can be seen as a cross section through the family of plots in graph (b) at the point on the X axis of (b) where $V_{ce} = 5.0V$. (The base current I_b is directly related to the base emitter voltage V_{be}.) Thus as V_{be} changes the line in graph (b) tracks up and down in response. Notice that I_b is measured in microamps while I_c is measured in milliamps, and so it seems pretty clear that the goal of amplification is being realised here. (Recall the one line description from the chapter heading: transistors amplify, buffer, and switch electrical signals.)

Any more detailed analysis of the graphs is unnecessary at this point. Suffice it to say that if an amplifier or other circuit design is being developed, reference to these and other plots provided in the data sheets of the devices to be used would provide the information necessary in order to develop a circuit design to the required performance specifications.

In addition to numerous graphs, transistor data sheets also contain a wide range of numerical data further specifying the behaviour and performance of the particular device. Some of these numbers are more useful and more important than others. Table 17.1 provides examples of some of the more commonly encountered numerical specifications, given for a range of typical devices.

Table 17.1 Some typical BJT specifications

Part	Type[a]	V_{ceo}	I_c	h_{fe}	P_{tot}	Package
2N3904	NPN	40V	200mA	100–300	625mW	TO-92
2N3906	PNP	40V	200mA	100–300	625mW	TO-92
2N5088	NPN	30V	50mA	300–900	625mW	TO-92
BC547A	NPN	45V	100mA	110–220	625mW	TO-92
BC547B	"	"	"	200–450	"	"
BC547C	"	"	"	420–800	"	"
BC557A	PNP	45V	100mA	120–220	625mW	TO-92
BC557B	"	"	"	180–460	"	"
BC557C	"	"	"	420–800	"	"
AC125	PNP Ge	32V	100mA	50–	500mW	TO-1
MPSA13	NPN Dar.	30V	500mA	5000–	625mW	TO-92
BD437	NPN	45V	4.0A	85–375	36.0W	TO-126
BUF725D	NPN	400V	5.0A	10–	40.0W	TO-220

a. Ge = germanium, Dar. = Darlington.

The columns in Table 17.1 cover a number of important metrics for BJTs as described under the headings below:

Part – the part number uniquely identifies the device to be used. Search on this number to find a data sheet for the device, as well as other resources, sellers, and discussion threads related to its use.

Type – in BJTs this is primarily either NPN or PNP, specifying the device's polarity. Germanium and Darlington devices have also been indicated here.

V_{ceo} – this is the maximum allowable voltage between collector and emitter. Exceeding this limit is likely to damage the device

I_c – similarly, this is the maximum current which should be allowed to flow in the collector

h_{fe} – the DC current gain of the device. A range from minimum to maximum is usually provided, although occasionally only a guaranteed minimum is quoted.

P_{tot} – specifies the maximum total device power dissipation, usually measured in milliwatts for small signal transistors, and running into watts for larger devices.

Package – indicates the package outline which the device comes in. Common package outlines are described earlier in this chapter.

Looking at the example devices covered in Table 17.1, top of the list is the 2N3904 encountered earlier. Next comes the 2N3906 which is the PNP counterpart to the 2N3904. Transistors are often used in complementary pairs involving two devices with similar specs but opposite polarities. The 2N5088 is another common NPN device, with a higher gain that the 2N3904. The next six devices can be seen as a family, with three each NPN and PNP. The A, B, and C variants are a common way of specifying low, medium, and high gain devices with otherwise similar specs.

The first nine devices listed are typical NPN and PNP small signal devices, all quite similar. The last four entries in the table provide a bit of variation. The AC125 is a PNP germanium device, where all the rest are silicon based. The MPSA13 is an example of a Darlington device, with its commensurately high minimum gain factor of 5000, and the last two entries are power devices (notice their package outlines and P_{tot} figures in particular). While small signal transistors are fine in preamplifier and other low power circuits, these last two devices are ideal for building power amplifiers and the like.

Substituting BJTs

Since there are so many different BJTs available, the situation will soon arise where a circuit is encountered using BJTs which can not easily be found. It then becomes necessary to figure out what readily available device might successfully be substituted for the unavailable one specified in the circuit design. In the case of small signal devices a simple checklist should yield a suitable substitute.

Step 1: Check the polarity. NPN and PNP devices can not be interchanged.

Step 2: It must match in general type. Is it Si or Ge? Is it a Darlington device?

Step 3: The next criteria to consider is the gain (h_{fe}). Transistor gains vary a great deal (observe the wide ranges of values for each type in Table 17.1). Generally it is sufficient to match gain as being low, medium, or high.

Step 4: Finally make a quick check to see that maximum voltage, current, and power levels are close to the original, or look reasonable for the circuit in question. For instance in a circuit to be powered from a 9V battery, it should be fine to use a device with $V_{ceo} = 30V$ where one with $V_{ceo} = 45V$ was specified, but if the circuit voltages are closer to the limit then care must be taken. Using substitutes with higher values here are fine, but values lower than the original need to be checked.

For substituting non small signal BJTs such as those in a power amp, the same general principles apply but a closer match might be sought, and particular attention must be paid to figures such as I_c and P_{tot}.

Matched Pairs

Occasionally a circuit design calls for a closely match pair of NPN and PNP transistors in order to achieve best performance. As mentioned above, h_{fe} can vary quite significantly between different samples of the same device, so once the types to be used for the matched pair have been selected, it can be helpful to seek a close gain match. This is done by measuring the gain of a number of devices of each type and selecting a pair whose gain values are as close together as possible. Usually gain measurement is a simple case of plugging each device into the socket provided in many hand-held digital multimeters and reading off the gain value (see Figure 17.7). Care must be taken to correctly identify both the polarity and the three pins of the device to be measured, and to insert it into the correct set of terminals on the DMM.

Figure 17.7 Using a DMM to measure the gain of a BJT.

BJT Circuits

This section presents a number of simple circuits to build and experiment with, in order to investigate the use and the operating characteristics of BJTs. A vastly oversimplified, but nonetheless useful, description of how an NPN BJT operates is presented in Table 17.2. Basically what it is saying is that if the base potential is low (close to or below that of the emitter), then the impedance between collector and emitter is high and the transistor is considered to be in its off state. Conversely, if the base potential is high (above that of the emitter, typically by about 0.8V or more), then the impedance between collector and emitter is low and the transistor is considered to be in its on state. Between these two extremes Z_{ce} travels down as V_{be} rises, and up as V_{be} falls.

Thinking about a transistor's behaviour within a circuit in these terms is often quite sufficient to determine in a broad sense the way in which the circuit will respond to varying inputs (like for instance a sine wave or an audio signal). By figuring out what

Table 17.2 Simplified behaviour of an NPN BJT

Input		Result	Transistor state
V_{be} low	$\Longrightarrow$	Z_{ce} high	Off
V_{be} intermediate	$\Longrightarrow$	Z_{ce} intermediate	Switching/amplifying
V_{be} high	$\Longrightarrow$	Z_{ce} low	On

the impedance between collector and emitter will be doing it is often possible to track the workings of a circuit.

Figure 17.8 illustrates how a BJT might be used as a switch. When a finger is touched to the finger contact points the base voltage is pulled positive relative to the emitter, the transistor is switched on, current flows from collector to emitter, and the LED lights up. Without a finger on the contact points the base emitter voltage returns to zero and the BJT is switched off. In this way a very small current flowing in the base controls a much larger current in the collector.

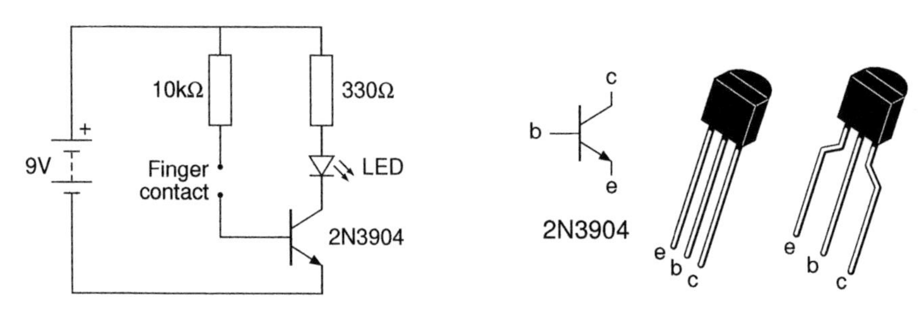

Figure 17.8 A simple BJT circuit. Touching the finger contacts lights the LED.

Using the graphs in Figure 17.6, the behaviour described here can be interpreted. When $V_{be} = 0V$, $I_c = 0mA$, and the transistor is in its off state. Touching the contact points pulls V_{be} positive. When it reaches about 0.5V on the graph, I_c starts to increase from zero. This collector current is flowing through the resistor and LED as well as along the collector-emitter path through the BJT. The resistor limits the amount of current which can flow, and prevents things from overheating and burning out.

Learning by Doing 17.1

BJT Finger Switch

Build the circuit shown in Figure 17.8 on a breadboard and confirm that it operates as described. Even for a circuit as simple as this it is worth designing a breadboard layout before you start plugging components into the breadboard, especially if this is one of the first circuits you have built.

For the finger contact points use two small U shaped pieces of bare wire plugged into the breadboard close to one another, but make sure they don't connect or touch.

The diagram on the right of Figure 17.8 shows the pinout for the 2N3904 transistor. If you are using an different device, remember to look up its pinout.

Replace the BJT with a higher gain device like a 2N5088. Can you see any difference in the performance of the circuit. Its always a good idea to remove power from a circuit when making any changes. Remember when switching devices that the pinout of the new device must be checked, and any adjustments needed to the circuit made before power is reconnected.

If the LED is not very bright, replace the resistance with a 220Ω one. Don't make the resistor too low or the LED may burn out.

Moisten your finger and try again. Does the LED light brighter? Pressing the contact points gently or hard will also change the brightness of the LED. In both cases we are changing the resistance between the base and the 9V supply, allowing more or less current I_b to flow, and as a result controlling the level of current I_c flowing, which in turn controls the brightness of the LED.

The next circuit, shown in Figure 17.9, implements a straightforward booster or preamplifier, ideal for adding a bit of drive to an electric guitar or other audio signal. Using just the simplified behavioural description of Table 17.2 and the voltage divider rule, this circuit can be analysed in basic terms.

A good way to start thinking about analysing any circuit is to initially assume no input, everything is at zero volts, and then switch on the power. Let's also start with VR1 turned all the way down, so there are matched 10kΩ resistances above and below Q1. This means once power is applied and things start to happen, the voltages at collector and emitter will mirror one another. When Z_{ce} is high, the collector is close to 9V and the emitter is close to 0V, and as Z_{ce} falls these two voltages move towards each other.

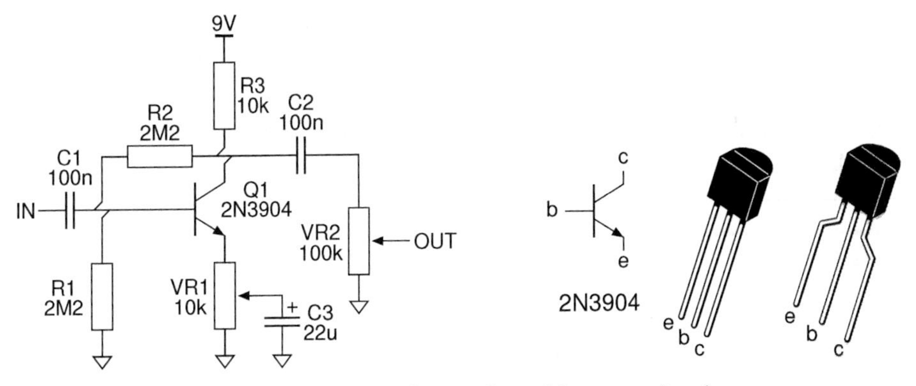

Figure 17.9 A simple BJT based booster circuit.

Initially Q1 is off and high impedance. With 9V applied at the top, R1 and R2 form a voltage divider. R3 is small enough to ignore for now, so the base of Q1, at the midpoint of R1 and R2 heads up from 0V towards 4.5V. Q1, which was initially off and therefore high impedance, starts to switch on as the base voltage rises, and as it does, its impedance drops. This causes the voltages at collector and emitter to head towards the middle too. An equilibrium will be found where any further decrease in Z_{ce} will tend to start switching Q1 back off again, thus increasing Z_{ce}.

And there it sits until a signal comes along. This is called the quiescent condition. The equilibrium is due to the action of R2. If Q1 tries to move more off, the voltage at the collector rises. This voltage increase feeds back through R2 to the base, forcing the transistor to move back in the on direction, and the opposite effect results if Q1 tries to move more in the on direction. The voltage falls, it feeds back through R2 and switches pulls Q1 back to its equilibrium point.

When a signal comes in, it modulates the voltage at the base up and down. As V_{in} at the base goes up, the transistor is pushed further on, and V_{out} at the collector goes down, and vice versa. Again the negative feedback influence of R2 keeps the whole process stable.

That is as much analysis as will be presented here, but it is well worth going through the process several times, to get a good feel for what is going on. The next step is to build the circuit and see how it performs.

LEARNING BY DOING 17.2

BJT BOOSTER

Design a breadboard layout for the circuit in Figure 17.9 (or use the one provided in Appendix D). As before, build the circuit, test it with a signal generator and oscilloscope, and then send some audio through it and see how it sounds. In this case it is supposed to be a clean boost, so hopefully the output sounds not too different to the input signal, just louder. Of course, as with any amplifier, if it is driven too hard it will start to clip and distort.

VR1 is a gain pot. How does it affect the level? VR2 just controls the output level, and shouldn't affect the quality or character of the sound.

Again, a stripboard layout is also included in Appendix D, so if you want to have a go at building a permanent version, that's a good place to start.

LEARNING BY DOING 17.3

ONE TRANSISTOR FUZZ CIRCUIT

A very simple fuzz effect can be made using little more than a diode and a transistor. Design a breadboard layout for the circuit shown (or use the one provided in Appendix D). There is lots of scope for experimentation with this circuit. Different types of diode and BJT can be tried, along with alternate values for the input and output capacitors, for a decent variety of tones and levels of fuzz.

It is always worth sending a test signal through a circuit like this. Feed a 1kHz sine wave into the input and view input and output on a two channel scope. Dial the sine wave amplitude up from nothing to a couple of volts and observe how the signal is distorted.

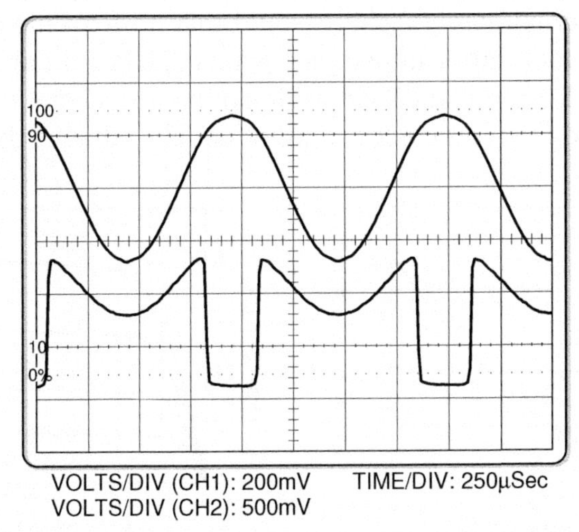

The incoming signal is processed in two modes. When the input is low, the transistor is off, the diode is forward biased, and the output tracks the input, remaining one diode drop above it. When the input rises enough to switch the transistor on, the diode is reverse biased and the output drops sharply.

Junction Field Effect Transistors (JFETs)

JFETs are made from a strip of one type of semiconductor material (the channel) with a terminal at each end (the drain and the source). A region of semiconductor material of the opposite polarity surrounds the channel, and is connected to the third terminal (the gate). The gate is the terminal that modulates the conductivity of a JFET (roughly equivalent to the base on a BJT). The drain and the source connected at either end of the channel correspond to the collector and emitter respectively on a BJT.

One of the main differences between BJTs and JFETs is that while in a BJT a small but significant current flows through the base to control a larger current between collector and emitter, in an JFET virtually no current flows through the gate. Thus JFETs have very high input impedances, and as such are ideal for building high impedance (hi-z) input buffers which are useful on guitar effects pedals and amplifiers.

JFET Characteristics

The first thing to become apparent from the graphs in Figure 17.10 is that JFETs are depletion mode devices, also known as normally on devices. This can be seen from the transfer characteristic in graph (a), where I_d, the drain current, is at its greatest when the gate source voltage, V_{gs} is zero, and V_{gs} must be taken several volts negative in order to stop current flowing in the channel. Because JFETs are formed from a p-n junction (exactly like a diode), care must be taken not to let V_{gs} go positive. At least, not more than the approximately 0.6V it would take to forward bias the p-n junction into conduction. The high input impedance which JFETs exhibit only persists as long as this rule is observed, so in a JFET circuit it can be instructive to see how the gate is going to be maintained at a voltage less than or equal to that of the source.

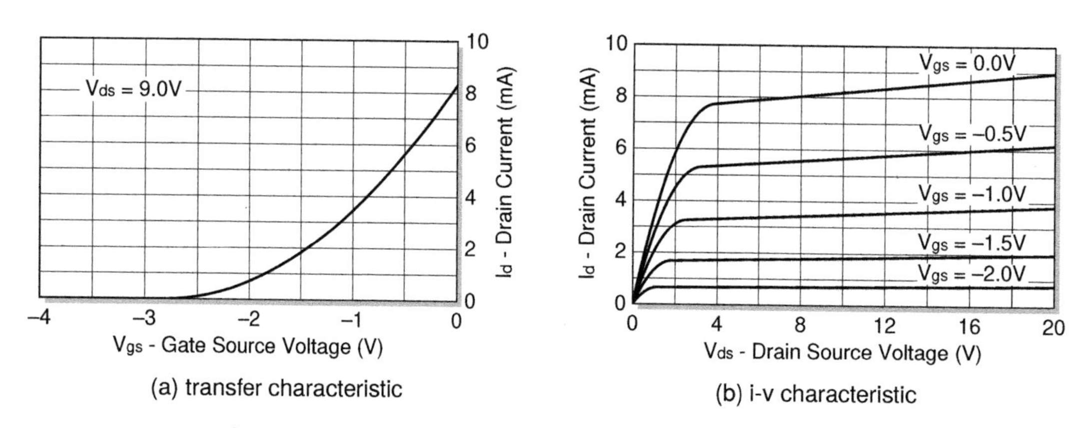

Figure 17.10 Typical n-channel JFET characteristic curves.

The amount of current which flows when $V_{gs} = 0V$ is called the device's saturation current. It is quoted for a particular drain source voltage; notice the corner of graph (a) where it says $V_{ds} = 9.0V$. In the i-v characteristic in graph (b), the topmost plot is labelled $V_{gs} = 0.0V$, and so it should be clear that this plot shows how the saturation current varies with V_{ds}. Reading off I_d from this plot for 9V yields a result just over 8mA. Returning to graph (a) it can be seen that this is exactly where the plot intersects the Y axis. In fact the plot in graph (a) is simply a cross section through all the plots in graph (b), taken along the vertical line where $V_{ds} = 9.0V$.

The saturation current just described (which is indicated I_{DSS}) is one of two important characteristics for a JFET which can be read from these graphs. The second, the pinch off voltage (V_p or $V_{GS(off)}$), can also be estimated from graph (a). It is defined as the gate source voltage at which the drain current reaches zero (again measured at a particular V_{ds}). In this case it looks to be a little past −2.5V

Table 17.3 lists the specified values of these two parameters for a number of common JFETs. A min-max range is provided rather than a single value, and it should be clear looking at the numbers, that these parameters can vary significantly between different examples of the same device. While circuit designs can often accommodate these wide levels of variation, it can sometimes be desirable to quantify these parameters more precisely for a particular component, either to allow for more precise design of a particular circuit, or in order to more closely match multiple components for use in a symmetrical design of some kind.

Table 17.3 Some typical JFET specifications

Part	Type	$V_{GS(off)}$ min	$V_{GS(off)}$ max	I_{DSS} min	I_{DSS} max	$r_{DS(on)}$ max[a]	Applications
2N5458	N-Ch	−1.00V	−7.0V	2.0mA	9.0mA	–	General purpose amp
2N5485	N-Ch	−0.50V	−4.0V	4.0mA	10.0mA	–	RF amp
BF245C	N-Ch	−0.25V	−8.0V	12.0mA	25.0mA	–	LF, HF and DC amp
J111	N-Ch	−3.00V	−10.0V	20.0mA	–	30Ω	Electronic switching
J175	P-Ch	3.00V	6.0V	−7.0mA	−60.0mA	125Ω	Electronic switching

a. The maximum on resistance ($r_{DS(on)}$ max) is usually only quoted for JFETs specifically intended for use in switching applications.

The final parameter listed on the Table 17.3, $r_{DS(on)}$, indicates the maximum resistance of the drain-source channel when the device is biased fully on (at V_{gs} = 0.0V). This parameter is usually only quoted for JFETs which are specifically designed for use in switching applications. In general, JFETs are designed with particular applications in mind, and their performance characteristics are optimised accordingly. Most JFETs are aimed at either various applications in amplification, or at switching type applications. For switching applications, the maximum resistance to expect when the switch is closed is an important consideration.

Measuring Pinch-Off Voltage and Saturation Current

It can be useful to be able to make a quick measurement of the two key design parameters for a JFET, $V_{GS(off)}$ and I_{DSS}. The two simple circuits shown in Figure 17.11 allow rough values for these parameters to be measured very easily.

In circuit (a) the gate and source are tied together, and so V_{gs} must equal zero. Recall that an ammeter is designed to have a very low impedance so that it affects the circuit it is measuring as little as possible. It is therefore reasonable to say that virtually all of the 9V of the supply will appear between the JFET's drain and source, and so V_{DS}, the drain-source voltage is equal to the full 9V of the supply. Based on the description of I_{DSS} given above, it should be clear that the ammeter will therefore be measuring the saturation current (for $V_{DS} = 9.0V$), as required.

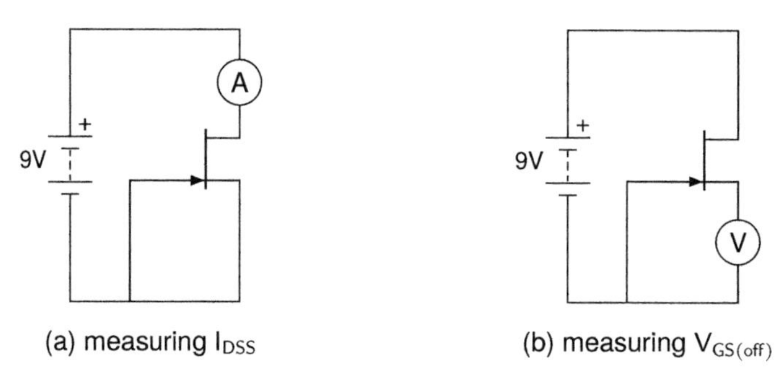

(a) measuring I_{DSS} (b) measuring $V_{GS(off)}$

Figure 17.11 Measuring I_{DSS} and $V_{GS(off)}$ for a JFET.

If instead of shorting together the gate and source, a resistance is placed between them, then any current flowing through the JFET's channel must also get through the resistor. Recall that a resistor drops voltage in proportion to the current flowing through it (Ohm's law), and so V_{gs}, the voltage between gate and source (and therefore across the resistor) will move away from zero. As the current is reduced and V_{gs} becomes more negative, the JFET starts to switch off (as per the graphs in Figure 17.10).

In order to measure the pinch off voltage, the current must be brought close to zero. A large resistor (say 100kΩ+) could be placed between source and gate to achieve this. However, a voltmeter has a very large resistance between its terminals, so instead of using it in the familiar fashion across the suggested 100k resistor, it can be placed directly into the circuit so that its own resistance is used to force the JFET into pinch off. This is the arrangement used in circuit (b), thus the pinch off voltage may be read directly off the voltmeter in this arrangement.

LEARNING BY DOING 17.4

MEASURING A JFET'S PINCH-OFF VOLTAGE AND SATURATION CURRENT

The procedure for measuring $V_{GS(off)}$ and I_{DSS} is described above, using two simple circuits as shown in Figure 17.11. In fact the two tests can be combined into one circuit as here, with $V_{GS(off)}$ displayed by default, and I_{DSS} measured when the momentary push switch is depressed. The JFET under test heats up fairly quickly as the relatively large saturation current flows. This both changes the value read, and can stress the device if maintained for too long, so a quick press-read-release action is best for measuring I_{DSS}.

Built on a breadboard, this circuit can test a large number of devices in a short period of time, allowing for a good pool of devices to choose from for the best set of characteristics in any given application. Measure a dozen devices of the same type and see how much of a spread appears in the readings. Are all devices within spec.?

Check the device's data sheet. This circuit is set up for testing n-channel JFETs. How would it need to be changed in order to test p-channel devices?

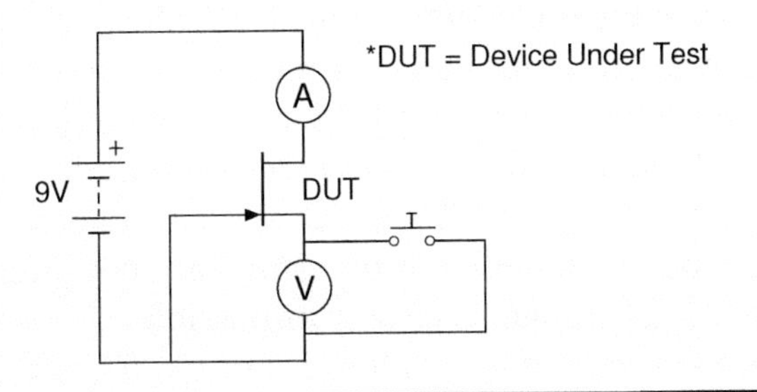

Reversible JFET Source and Drain

The standard convention for drawing FET circuit diagram symbols (both JFETs and MOSFETs) is that the arrow in the symbol appears on that side of the symbol which is closer to the source than the drain. This allows source and drain to be easily identified. However symbols like the second pair in Figure 17.12, with the gate connection bearing the arrow placed centrally between source and drain, are often encountered.

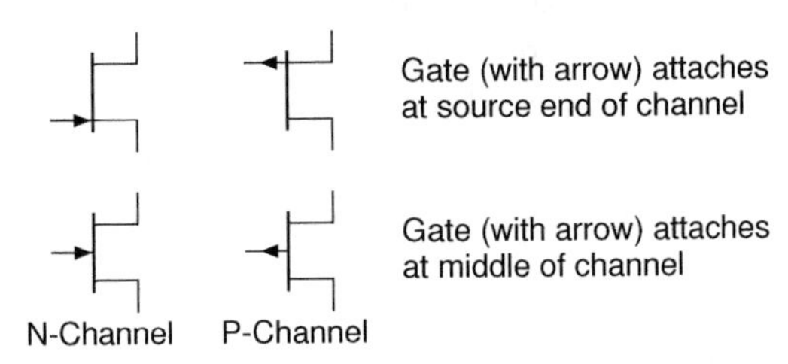

Figure 17.12 JFETs are often indicated using two slightly different symbol styles.

Source and drain now become indistinguishable, which would seem to be an undesirable state of affairs from the point of view of unambiguous interpretation of the circuit diagram. The reason that these symbols are used is that the design of JFETs is such that they are by and large reversible devices, the source and drain connections can often be interchanged and the device will still work. Some JFET manufacturing processes are designed to be entirely symmetrical, such that the behaviour will be indistinguishable between these two connection. A JFET's data sheet will often include in the device's list of features something like 'Source and drain are interchangeable'. It is a good idea to stick to the specified orientation where one is given, but being aware of the fact that some devices are reversible is useful.

Substituting JFETs

JFETs do not come in the same dizzying array of alternatives that is encountered with BJTs. Indeed with JFETs the problem is somewhat the reverse. Discrete JFETs are a far less popular device of choice than they used to be, and as such many previously popular devices are obsolete and are now difficult if not impossible to source. As such it is very often the case that when building a circuit which involves JFETs, it will be necessary to identify a suitable replacement part from those which are available to you. Sometimes a circuit may even require a bit of redesign in order to accommodate what ever is the best available choice for a substitution. In common with the checklist given earlier for substituting BJTs, a number of easy steps should be followed when looking for a suitable replacement part for a JFET which is no available.

Step 1: Check the type, n-channel and p-channel are not interchangeable

Step 2: Match the $V_{GS(off)}$ range as closely as possible

Step 3: Make sure the range of I_{DSS} values is not too far out

It is likely that a close match in both parameters will not be found. Given the wade range of values which a design can usually accommodate, this may not be a problem. Sometimes it can be easier to just try one or two parts, rather than attempting a detailed analysis and/or redesign of the circuit.

Learning by Doing 17.5

JFET Buffer

The section on JFETs is brought to a close with a simple JFET buffer circuit. JFETs have a much higher input impedance than BJTs. So high in fact that the input impedance of this circuit is really set by the resistors on the input. As such this circuit is ideal for interfacing to an electric guitar pickup or even a piezo pickup. These both exhibit high output impedances and so are best matched to a very high impedance input.

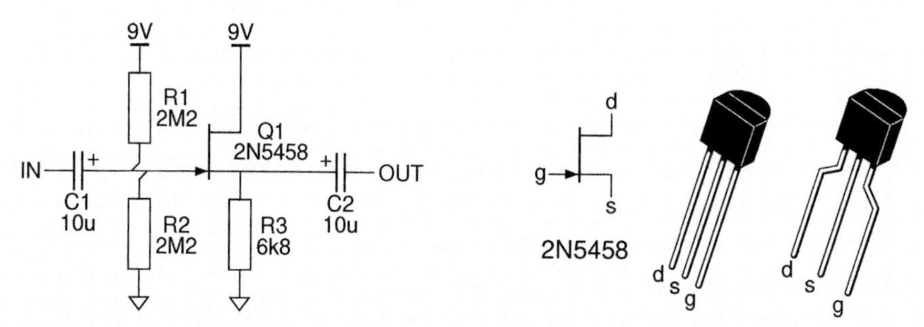

The circuit can be seen as two voltage dividers, the only wrinkle being that the second voltage divider, comprised of Q1 and R3 involves a variable impedance arm in the form of Q1, whose impedance is governed by V_{gs}.

As before, to analyse the circuit, imagine turning on the power and see what happens. Clearly the first voltage divider wants to bring the gate voltage up to 4.5V. From the JFET graphs it is clear that V_{gs} at or above zero keeps the JFET fully on (recall negative voltages switch it off). Hence the impedance of Q1 falls to below the 6k8 it is paired with, finding an equilibrium point which holds V_{gs} a little below zero.

If V_{gs} moves lower, Q1 starts to switch off, raising its impedance and thus pulling V_{gs} back up. If V_{gs} moves higher (towards zero), Q1 starts to switch on, lowering its impedance and thus once again pulling V_{gs} back to its equilibrium point.

If an input signal modulates the gate voltage, the action described above forces the output voltage at the source to track up and down along with it.

Metal Oxide Semiconductor Field Effect Transistors (MOSFETs)

The differences between MOSFETs and JFETs come mainly from the different ways in which the gate terminal is interfaced into the channel between drain and source. In JFETs this interface consists of an actual p-n junction, whereas in MOSFETs the gate is completely insulated from the channel by the oxide layer referenced in its name. This has two primary implications for the different ways in which these two types of devices operate. Firstly, it means that V_{gs} in a MOSFET can swing both negative and positive, whereas in a JFET the p-n junction must be kept reverse biased or significant current will be able flow through the gate terminal. Secondly, the p-n junction of the JFET allows a very small but not completely negligible working current to flow through the gate even in its normal reverse biased state. This is the same thing as the standard reverse bias leakage current described for diodes in Chapter 16. In MOSFETs on the other hand, with their gate electrically insulated from the channel, no appreciable gate current flows at all (values in the nanoamp range are typically quoted).

MOSFETs appear widely in applications such as power amplifier and switched mode PSU circuit designs but are much less commonly found in the kind of small signal audio circuits which are the focus of this book. Opinions differ on the subject of whether BJTs or MOSFETs make for superior power amplifier designs. Sloane expresses his preference for MOSFETs (Slone, 2002, p. 166) ('I make it no secret that L-MOSFETs are my favourite output device types...') while Self takes the contrary view extolling the merits of BJT based designs (Self, 2010, p. 128) where he presents his view that BJTs are 'the best choice for all three stages of a generic power amplifier'. As both authors point out there are advantages and disadvantages to both, and both authors agree that good designs can certainly be developed using each.

MOSFET Characteristics

Due to the restrictions their inherent p-n junction imposes, JFETs are exclusively depletion mode devices. That is to say they are normally on and require a voltage difference between gate and source in order to switch them off. MOSFETs on the other hand come in both depletion mode (normally on) and enhancement mode (normally off) varieties. That being said, the enhancement mode variants are far more commonly encountered, and are the only type considered here. Comparing the graphs of Figure 17.10 (for JFETs) with those of Figure 17.13 (for MOSFETs). The negative range of values for

V_{gs} in the former is replaced with a positive range in the latter. Thus at $V_{gs} = 0.0V$, unlike the JFET, the enhancement mode MOSFET is firmly off. Increasing V_{gs} brings the channel into conduction, allowing the drain current I_d to climb.

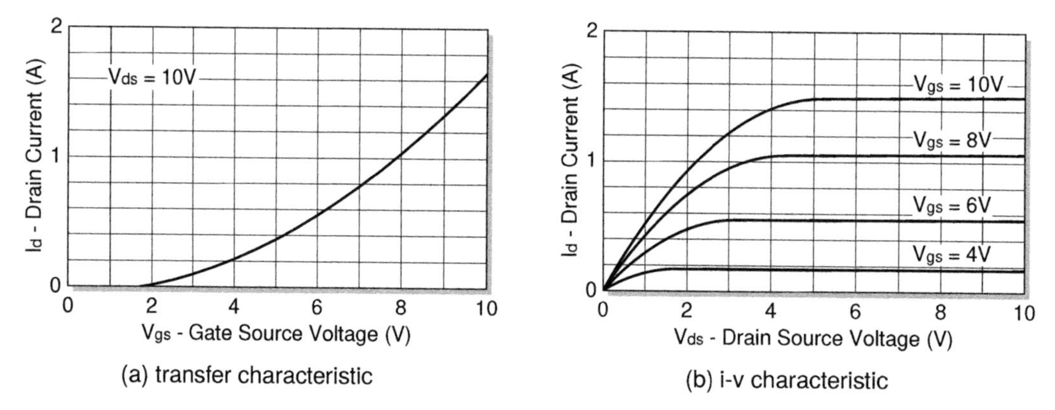

Figure 17.13 Typical n-channel enhancement MOSFET characteristic curves.

In Table 17.4 some key parameters for a number of enhancement MOSFETs, both n-channel and p-channel, are listed. The first two devices shown, the 2N7000 and the BS170, are the MOSFETs most likely to be encountered in the small signal audio circuits of interest here, while the BS250 is probably the most commonly found p-channel device. The VP3203 is another p-channel device, notice in particular its quite low $R_{DS(on)}$ value as compared to the three previous devices. The final three, with 'IR' prefixes, exhibit much lower values for $R_{DS(on)}$, along with higher current handling, and associated larger package outlines.

Table 17.4 Some example MOSFET specifications

Part	Type[a]	V_{DSS}	I_D	$R_{DS(on)}$	Package
2N7000	N-Ch enh.	60V	0.20A	5.00Ω	TO-92
BS170	N-Ch enh.	60V	0.50A	5.00Ω	TO-92
BS250	P-Ch enh.	−45V	−0.23A	14.00Ω	TO-92
VP3203	P-Ch enh.	−30V	−4.00A	0.60Ω	TO-92
IRL2703	N-Ch enh.	30V	24.00A	0.04Ω	TO-220
IRF520N	N-Ch enh.	100V	9.20A	0.27Ω	TO-220
IRF644N	N-Ch enh.	250V	14.00A	0.24Ω	TO-220

a. The majority of MOSFETs encountered are n-channel enhancement, with p-channel enhancement types coming in a distant second. Depletion mode devices are relatively rare and are not addressed here.

The three numerical parameters listed in Table 17.4 are the most commonly examined metrics from the typical large set of specifications given in device data sheets.

Maximum drain source voltage V_{DSS} – indicates the absolute maximum voltage which the device is capable of accommodating between its drain and source terminals. Generally this is high enough to be of no concern in the applications considered here.

Maximum continuous drain current I_D – says what level of constant current the device can handle before it is liable to be damaged. A few hundred milliamps for smaller devices, tens of amps for larger devices. Devices will be able to handle short pulses of higher current but this is the best figure of first resort.

Drain-source on resistance $R_{DS(on)}$ – the effective channel resistance when the device is fully switched on, and the channel is in its lowest resistance state. The very low values in some of the devices listed makes them good candidates for low drop out applications, where minimal voltage drop is desired when the device is in its on condition.

MOSFET Circuits

MOSFETs are not encountered too regularly in the kinds of small scale audio circuits of particular interest here. It is however well worth having a look at a selection of simple applications, illustrating the kinds of places where they might be found.

LEARNING BY DOING 17.6

SIMPLE MOSFET DIMMER CIRCUIT

A first simple application uses a MOSFET to implement a basic dimmer or throttle arrangement. With the variable resistor turned all the way down, V_{gs} is zero, and the MOSFET is off, so no current flows through the load. From the transfer characteristic in Figure 17.13a, we can see that as V_{gs} rises above about 2V the drain current starts to flow. Comparing V_{ds} of 6V and 10V on graph (b) indicates that the transfer characteristic, shown here for $V_{ds} = 10V$ should look very similar at 6V.

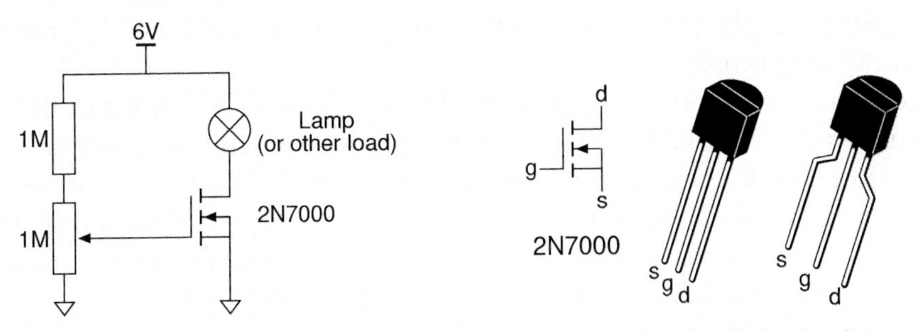

With the arrangement shown, based on a supply voltage of 6V, the maximum value which V_{gs} can take is 3V, with the variable resistor turned all the way up. On the graph this corresponds to a maximum of about 100mA flowing in the drain, and hence through the load. From Table 17.4, a 2N7000 can handle up to 200mA continuous drain current, so this should pose the MOSFET no problems. The current may or may not be further limited depending on the load being driven. So long as the load is happy to see up to 6V across it all will be well.

Build this circuit and assess its performance. Add an ammeter to the circuit and monitor the current flowing as the pot is turned. How could the design be modified so that current starts to rise earlier through the rotation of the pot? If the lamp were replaced by an LED, a current limiting resistor would be needed. Use the calculation introduced in Chapter 16 to select a suitable resistor, modify the circuit accordingly, and repeat your tests.

An application where the low $R_{DS(on)}$ values seen in the last three entries in Table 17.4 can be of particular benefit is when using a MOSFET in a reverse polarity protection circuit, as shown in Figure 17.14. Either an n-channel or a p-channel device can be used, as in Figure 17.14a and Figure 17.14b respectively.

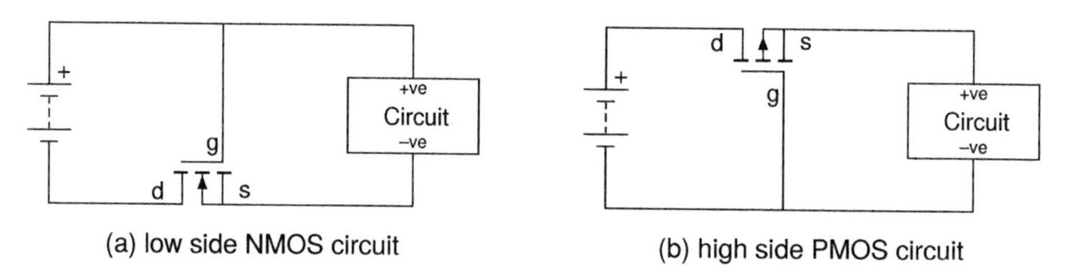

Figure 17.14 Reverse polarity protection using either a low-side n-channel MOSFET or a high-side p-channel MOSFET.

The operation of both circuits is fairly straightforward. Taking the NMOS[a] version, if the battery is correctly inserted, V_{gs} is positive, the MOSFET is switched on, and the circuit is powered. Reversing the battery makes the gate negative relative to the source, the MOSFET is switched off, and no current is allowed to flow, protecting the circuit from potentially damaging reverse powering. The main design requirement is that the battery or power supply must be of a sufficient voltage to fully switch on the MOSFET when correctly connected.

Another important requirement is a low $R_{DS(on)}$. This is important so that the voltage drop across the MOSFET is kept to a minimum. Any voltage dropped across the MOSFET is lost to the circuit being protected. If a MOSFET with a relatively high on resistance, such as a 2N7000 were used, significant voltage level would be lost. Consider a circuit which draws 100mA. At an $R_{DS(on)} = 5\Omega$, Ohm's law indicates that the voltage drop across the 2N7000 would be $V = I \times R = 100m \times 5 = 500m = 0.5V$. Half a volt may not seem like much but, especially in a battery powered application, it can make a great deal of difference. Using a device with $R_{DS(on)}$ equal to say 0.24Ω, that voltage drop is reduced to $100m \times 0.24 = 24mV$, which is much more sustainable.

a. In this context NMOS and PMOS can be considered shorthand for n-channel MOSFET and p-channel MOSFET respectively.

It may on first analysis seem that the devices in these two diagrams have been inserted the wrong way round. Generally in any FET current flows from drain to source, but these devices are reversed. The reason is down to a peculiarity of MOSFETs which usually does not impact upon the operation of a circuit. MOSFETs have an intrinsic body diode which is oriented from source to drain. In most applications this remains reverse biased and does not come into play. In this instance things are not so simple. If the MOSFETs were inserted the other way round, a correct power connection would still work, switching the device on and powering the circuit. However reverse polarity connection would not now be blocked. The gate source voltage would switch the main channel of the MOSFET off as intended. Unfortunately, the body diode would bypass this and allow current to flow.

Basically a MOSFET, even when switched off can conduct in the source-drain direction. This fact is actually used to advantage in the exercise which follows.

LEARNING BY DOING 17.7

WIRING A MOSFET AS A DIODE

One of the most fertile grounds for experimentation in audio electronics is in the area of distortion circuits. As is examined in Chapter 16, diode clipping is the commonest route to simple distortion circuits. The ubiquitous back to back diode pair lends itself to a multitude of alterations and variations. One branch of these is the MOSFET clipper variations described here.

As has just been discussed above, MOSFETs include a body diode in their basic makeup. This leads to two possible approaches to the clipping task, either keep the transistor off and just use the diode, or allow the transistor to conduct and use both the MOSFET's channel, and the body diode. The first diagram illustrates the former, while the second shows the latter.

In the first case the gate is tied to the source, so V_{gs} is held at zero volts, keeping the transistor firmly switched off. In this case just the body diode gets involved and the transistor behaves much like a regular diode. The shape of its i-v curve will differ from a standard diode however, and as such the tone it produces should be different, if perhaps subtly so.

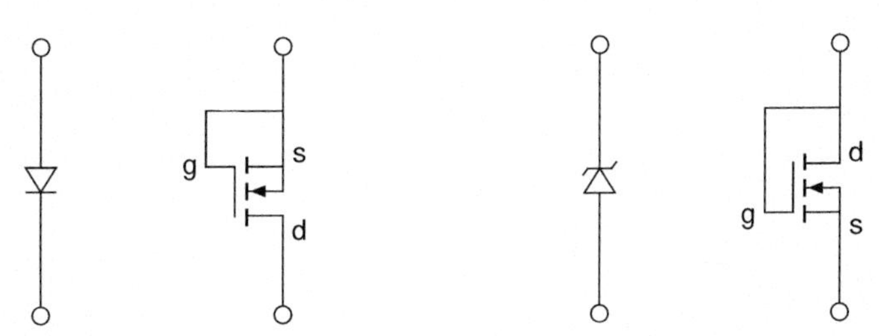

In the second case, gate is tied to drain. Now the body diode conducts in one direction, and the channel conducts in the other. The diode starts to conduct at the usual si diode forward bias voltage of round about 0.7V, and in the other direction the channel starts to conduct (as per Figure 17.13a) at about 2.0V.

Both series and parallel combinations of devices are possible giving various different effects in a standard distortion circuit.

Wire up a 2N7000 in each configuration, and use an oscilloscope in X-Y mode, as described in Learning by Doing 16.2 (p. 281) to view the i-v characteristic achieved.

Also try the clipping experiments in Learning by Doing 16.3 (p. 288) to compare how MOSFET clipping compares to the regular diode variety.

References

A. Evans et al. *Designing With Field-Effect Transistors*. McGraw-Hill, 1981.

E. Evans, editor. *Field Effect Transistors*. Mullard, 1972.

Fairchild. *AN-6609: Selecting the Best JFET for Your Application*. Fairchild Semiconductor Corporation, 1977.

J. Falin. *Reverse Current/Battery Protection Circuits*. Texas Instruments, 2003.

Maxim. *AN 636: Reverse Current Circuitry Protection*. Maxim Integrated Products, 2001.

C. Motchenbacher and J. Connelly. *Mullard Technical Handbook, Book 1 (Part 1): Semiconductor Devices, Transistors AC127 to BF451*. Mullard, 1974.

D. Self. *Audio Power Amplifier Design*. Focal Press, 6th edition, 2010.

R. Severns, editor. *MOSPOWER Applications Handbook*. Siliconix Incorporated, 1984.

G. Slone. *The Audiophile's Project Sourcebook*. McGraw-Hill, 2002.

H. Zumbahlen. *Linear Circuit Design Handbook*. Newnes, 2008.

18 | Integrated Circuits

Integrated circuits provide complete circuits in a prefabricated package

Integrated circuits (ICs) are just circuits pre-wired on a silicon chip. They can come in a lot of shapes and sizes, but the ones used here are almost all in DIL (dual in-line) packages, Figure 18.1b. The name refers to the fact that there are two rows of legs for making connections. This is a through hole format; the legs are designed to be inserted through holes on the circuit board and be soldered on the other side, as opposed to surface mount packages which are also common, but which are not discussed here. Related commonly seen abbreviations for the same package style are of the form DIP8 and DIP14. The DIP stands for dual in-line package, and the number refers to the number of legs, so for instance the 386 power amplifier chip encountered a few times already would be described as coming in a DIP8 format. An obvious variation on this theme is the SIL or single in-line package. An example of this style is the three pin LM317 voltage regulator, which is introduced later. Obviously most transistors could also be referred to as being SIL packages. SIL packages with a leg count over three are less common but by no means unheard of.

(a) SIL (b) DIL

Figure 18.1 Typical integrated circuits in (a) SIL and (b) DIL packages.

When using ICs remember that it is almost always necessary to provide power to the chip, even if these connections are not shown in the circuit diagram for the circuit being built. If not all elements on a chip are being used in the circuit, it is often a good idea to tie unused inputs either to ground or sometimes to the positive voltage rail or some other available reference voltage, in order to minimise noise and interference. See also Chapter 7 for some additional notes on this topic.

Selected ICs

Table 18.1 lists some commonly encountered ICs. Some of these are examined in more detail in the sections which follow. All are worth getting familiar with. As usual a good place to start is with the manufacturer's data sheets for the devices. A quick search online will yield the relevant documents (and can quickly lead down the rabbit hole of online forum discussion threads and similar valuable, though sometimes variable quality, resources).

Table 18.1 Selected ICs commonly found in audio circuits

Category	Example part #	Notes
Opamp (single)	LM741	Very common (though now very old) opamp
"	LF356	JFET inputs gives high Z_{in}
"	NE5534	Low noise opamp
Opamp (dual)	TL072	Popular dual opamp with JFET inputs
"	LM358	Works well close to power rails
"	NE5532	Dual version of the NE5534 low noise opamp
Opamp (quad)	LM324	Four opamps in a single fourteen pin package
OTA	NJM13700	One of very few OTAs still available
Power amp	LM386	Common, easy to use low power audio amplifier
"	LM380	For a little more power than the 386 can provide
Voltage regulator	LM317	Output voltage adjustable from 1.25V to 37V
"	LM337	Negative power rail equivalent of the LM317
Precision timer	NE555	Popular timer chip, found in many audio projects
"	NE556	Two 555 timers on one chip
Digital delay/echo	PT2399	Single chip digital delay/echo processor
Tone decoder	LM567	Re-purposed by DIYers to make interesting noises
4000 series	various	See Table 18.4 for some examples
Optocoupler	NSL-32	LED/LDR optocoupler

Opamps

The operational amplifier (opamp) is the most important and most commonly encountered standard circuit building block in audio electronics, and as such a significant portion of this chapter is dedicated to understanding and using opamps effectively. Opamps come in two forms, either prefabricated as ICs, or built from the ground up from discrete components. IC opamps make life a great deal easier for the audio circuit builder, and here only the IC approach to opamp circuit construction is considered. Opamps are one of the most fundamental building blocks of audio circuits, so it is important to have a good grasp of how they are used.

The practical operation of an opamp is very straightforward; the output is just a greatly amplified version of the difference between the two inputs. This relationship is expressed in Eq. 18.1. The triangle symbol in Figure 18.2 is the standard representation in electronic circuit diagrams for an amplifier in general and for an opamp in particular. In this case the *G* inside the triangle represents the opamp's open-loop gain. The only important thing to remember about this is that it is a very big number. So, by Eq. 18.1, V_{out} is equal to the difference between the two inputs multiplied by a very big number. This rule never changes and is key to understanding how opamp circuits work.

$$V_{out} = G(V_+ - V_-) \tag{18.1}$$

The other important rule to remember about opamps is that the inputs generally have very high input impedances. This means that their effect can usually be ignored when analysing how a circuit is going to behave. These two rules, very high gain and very high input impedance, are all the knowledge about opamps that is really needed to analyse basic opamp configurations.

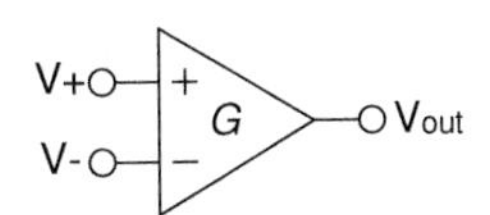

Figure 18.2 Basic characteristics of an opamp.

Negative Feedback

In fact this open loop gain *G* is far too high to be used directly as the amplification factor in most analog circuits; even the tiniest input signal would appear at the output too large to be useful. A simple trick allows this gain to be brought under control, and set and varied exactly as required in a typical amplifier application. That simple trick is called negative feedback (NFB). Most opamp circuits which will be encountered employ negative feedback in order to achieve signal control. The thing to look for in order to identify the application of negative feedback is a pathway (usually involving a resistor) from the opamp's output back around to its negative (or inverting) input (Figure 18.3). The inverting input is usually labelled with a minus sign, with the noninverting input labelled with a plus, as illustrated in the Figure 18.2.

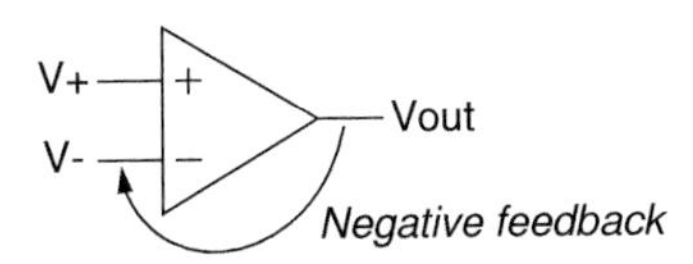

Figure 18.3 Negative feedback from output to inverting input.

Negative feedback allows the excessive open-loop gain *G* to be turned into a smaller (and controllable) closed-loop gain, represented as g. So the transfer function for an opamp circuit employing negative feedback is as in Eq. 18.2. (Recall that a transfer function is simply an equation defining the output in terms of the input.) So the output is just the input multiplied by this new lower gain factor, g, as expected.

$$V_{out} = g \times V_{in} \tag{18.2}$$

Negative feed back leads to a simple rule for analysing the behaviour of any opamp circuit which uses it:

Negative feedback forces an opamp's inputs to remain locked together

To understand why this is so, consider what happens if either input is moved away from the other in either the positive or the negative direction. Always remember that the fundamental, unchangeable behaviour of an opamp is described in Eq. 18.1.

Referring to the inputs as V_+ for the noninverting input and V_- for the inverting input, and the output as V_{out}, if V_+ moves a bit above V_-, then V_{out} wants to move up a lot (by Eq. 18.1). However, as V_{out} starts to move up, its negative feedback connection to V_- pulls V_- up along with it. This brings V_- back level with V_+ again, and so V_{out} no longer wants to move, since what it amplifies is the difference between V_+ and V_-.

Exactly the same thing happens if V_+ moves below V_-. This time V_{out} moves down, again pulling V_- with it, and again both settle at a new equilibrium. And so the negative feedback combined with the very large open-loop gain G prevents the two inputs from diverging. If they try, the negative feedback pulls them back together.

It is then a question of how to set the value for the closed-loop gain, g, and the operation of the negative feedback opamp circuit is known. The setting of g is down to the configuration of the circuit involved. Two standard circuit configurations are examined below. They are called the inverting and the noninverting configurations. The names indicate which of the opamp's two inputs has the input signal connected to it, and as a result, whether or not the output signal is inverted relative to the input.

Inverting Configuration

Figure 18.4 illustrates the standard inverting configuration for an opamp. The noninverting pin (+) is tied to ground, and the input signal is applied to R2, which feeds into the inverting input (−). R1 implements the negative feedback. The gain which this configuration achieves is defined by Eq. 18.3. The gain factor is simply the ratio of the resistors used. The minus sign in Eq. 18.3 indicates that this circuit will invert the polarity of the input signal when it appears at the output.

$$g = -\frac{R_1}{R_2} \tag{18.3}$$

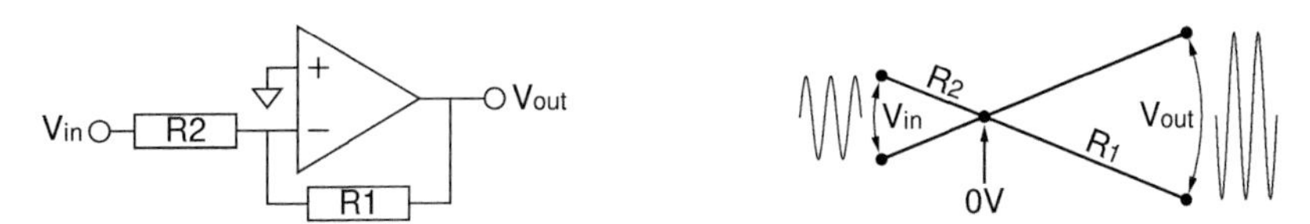

Figure 18.4 Opamp inverting configuration.

It thus becomes a simple matter to design a circuit which will (for instance) double the size of a signal. (The signal will also always be inverted using this circuit. The next section presents the noninverting option.) If the gain is to be −2, then by Eq. 18.3 R_1/R_2 must equal 2, and so R_1 must be twice the size of R_2. Clearly many pairs of values satisfy this requirement. One value must be chosen and then the second can be calculated. There are tradeoffs favouring smaller and larger resistor values in such circuits, but values around the 10k to 100k range are often a reasonable place to start. So if $R_1 = 50k$ were chosen, then the second value would be calculated as $R_2 = R_1/2 = 25k$.

The behaviour of this circuit is illustrated on the right in Figure 18.4, where the lengths of the lines are proportional to the resistor values R_1 and R_2. The lines start at V_{in} and end at V_{out}, with the crossover point locked in at zero volts where R_1 and R_2 meet. In this case R_1 is twice the size of R_2, and so the end of the line at V_{out} moves twice as far as that at V_{in}. Changing the relative lengths of R_1 and R_2 changes the relationship between V_{in} and V_{out}, precisely as specified in Eq. 18.3. Notice that this is the inverting configuration so that when V_{in} goes up V_{out} goes down, and vice versa.

Noninverting Configuration

Figure 18.5 shows the noninverting opamp configuration. In this case R2 is tied to ground, and the input signal V_{in} is applied to the noninverting opamp input (+). As always, the golden rule of negative feedback applies, and the minus and plus inputs stay locked together. This is illustrated on the right of Figure 18.5 where the point between R_1 and R_2 (connected to minus) is shown to track V_{in}, which is the signal at plus.

Figure 18.5 Opamp noninverting configuration.

The analysis of the noninverting configuration proceeds along the same lines as the inverting case. The voltage at the minus input tracks V_{in} due the negative feedback provided by R1. This time V_{in} and V_{out} always move in the same direction, with the far end of R2 tied to 0V. Thus the result is a non-inverted output.

$$g = 1 + \frac{R_1}{R_2} \tag{18.4}$$

Notice that if R_1 becomes zero (a short circuit), then the gain equals one. In fact, the configuration shown in Figure 18.6 can be seen as a special case of the general noninverting opamp configuration where $R_1 = 0$ and $R_2 = \infty$ (i.e. a short circuit and an open circuit respectively). This arrangement is very commonly encountered, and is referred to as a voltage follower or unity gain buffer. With a gain of one, output simply equals input at all times (Eq. 18.5).

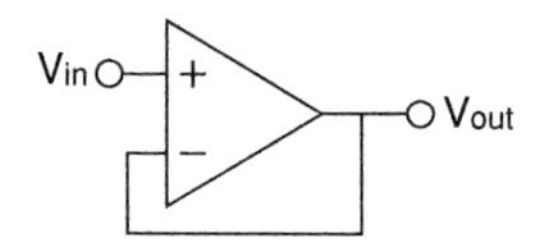

Figure 18.6 Opamp voltage follower (unity gain buffer) configuration.

$$V_{out} = V_{in} \tag{18.5}$$

The voltage follower is often encountered inserted between sections of a circuit. It would appear on first analysis to play no role in the operation of the wider circuit, as the output simply equals the input. What it does is to buffer one part of a circuit from another, while still passing signal from one side to the other. This isolation can be very useful, preventing unwanted interactions between the sides.

Single, Dual, and Quad Opamp Packages

Opamps come in a number of standard IC packages, containing one, two, or four opamp circuits on a single chip. While it is always good practice to check the pinout for a particular device, the arrangements shown in Figure 18.7 are unlikely to be deviated from. One consequence of these standardised footprints is that it facilitates drop-in replacements. While not always the case, it is often possible to swap one device for another of the same configuration. This can be done in order to achieve lower noise or better linearity for instance. A TL072 might be replaced by an NE5532 for example. These are both dual opamps (see Table 18.1), and thus share a common pinout.

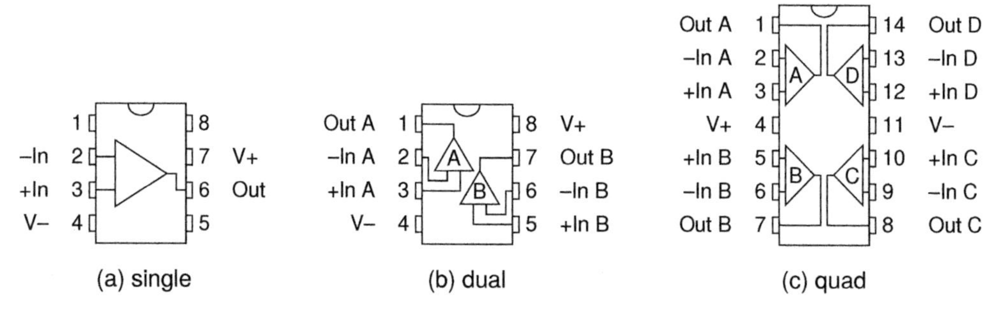

Figure 18.7 Standard pinouts for single, dual, and quad opamp ICs.

18.1 – Inverting Opamp Gain Equation

The inverting opamp gain formula shown in Eq. 18.3 can be derived quite simply. Consider R_1 and R_2 from Figure 18.4 as a voltage divider circuit, and recall that the high input impedance of the opamp means that it can be ignored in this analysis. The general case of the voltage divider rule or VDR (Eq. 12.4 from Chapter 12) is all the additional information that is needed. The steps required are as follows:

Identify the corresponding quantities in the VDR and the inverting opamp circuit (see table below) substitute the appropriate values into Eq. 12.4, and rearrange the resulting equation in order to arrive at the desired result. From Eq. 18.2 we know that $g = V_{out}/V_{in}$. We have expressions involving V_{in} and V_{out}, and we want to find g, so something equal to V_{out}/V_{in} is the expression to aim for.

Table 18.2: Inverting parameters

VDR		Opamp
R_1	$\Longrightarrow$	R_1
R_2	$\Longrightarrow$	R_2
V_a	$\Longrightarrow$	V_{out}
V_b	$\Longrightarrow$	V_{in}
V_{out}	$\Longrightarrow$	0V

$$V_{out} = (V_a - V_b)\frac{R_2}{R_1 + R_2} + V_b \qquad \text{Eq. 12.4 – VDR (general form)}$$

$$\Rightarrow \quad 0 = (V_{out} - V_{in})\frac{R_2}{R_1 + R_2} + V_{in} \qquad \text{Substitute from table}$$

$$\Rightarrow \quad \frac{V_{out}}{V_{in}} = \frac{\left(\frac{R_2}{R_1+R_2} - 1\right)}{\frac{R_2}{R_1+R_2}} \qquad \text{Rearrange to make } V_{out}/V_{in} \text{ subject}$$

$$= -\frac{R_1}{R_2} \qquad \text{Simplify}$$

$$\Rightarrow \quad g = -\frac{R_1}{R_2} \qquad \text{Eq. 18.3 as required}$$

A full expansion of this derivation would be a bit longwinded for the current brief examination. If your maths is rusty, the 'rearrange and simplify' steps may seem a little tricky. Just be methodical in multiplying out and simplifying fractions, and you will get there.

18.2 – Noninverting Opamp Gain Equation

The noninverting opamp gain formula shown in Eq. 18.4 can be derived in a similar fashion to the inverting version. Once again consider R_1 and R_2 as a voltage divider circuit, and apply the general case of the voltage divider rule or VDR (Eq. 12.4 from Chapter 12).

Identify the corresponding quantities in the VDR and the inverting opamp circuit (see table below) substitute the appropriate values into Eq. 12.4, and rearrange the resulting equation in order to arrive at the desired result. From Eq. 18.2 we know that $g = V_{out}/V_{in}$, so this is the form to aim for.

Table 18.3: Noninverting parameters

VDR		Opamp
R_1	$\Longrightarrow$	R_1
R_2	$\Longrightarrow$	R_2
V_a	$\Longrightarrow$	V_{out}
V_b	$\Longrightarrow$	0V
V_{out}	$\Longrightarrow$	V_{in}

$$V_{out} = (V_a - V_b)\frac{R_2}{R_1 + R_2} + V_b \qquad \text{Eq. 12.4 – VDR (general form)}$$

$$\Rightarrow \quad V_{in} = (V_{out} - 0)\frac{R_2}{R_1 + R_2} + 0 \qquad \text{Substitute from table}$$

$$\Rightarrow \quad \frac{V_{out}}{V_{in}} = \frac{R_1 + R_2}{R_2} \qquad \text{Rearrange to make } V_{out}/V_{in} \text{ subject}$$

$$= 1 + \frac{R_1}{R_2} \qquad \text{Simplify}$$

$$\Rightarrow \quad g = 1 + \frac{R_1}{R_2} \qquad \text{Eq. 18.4 as required}$$

Some circuits rely on specific characteristics of the opamps they use, which may not be shared by potential replacement parts, so this procedure will not always work, but it is a common practice in many circuits, where better (or just different) performance is sought.

The unlabelled pins on the single opamp pinout (1, 5, and 8) are used in different ways on different devices, but are often left unconnected, and so drop-in replacements may still work, but it is a good idea to examine these pins in order to spot potential pitfalls before attempting a swap.

LEARNING BY DOING 18.1

OPAMP BASED PREAMP CIRCUIT

A simple but effective preamp circuit is shown below, built around a single opamp IC. The part shown is the LF356N but any standard opamp will work here. This design represents a useful building block in a wide range of audio circuits.

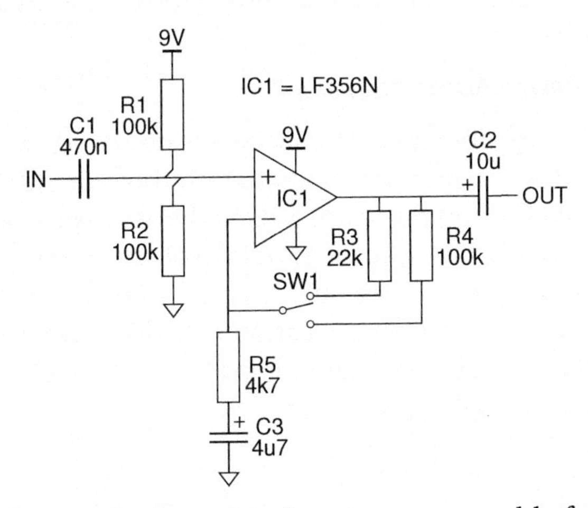

The voltage divider on the input has been encountered before. The JFET buffer from Chapter 17 utilises very much the same input circuit. The arrangement biases the input up to half the supply voltage, so as to accommodate an input signal swing above and below the centreline.

The circuit adopts the noninverting configuration presented earlier. The spdt switch SW1 provides for two different gain levels. The noninverting gain equation, $g = 1 + R1/R2$, tells us the gain levels available with the switch in each position.

$$g = 1 + \frac{22k}{4k7} \approx 5.7 \qquad \text{(which in decibels equals: } 20\log(5.7) = 15\text{dB)}$$

$$g = 1 + \frac{100k}{4k7} \approx 22.3 \qquad \text{(which in decibels equals: } 20\log(22.3) = 27\text{dB)}$$

All three capacitors primary functions are to accommodate the single rail opamp power supply setup. With a split supply (±9V) ground becomes the midpoint, and no offset biassing is needed, but with the opamp's negative rail tied to ground, its operating point need to be elevated to the midpoint of the power supply. Thus the input and output capacitors C1 and C2 are needed to shift the levels of signals as they enter and leave, and C3 allows the gain network to float up to the bias point set by R1 and R2. In general, a split supply simplifies a circuit but providing the split supply itself is more complex, and so often a single rail supply is the arrangement chosen.

Design a layout (or refer to Appendix D), gather the necessary components, build and test. As ever a function generator and an oscilloscope are a good way to test such a circuit before attaching it to a power amp and sending audio through it.

IC Power Amps

Many IC power amps exist, but the most commonly encountered for low power audio work has got to be the 386. Others are well worth investigating if a little more power is required. The range of 386 chips available top out at about 1½W, and distortion levels are getting pretty high by then. As a simple lo-fi utility amp however, the 386 is hard to beat. As such it is the device considered here, and the one used in numerous other places throughout this book.

LM386/JRC386 Low-Power Audio Amplifier

The 386 low-power audio amplifier Ic can be encountered in a number of places throughout this book. It is a workhorse of audio electronics builders and very many circuits have been designed around this chip. Several have already been encountered and Figure 18.8 adds to the possibilities, presenting two of the simplest possible power amp circuits. The bare-bones amp in (a) provides a signal gain of 20, which can be plenty in many situations, but for a bit more amplification (b) adds a gain switch between pins 1 and 8. When the switch is closed, the gain jumps from 20 to 200. The capacitor on pin 7 helps stabilise the amp at higher gains.

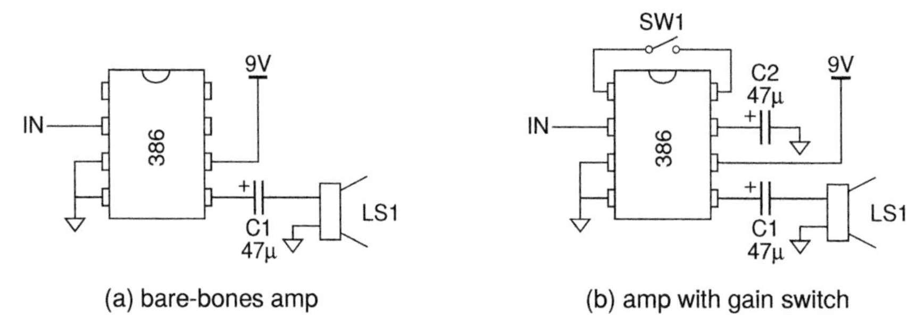

Figure 18.8 Two very simple 386 based amplifier circuits.

LM317 Voltage Regulator

Using a voltage regulator provides a number of advantages over an unregulated power supply. Voltage regulators can eliminate noise, and provide output voltages which are both precise and stable, even with a poorly defined, noisy input voltage prone to drift. There are many options available, both fixed and programmable. Fixed variants are simple to use but provide only one possible regulated output voltage.

Here the LM317 programmable voltage regulator is considered. For good regulation the input voltage needs to be at least 3V above the output. In other word, if a 9V input is used, a regulated output up to 6V can be reliably generated. (The minimum output level is 1.25V, regardless of input.) The input voltage can be as high as 40V, which means regulated outputs between 1.25V and 37V are possible. The output current can be as high as 1.5A, but this kind of level will certainly require a heat sink. Lighter loads may not require heat sinking – if the power dissipated by the regulator ($(V_{in} - V_{out}) \times I_{out}$) does not exceed 1/4W then no heat sink is required.

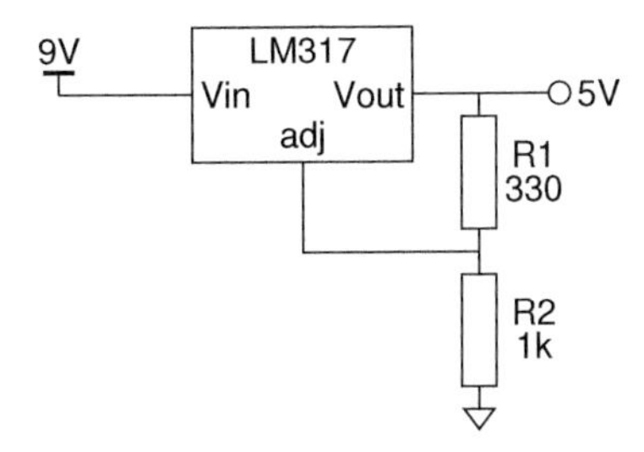

Figure 18.9 How to configure an LM317 voltage regulator.

Figure 18.9 shows the basic configuration used to set the output voltage of an LM317. R1 is typically kept between about 120Ω and 1k2Ω, with 240Ω being a commonly used value. Once R1 has been selected, R2 can be calculated using the expression shown in Eq. 18.6.

$$V_{out} = 1.25 \times \left(1 + \frac{R2}{R1}\right) \tag{18.6}$$

Since usually only a limited number of resistor values will be available, it can take a bit of work to find a pair of available values which give a result close enough to the target level. The values shown here in Figure 18.9 (R1 = 330 and R2 = 1k) are very commonly available resistor values, and give a result very close to 5V. If a very particular output voltage is needed, R2 can be replaced with a trim pot, or alternatively a regular pot can be used to provide a user variable output voltage level.

Learning by Doing 18.2

Voltage Regulator

This exercise presents a useful building block circuit expanding on the basic configuration shown in Figure 18.9. The added capacitors can help further smooth and regulate the output. Various different capacitor combinations can be encountered in different applications. The LM317 data sheet provides useful guidance as to when the various capacitors might be useful.

Build the circuit shown (with or without capacitors), and measure the output voltage level. It should be very close to 5V with the resistor values shown. Increase the input voltage to 15V. The output level should remain steady.

With $V_{in} = 15V$, the output can be set at anything up to 12V, as described above. Let's say we want an output of 10V, and we have a full range of E6 resistor values available (multiples of: 10, 15, 22, 33, 47, 68). If we decide to limit the value of R1 to be one of 220, 330, or 470 in line with the notes above, we can calculate the best single resistor value for R2 in each case, and determine which resulting pair provides the best match.

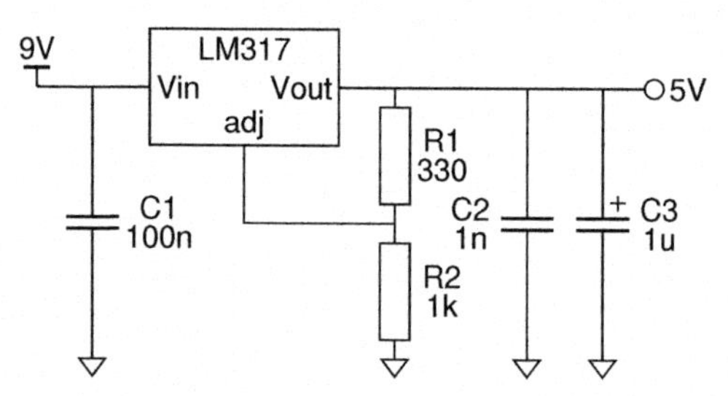

Step 1: Rearrange the LM317 equation to make R2 the subject, and insert the required output voltage level of 10V

$$
\begin{aligned}
V_{out} &= 1.25 \times \left(1 + \frac{R2}{R1}\right) && \text{LM317 equation} \\
\Rightarrow \quad R2 &= R1 \times \left(\frac{V_{out}}{1.25} - 1\right) && \text{Rearrange} \\
&= R1 \times \left(\frac{10}{1.25} - 1\right) && \text{Insert required voltage} \\
&= R1 \times 7 && \text{Simplify}
\end{aligned}
$$

Step 2: Calculate exact values for R2, for each possible R1

$R2 = R1 \times 7 = 220 \times 7 = 1540$

$R2 = R1 \times 7 = 330 \times 7 = 2310$

$R2 = R1 \times 7 = 470 \times 7 = 3290$

Step 3: Select closest E6 values for R2

$R2 = 1k5\Omega$

$R2 = 2k2\Omega$

$R2 = 3k3\Omega$

Step 4: Calculate resulting V_{out} values and select the best

$$V_{out} = 1.25 \times \left(1 + \frac{R2}{R1}\right) = 1.25 \times \left(1 + \frac{1k5}{220}\right) = 9.77V$$

$$V_{out} = 1.25 \times \left(1 + \frac{R2}{R1}\right) = 1.25 \times \left(1 + \frac{2k2}{330}\right) = 9.58V$$

$$V_{out} = 1.25 \times \left(1 + \frac{R2}{R1}\right) = 1.25 \times \left(1 + \frac{3k3}{470}\right) = 10.03V$$

Clearly the final pair gives the result closest to the target value of 10V (just 0.03V out), and so R1 = 470 and R2 = 3k3 looks like a good pair in this case. Replace the resistors in your circuit and measure the new output voltage. It should match the calculated value closely.

Now follow the same procedure to come up with the best pair of resistors for V_{out} = 6.3V (this happens to be a value commonly used for vacuum tube heaters). Again, replace the resistors and confirm that the output level is as expected.

If the heaters connected to such a 6.3V supply draw 300mA, and V_{in} = 9V is used as the voltage regulator's input voltage, will a heat sink be needed on the LM317? (The necessary equation is given in the main text.)

NE555 Precision Timer

The 555 timer has long been a favourite with audio circuit builders. This chip can be used to make simple and versatile oscillator circuits. In Figure 18.10 the pinout for the 55 can be seen, alongside its big brother, the 556. This second chip is just two 555 timer circuits packaged up into one IC.

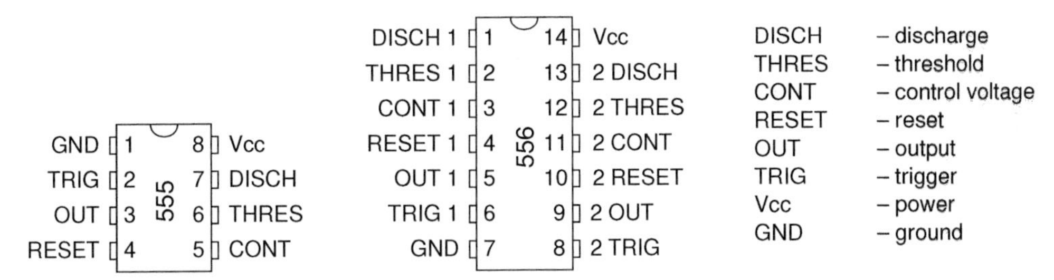

Figure 18.10 Pinouts for the 555 timer and the 556 dual timer.

A wide range of circuits can be found in Mims (1996) including the stepped-tone generator, a variation of which can be seen in Figure 18.11. This circuit is examined in detail in one of the projects presented in Chapter 11.

4000 Series ICs

The 4000 series consists of a large collection of utility ICs, mainly intended for building digital logic control oriented circuitry. Many have been pressed into service in audio circuits, in various imaginative ways. Often the first to be encountered in this context is the CD40106 hex Schmitt trigger inverter, which with the addition of only a resistor and a capacitor, can be used to build an oscillator (six oscillators potentially, since the 'hex' in the name indicates that the IC contains six inverters). The practical exercise below explores a similar circuit which uses NAND gates (courtesy of the CD4093) rather than inverters. An excellent source of ideas for 4000 series circuits is Collins (2009).

Table 18.4 lists a small selection of 4000 series ICs which can be particularly useful in making interesting audio circuits. A brief note on some potential uses for each are provided below, but the possibilities are endless.

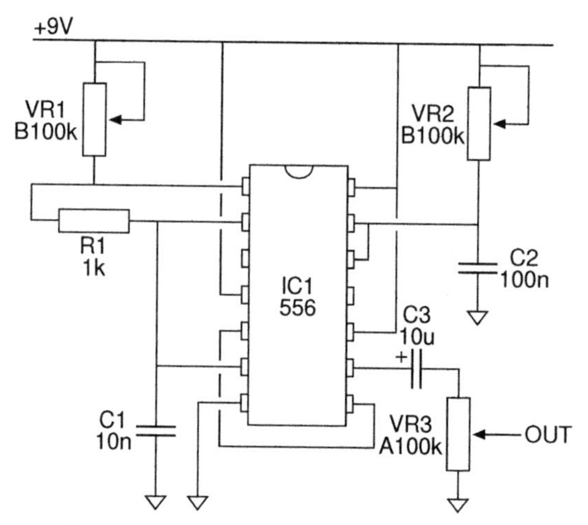

Figure 18.11 The stepped-tone generator circuit from Chapter 11 is a very popular 555 based circuit. Here it is built using a 556, but two 555s can also be used.

Table 18.4 A small selection of interesting 4000 series ICs

Part	Package	Description
CD4040B	DIP16	12 bit binary counter
CD4051B	DIP16	single 8 channel multiplexer
CD4053B	DIP16	triple 2 channel multiplexer
CD4066B	DIP14	quad bilateral (spst) switch
CD4069UB	DIP14	hex inverter (unbuffered)
CD4070B	DIP14	quad 2 input XOR (exclusive OR)
CD4093B	DIP14	quad Schmitt trigger 2 input NAND
CD40106B	DIP14	hex Schmitt trigger inverter

The **CD4040** is a 12 bit binary counter. Each time the counter's input goes from high to low, the 12 bit binary count goes up by one. This doesn't sound much use until you realise that a 12 bit binary number counting up looks like twelve square wave, each with a frequency half that of the one before. So putting an LFO into the input generates twelve synchronised square waves on the twelve outputs. These can be used directly as audio signals, or they can be used as control signals. They are often used in this second way, fed into the control inputs of the 4051 multiplexer (see below).

The **CD4051** is a single 8 channel multiplexer. This is basically a digitally controlled single pole eight throw (sp8t) switch. Three binary digits let you count to eight, so three digital inputs control the switch, selecting which throw is currently connected. Pick three of the twelve square waves coming out of the 4040 above, and you get a nice repeating switching pattern jumping around the multiplexer channels.

The **CD4053** is a triple 2 channel multiplexer, three independent spdt switches, controlled by digital inputs. Where a standard spdt switch appears in a circuit, this chip might be used to replace it with an electronic switch, perhaps controlled from an arduino or similar microcontroller, or depending on what the switch does, driven from a simple square wave oscillator.

The **CD4066** is a quad bilateral (spst) switch. This one has been widely used for implementing electronically controlled bypass switching on effects circuits, but again, wherever an spst switch is called for one of these might be considered for some automated control.

The **CD4069** is an unbuffered hex inverter which can be found used in booster and distortion circuits. As with many of the applications suggested here, this is somewhat outside the intended application of this device. Several interesting applications can be found in the 'Circuit Snippets' collection which has been referenced several time throughout this book.

The **CD4070** is a quad two input XOR (exclusive OR). This gives a high output if and only if one input is high and the other is low; if both are low or both are high, the output is low. This chip is used to good effect to build a 'digital ring modulator' in Penfold (1986).

The **CD4093** is a quad Schmitt trigger two input NAND. This is the chip used here in Learning by Doing 18.3 to build a couple of cascaded oscillators. The Schmitt trigger bit of the name means that the switching of these gates does not happen symmetrically about the mid voltage level. Instead, inputs moving low to high need to go higher, and inputs moving high to low need to go lower, before switching happens. This is a common situation, used to eliminate unwanted switching with noisy inputs. This same property makes these Schmitt trigger variants (both this and the 40106 below) better options than non-Schmitt trigger types for building oscillators.

The **CD40106** is a hex Schmitt trigger inverter. Inverters are also called NOT gates; the output is NOT the input, i.e. it is inverted. The 40106 is very commonly encountered used to build simple oscillators. This is done in the same basic way as with the

4093, used in the exercise below, needing only a resistor and a capacitor in order to make a working oscillator.

LEARNING BY DOING 18.3

NAND GATE OSCILLATOR

Theory

This circuit is composed of two oscillators with the first modulating the second, in order to create a more interesting sound palette. Consider the first oscillator, built around IC1a. When power is applied to the circuit, input A (tied to 9V) goes high, while input B starts out low. The truth tables for AND and NAND gates are shown below, telling you what output to expect for any given combination of inputs. Our starting position is shown in row 3 (IN A = 1, IN B = 0), so initially the output of the first NAND gate is a high voltage.

Table 18.5: AND and NAND gate truth table

In A	In B	Out (AND)	Out (NAND)
0	0	0	1
0	1	0	1
1	0	0	1
1	1	1	0

However, notice that the variable resistor VR1 connects this high voltage back to input B. This causes the capacitor C1 to start charging up, and thus the voltage at input B starts to rise. When this voltage reaches the switching level IN B switches form 0 to 1, and from the truth table we can see that the output will then go low. Now IN B is at a higher voltage than OUT, and so the capacitor starts to discharge back through VR1 until once again IN B reaches its switching level and transitions from 1 back to 0, returning to the original situation and causing OUT to once again return to 1. And so the cycle repeats – an oscillator.

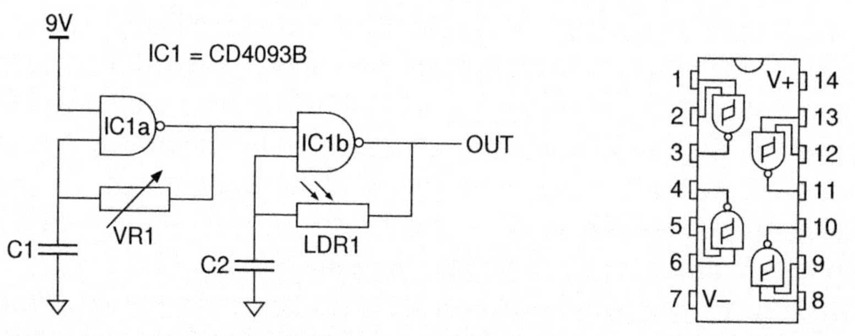

The frequency at which the oscillator oscillates is controlled by two factors: the size of the resistor (a larger resistor slows the charging process and vice versa), and the size of the capacitor (a larger capacitor takes more charge to fill it and likewise viceversa). Thus making the capacitor or the resistor larger reduces the frequency, and making either smaller increases the frequency.

Practice

First just build one oscillator. Experiment with different sizes for the pot and the capacitor, to get a good working frequency range (check the output on an oscilloscope and listen to it through an amp). In the final circuit you can make this oscillator run either at audio frequencies, or set it lower so that it acts as an LFO. Both options can produce interesting results.

Once oscillator one is up and running, cascade in number two. Different control mechanisms are possible. The circuit here shows an LDR as the resistive element in oscillator two, but a pot can be used here too. Another good option is touch control – clamp two coins with clip leads and push your fingers on them to make a variable resistance. You'll probably want a smaller capacitor for this final option to compensate for the high resistance.

Analog Optocouplers

Optocouplers (aka opto-isolators) come in a number of different forms, but usually involve an LED shining on a light sensitive device of some kind. In the two symbols illustrated in Figure 18.12 the sensors indicated are a photoresistor and a phototransistor respectively – other possibilities also exist. Optocouplers provide electrical isolation between their two sides; while this is their primary intended application, they can also add interesting distortion to an audio signal, as well as providing other novel control options.

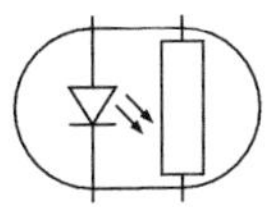

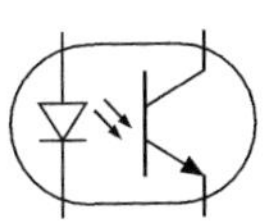

Figure 18.12 Symbols for photoresistor and phototransistor based optocouplers.

The two names given above highlight these devices primary characteristics; '-coupler' because they can pass a signal from one side to the other, '-isolator' because they provide electrical isolation between the two sides while doing so, and of course the 'opto-' part indicates that the way all this is achieved is through the use of light.

A signal entering on one side of the device causes the LED within to shine brighter and dimmer, in time with the variations in the signal. This varying light level causes the detector element on the other side (photoresistor, phototransistor, or whatever) to adjust its output in time with the varying light. Much of the distortion which is introduced (most especially in the photoresistor variety) is due to the slow response such devices have to changes in light level. It takes a little time for the resistance of a photoresistor to change in response to the changing intensity of the light falling upon it. This can for instance result in a satisfying softness in the response of compressors and limiters built using these devices.

Commercial optocouplers most often combine an LED with a phototransistor, and this configuration certainly merits experimentation also, but the easiest approach to a homemade optocoupler uses a photoresistor (or LDR) as the detector. This is the approach used in the exercise below.

LEARNING BY DOING 18.4

HOMEMADE OPTOCOUPLER

Materials

- LED
- LDR – light dependant resistor
- Heat shrink tubing

Theory

Many optocouplers are designed for use in digital systems, where binary 'hi-lo' rather than continuously varying analog signals are to be transmitted. As such, these devices are unlikely to provide much useful function in an analog audio circuit. Analog optocouplers are less common (although by no means unheard of), but in any case it is a simple matter to construct an optocoupler from an LED, an LDR, and a bit of heat shrink tubing. Simply place the two devices end to end as shown below, slide the heat shrink over the top and apply heat. Some extra narrow heat shrink around the component legs can prevent short circuits. Be sure that you can identify the anode and cathode of the diode. It is important to connect it up the right way round. The LDR leads can be connected in either order.

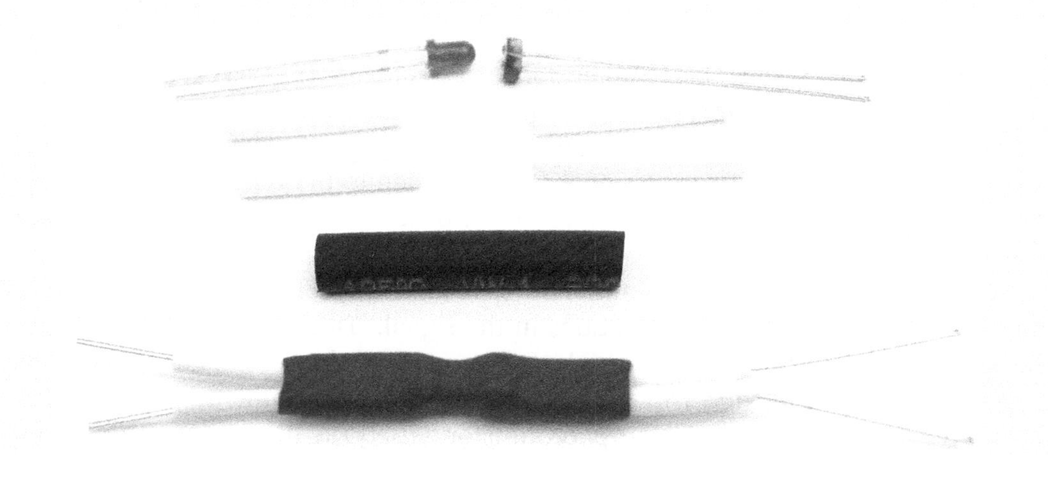

Practice

Once you have an optocoupler to play with there are a number of experiments which you might like to try. Some interesting applications of this device can be found in Collins (2009, chapter 21). The Uglyface from Escobedo's 'Circuit Snippets' collection provides another interesting application where this might be tried. The illustrations below outline two basic ways in which an optocoupler can be integrated into a circuit. In the scheme on the left, an input signal is used to modulate the LED and this signal transfers across to the other side by setting up a voltage divider where the top arm is the variable resistance of the LDR. More involved interfacing methods may be tried, but this simple approach produces some nice results.

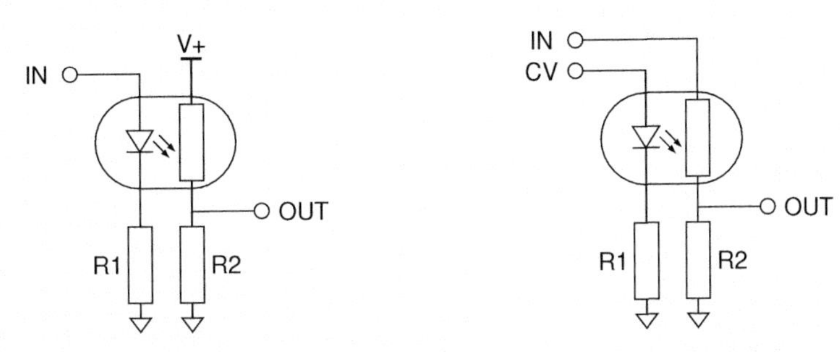

In the second scheme, illustrated on the right, two signals are used. A control voltage modulates the LED, and this in turn varies the amount of the input signal making it through the voltage divider to the output. Try attaching an LFO to the CV input (perhaps a 4093 or 40106 based oscillator, as described previously), to implement a kind of tremolo or chopper effect.

REFERENCES

G. Clayton and S. Winder. *Operational Amplifiers*. Newnes, 5th edition, 2003.

N. Collins. *Handmade Electronic Music*. Routledge, 2nd edition, 2009.

W. Jung, editor. *Op Amp Applications Handbook*. Newnes, 2005.

F. Mims. *Engineer's Mini-Notebook: 555 Timer IC Circuits*. Radio Shack, 3rd edition, 1996.

R. Penfold. *More Advanced Electronic Music Projects*. Bernard Babani, 1986.

R. Severns, editor. *MOSPOWER Applications Handbook*. Siliconix Incorporated, 1984.

TI. *The TTL Data Book*. Texas Instruments, 5th edition, 1982.

R. Widlar et al. *Linear Applications Handbook*. National Semiconductor, 1994.

19 | Vacuum Tubes

Tubes provide the same kind of functionality as diodes and transistors

Vacuum tubes (aka thermionic valves) were the forerunners of semiconductor devices such as diodes and transistors. There are quite a few different types of vacuum tubes but the discussion here is restricted to the commonest two variants, diode tubes, and triode tubes. These two types in particular correspond closely to the functionality of semiconductor diodes and transistors respectively.

The semiconductor alternatives have many advantages over their vacuum tube forebears – size, robustness, reliability, power consumption, and linearity of operation come to mind. However, for some applications – and to some ears – the different character which vacuum tubes can impart to an audio signal makes their sound a valuable tool in the audio electronics arsenal.

Principles of Operation

The goal of any electronic component is to exercise some control over how electricity flows in a circuit. Vacuum tubes achieve this control by virtue of the fact that heat can encourage electrons to leave the surface of a piece of metal and fly around in the surrounding space.

The basic principles upon which vacuum tubes operate are quite straightforward. These principles are illustrated in Figure 19.1. As a piece of metal is heated up, the electrons in the metal are given more energy. As such they fly around more vigorously, and as a result some electrons are able leave the surface of the metal and fly into space.

If it is just a piece of metal sitting on the desk, they do not fly far before returning (various factors keeping them close by). If this heated piece of metal is enclosed inside an evacuated glass envelope (i.e. a vacuum tube), then one of the factors keeping the electrons close – the surrounding air molecules – is removed. The electrons can then stray a little farther, but still tend to return to base on account of their negative charge and the positive charge they leave behind – recall that opposites attract.

In order to make a diode tube from this starting point all that is needed is to insert a second piece of metal (aka electrode) into the evacuated glass envelope, and provide

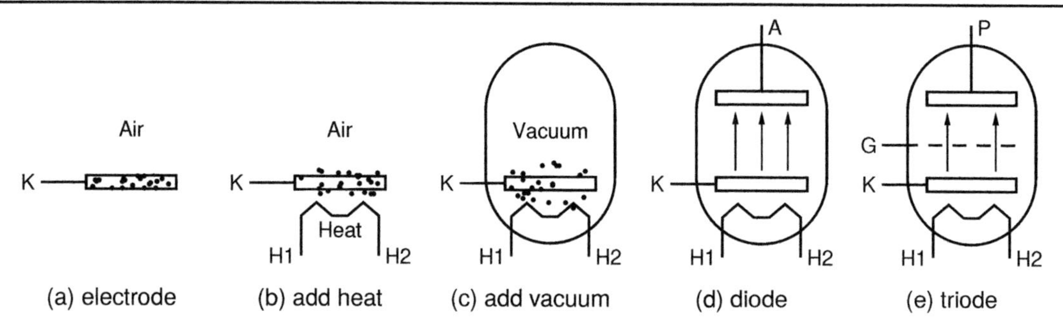

Figure 19.1 How vacuum tubes work. Start with an electrode (the cathode – K). Heat it with a heater (H1 and H2), and place it in an evacuated glass envelope (the vacuum tube). Add a second electrode (the anode – A, or plate – P) to make a diode. Add a third electrode for control (the grid – G) to make a triode.

electrical connections to the two metal electrodes (Figure 19.1d). Now if a positive charge is applied to the second electrode then the electrons flying around in the vacuum see this positive charge as an inviting alternative destination, and some of them move across the gap between the two pieces of metal and fly headlong into the second electrode. This movement of electrons constitutes an electric current.

Note that it is only a one way process, because the second piece of metal is not being heated like the first, and so electrons are not nearly so likely to fly off the surface and into the vacuum between electrodes. Therefore if the second electrode is made more negative than the first, instead of being more positive, then no electrons cross the gap and no current flows.

This is exactly the action of a diode as described in Chapter 16 – diodes let current flow in one direction but not the other. The two terminals of a diode are commonly referred to as the cathode (the heated electrode where electrons emerge into the vacuum), and the anode (which can absorb electrons coming its way, but does not emit any).

The diode (or rectifier) is a very useful electronic component, but in order to arrive at a device in which one signal can be used in order to control another, a third electrode must be added to the mix, as in Figure 19.1e. This 'control' terminal is referred to as the grid, and with its addition the diode becomes a triode. The grid can be thought of as the means by which the valve is opened and closed. The valve is still only a one way street as before, but now the magnitude of the current flow in that forward direction can be controlled. This should be a very familiar basic scenario from the description of the action of transistors in Chapter 17.

The discussion presented above clearly illustrates the origin of the two general names for devices of this type: 'vacuum tube' because there is a glass tube with a vacuum inside, and 'thermionic valve' because heat is used to facilitate current flow.

Diode Tubes

The same one-line description which is used in Chapter 16 to characterise the semiconductor diode can also be seen as a valid description of the tube version of this component – diodes let current flow in one direction but not the other. When talking about semiconductor diodes, the next most important aspect of their behaviour is the forward bias voltage drop. Recall that a typical silicon diode requires approximately 0.6V to 0.7V across its terminals before it will start to conduct. This fact is used to good effect in the diode clipping circuits discussed in Chapter 16.

This forward bias voltage is required in order to overcome the depletion layer which develops at the p-n junction within these devices, as described in Chapter 3. One important difference between semiconductor and tube based diodes is that the tube diode requires no such bias voltage before it starts to conduct. As illustrated in the graph in Figure 19.2, the tube diode's forward current I_d starts to flow as soon as the voltage V_d rises above zero. As should be obvious from the graph, the relationship, although not very far from a straight line, is still decidedly nonlinear. As such a tube diode will impart some level of distortion onto any signal passing through it.

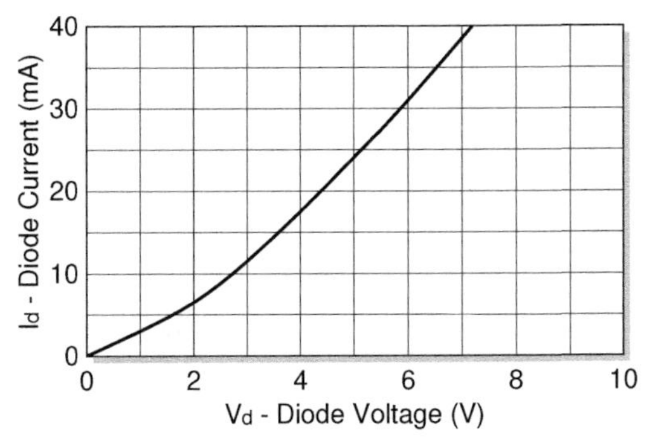

Figure 19.2 Typical diode valve i-v characteristic curve. This curve is for the EB91 dual diode, as pictured in Figure 19.6a, but is representative of the behaviour which can be expected from most diode tubes.

In common with their semiconductor based brethren, tube diodes are fundamentally two terminals devices, with an anode and a cathode at their positive and negative ends respectively. That being said, unlike semiconductor diodes, practical tube diodes have some additional connections. Typical circuit symbols used to indicate a diode tube in a schematic diagram are shown in Figure 19.3. As is often the case with such symbols, variations on the style illustrated here are regularly encountered, but the basic format is usually fairly close.

As described above, all tubes require heating, and the heater connections may or may not be indicated on any given symbol, as is highlighted in the two variants shown here. The extra connections shown on the bottom of the symbol on the left represent the heater. See the discussion around Figure 7.6 (p. 99) for more on circuit diagram conventions surrounding the inclusion of vacuum tube heater connections.

Figure 19.3 Circuit symbols for a diode (aka rectifier) valve.

When using tubes it is quite important to observe the correct supply voltage levels for heater connections. Running the heaters too low is likely to compromise the effective operation of the valves, although is generally not a major problem. On the other hand, supplying the heaters with a voltage in excess of that in the manufacturer's specifications is likely to significantly shorten a tube's working life. Notice the very specific heater voltages quoted in Figure 19.4 and Figure 19.5. The level specified in both cases here (6.3V) is a common, but by no means universal heater voltage level.

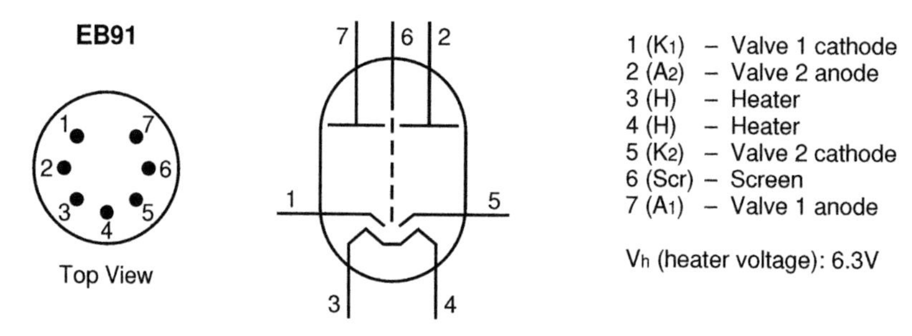

Figure 19.4 Pinout for an EB91 dual diode vacuum tube.

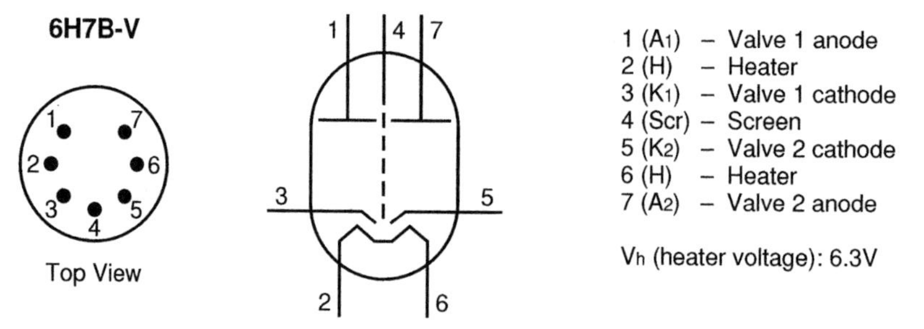

Figure 19.5 Pinout for a 6H7B-V dual diode vacuum tube.

The two devices whose details are provided in Figure 19.4 and Figure 19.5 are both what are referred to as dual diode vacuum tubes. This means that there are in fact two separate diode valves contained in the one device. as is often the case with such devices, a screen connection is provided, which is designed to isolate the two diodes from one another, so as to minimise the chances of crosstalk contamination allowing the signal in one diode to leak across into the other. This screen connection is usually bonded to the circuits ground plane.

The pair of connection points at the bottom of each diagram are for the heater, and the final two pairs represent the anode and cathode connections for each diode. Notice that the pin numbering is completely different between the two devices. They both have the same seven connections, but the ordering changes. It is always important to check the pinout for the specific device which you are using.

These two devices are also physically very different, as can be seen in Figure 19.6. The EB91 is a much broader tube and its pins are short and solid, designed to be plugged directly into a suitable tube socket. The 6H7B-V on the other hand is a very slim tube, and provides connections on much longer, flexible wires designed to be soldered directly into a circuit.

(a) EB91　　(b) 6H7B-V

Figure 19.6 Two commonly found dual diode vacuum tubes.

In terms of circuits to experiment with, these devices can usefully be tried in any of the circuits provided for semiconductor diodes in Chapter 16 and elsewhere. Extra circuitry is needed in order to accommodate the required heater supply, and the screen should probably be grounded as described above, but otherwise these devices offer an interesting alternative to the more commonplace semiconductor approach. In particular, anywhere that diodes have been used explicitly for their clipping and distortion attributes, such a substitution might provide an interesting alternative.

Triode Tubes

As is the case with transistors, triodes can be used to amplify, buffer, and switch. The example characteristic curves shown in Figure 19.7, although quite different from the equivalent transistor curves given previously in Chapter 17, do share many general characteristics in common with those transistor curves.

The transfer characteristic in Figure 19.7a is most reminiscent of that of the JFET with its negative gate-source voltage range. Here a negative grid voltage takes its place, with the anode current reducing the more negative the grid voltage becomes.

The curves of the i-v characteristic in Figure 19.7b rise in a rather different fashion to their transistor cousins, and do not level off in the same way. These differences notwithstanding, the general behaviour of the family of graphs plotted here does represent a

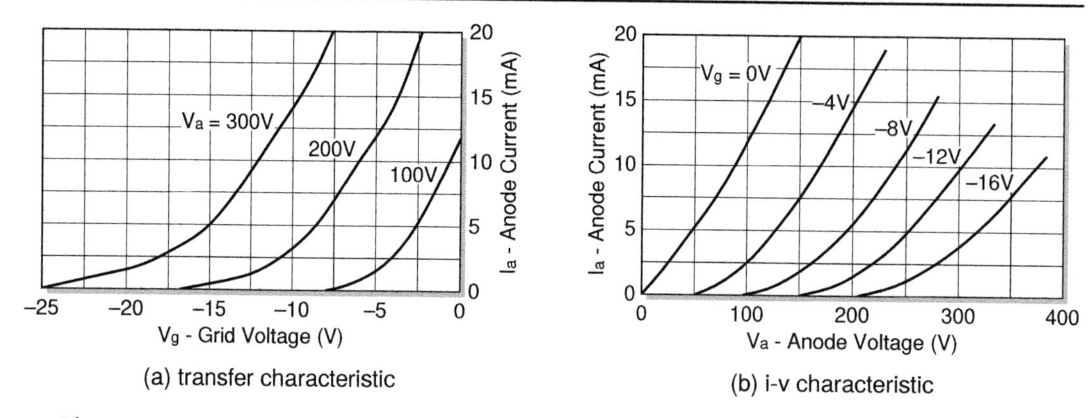

Figure 19.7 Typical triode valve characteristic curves. Those shown are for a 12AU7 dual triode valve (pictured in Figure 19.9).

kind of control over the signals passing through the devices similar to that provided by transistors.

In fact, the most striking way in which these curves depart from the transistor's characteristic curves is in the anode voltage levels quoted. While the maximum equivalent transistor voltages are typically in the low double digits, here the anode voltage is shown going up to 300 and even 400V, about twenty times the transistor maximums. This reflects a commonly recognised fact that tube amps tend to have some very high voltages knocking around inside them, which should be viewed with caution. This chapter steers clear of such hazards, and instead places its focus on a few low voltage possibilities for these devices, as described later.

Triode tubes are represented in schematic diagrams using symbols such as those shown in Figure 19.8. As was the case with the diode tube symbols introduced earlier, two variants are illustrated, one with and one without the requisite heater connections shown. As before, since many triodes come as a pair of valves in a single tube, and usually one set of heater connections can serve to heat both valves, it is often the case that one of each of the two symbols shown will appear in a schematic diagram, so as not to duplicate the required connections. The diagram for the tube booster circuit below illustrates this convention well.

Figure 19.8 Circuit symbol for a triode valve.

As mentioned, triode tubes most often come in the form of a dual triode device, with two independent triodes housed in the same glass envelope. This is the case with the 12AU7 (aka ECC82) which is the triode illustrated in Figure 19.9. As is the case with most dual triodes, this tube has nine pins. Six give the grid, anode (or plate), and cathode of the two triodes (corresponding to the base, collector, and emitter of a BJT). The final three pins connect to the heater.

Figure 19.9 The 12AU7 (aka ECC82) dual triode vacuum tube.

Tube heater voltages are usually quite low. As indicated in Figure 19.10, the 12AU7 heater is designed to operate at either 12.6V or 6.3V, depending on how it is wired. Pin nine is a centre tap, with pins four and five connecting to either end of the heater. To use 12.6V connections should be made between pins four and five. To use 6.3V connections should be made between pin nine and either or both of pins four and five. The side of the heater to pin four services one valve, while the side to pin five services the other, as shown in Figure 19.10. It is far preferable to under supply the heater than it is to over supply it (as over voltage will lead to premature failure). Thus if for instance a nine volt supply is to be used to run the heater it should be connected between pins four and five, leaving pin nine unconnected.

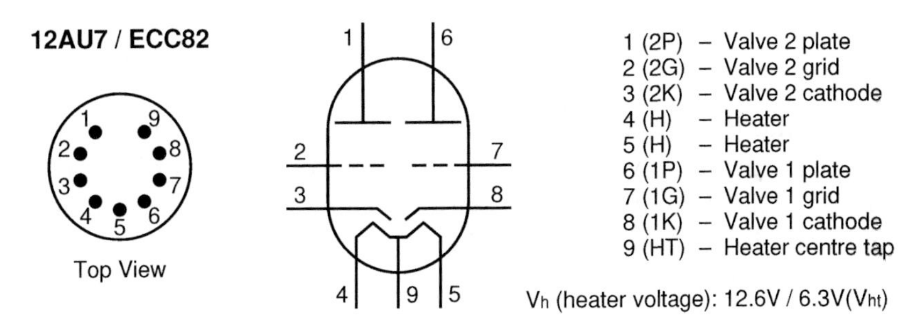

Figure 19.10 Pinout for a 12AU7 dual triode vacuum tube.

As referred to above, tubes themselves usually operate at very much higher plate voltages than the heater voltages just discussed, typically two or three hundred volts. It is however possible to squeeze some life out of a tube at much lower voltages. The material presented here deals only with these low voltage tube circuit designs. If venturing into higher voltage tube circuits tread carefully.

Low Voltage Tube Circuits

Standard (high voltage) tube circuits (most commonly seen today in guitar amps and other audio amplifiers) are a vast and complex subject area (see Jones, 2004; Pittman, 2004; Valley et al., 1948, for example). High voltages are involved, often in excess of the mains voltage powering the amp. Furthermore even the most basic tube amp kit is likely to cost at least several hundred euros, including tubes, power and output transformers, and various high voltage and high power handling components. In the spirit of the lo-fi, low cost audio electronics which this book embraces, such projects are left to others. It is however possible to have some fun with tubes running at low voltages too; definitely not where they were designed to operate but some life can still be coaxed out of them.

In the section on diode tubes above some suggestions are given as to experiments which can be conducted replacing semiconductor diodes in circuits with their vacuum tube counterparts. Perhaps no diode tubes are readily available but triode tubes such as the 12AU7 come easily to hand. If this is the case (and in similar fashion to the 2N7000 'MOSFET as a diode' hack presented in Chapter 17) it is a simple matter to re-task a triode tube as a diode (Figure 19.11).

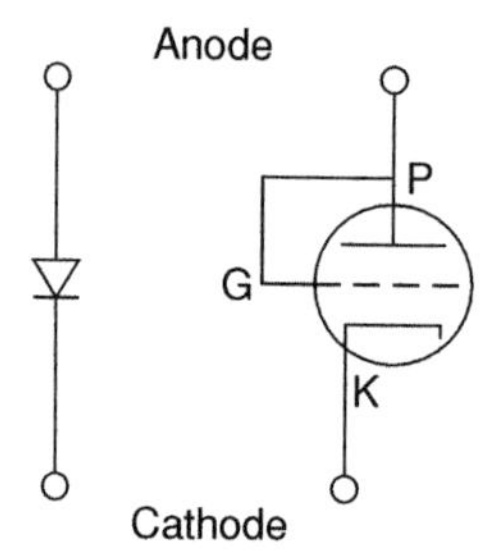

Figure 19.11 How to wire a triode valve to operate as a diode.

Clearly, the cathode must remain as the cathode, since the heater makes this the only source of electrons in the device. Various different connections of the remaining two terminals are possible (five in fact), but wiring together the grid and plate to form the anode of the diode is the best option. Other arrangements, such as, grid and cathode wired together can also be found suggested in various places, however the wiring shown in Figure 19.11 provides the best performance.

Learning by Doing 19.1

Tube Booster

To finish off this chapter on vacuum tubes a complete circuit is given here as a good starting point into the possibilities of building low voltage tube circuits. The circuit shown is a simple booster running off nine volts. Since the heater in particular draws quite a lot of current, this circuit will drain a nine volt battery relatively quickly. It still works quite well, but it is probably better to consider using a power supply derived from the mains. be sure to use a well filtered and regulated supply or noise is likely to get into the audio.

Using a PSU also opens up the possibility of increasing the supply voltage a bit, but be sure not to exceed 12.6V on the heater supply or your tubes won't last as long as they should. Altering the supply voltage (both to the main circuitry and to the heaters) can significantly alter the tone of circuits like this, so it can be well worth while to experiment a bit before settling on a final design. Even better, a variable supply (to heater and/or circuit) could be implemented giving the maximum possible flexibility and range of tones available. As always, it is also worth experimenting with capacitor values, to assess their impact on the overall tone of the circuit before settling on a final design.

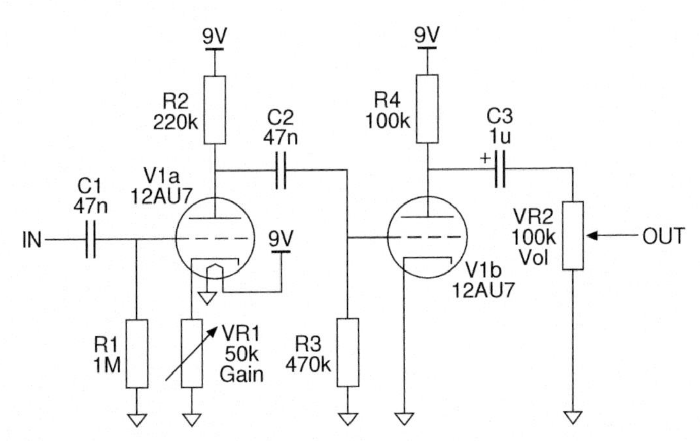

So pull together the parts you need, and spend some time with this circuit to see how much you can get out of it. To take it a bit further, have a look for circuits which add a 386 power amplifier chip to the end of this configuration, to turn it into a low power practice amp with a tube preamp for improved tone. This approach actually performs remarkably well in getting the best out of the tiny 386 chip. One such circuit is called the tube cricket, a variation on a popular 386 amp design called the noisy cricket.

References

J. Hood. *Valve and Transistor Audio Amplifiers*. Newnes, 1997.

D. Hunter. *The Guitar Amp Handbook*. Backbeat Books, 2005.

M. Jones. *Valve Amplifiers*. Newnes, 3rd edition, 2003.

M. Jones. *Building Valve Amplifiers*. Newnes, 2004.

A. Pittman. *The Tube Amp Book*. Backbeat Books, 2004.

RCA. *Receiving Tube Manual*. Radio Corporation of America, 1950.

R. Tomer. *Getting the Most out of Vacuum Tubes*. Sams, 1960.

G. Valley, H. Wallman, and H. Wenetsky, editors. *Vacuum Tube Amplifiers*. McGraw-Hill, 1948.

B. Vogel. *How to Gain Gain: A Reference Book on Triodes in Audio Pre-Amps*. Springer, 2008.

G. Weber. *Tube Amp Talk for the Guitarist and Tech*. Kendrick Books, 1997.

20 | Audio Transducers

Transducers transform signals from one form into another

Transducers come in a wide variety across many application areas. In the realm of audio electronics the two which come most readily to mind are microphones and loudspeakers. Others include coil pickups (standard electric guitar pickups) and piezo elements (which can operate as transducers in two directions – mechanical to electrical and vice versa). The standard spring reverb pan combines two transducers linked with springs. All these are examined in this chapter, along with some useful and interesting circuits using them.

MICROPHONES

Microphones come in a range of types with vastly differing principles of operation. The two commonest classes in pro audio are the dynamic mic and the condenser mic. A particularly simple and inexpensive variant of the condenser mic is the electret condenser. Electret condenser mic capsules like those shown in Figure 20.1 are widely and cheaply available, and can be built into surprisingly effective condenser microphones relatively easily – probably the most significant difference between such circuits and good quality commercial microphones, which sets them apart, is actually the mechanical and acoustical design of the enclosure.

Figure 20.1 Electret condenser microphone capsules.

On top of the electronics considerations addressed here, much creative experimentation can go into the mounting and finishing of a project such as a microphone. For most electronics projects working on the enclosure can be every bit as fun and rewarding as the electronics, but it is primarily just a question of the aesthetics, and to some extent the general robustness of the finished project. On the other hand, most transducer circuits can have their performance significantly altered by the manner in which they are finished, with a well thought out and carefully constructed enclosure potentially making a great difference to the final success of the project. This can be particularly true of microphone projects. Inspiration in this regard can be found in sources such as Hopkin (1996), which, although focused on musical instrument design, provides lots of food for thought when it comes to acoustically active enclosures.

Phantom Powered Electret Condenser Microphone

With all this in mind, the phantom powered electret microphone circuit in Figure 20.2 makes a good starting point for the development of a homemade condenser mic suitable for use in recording interesting sounds with an individual character and a unique twist. As with all the major circuits throughout this book, possible breadboard and stripboard layouts are provided in Appendix D, but as always, have a go at your own layouts before resorting to the suggestions provided there. In this instance it might be useful to aim for a long thin stripboard layout so as to fit the finished circuit inside a slender enclosure, to make a good looking microphone.

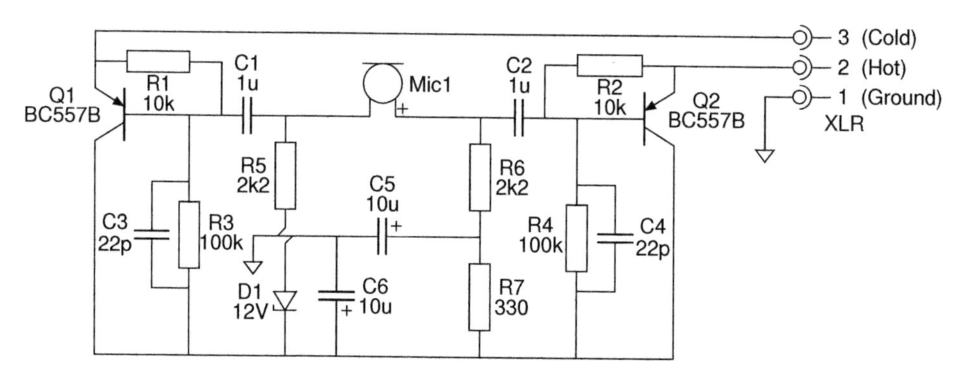

Figure 20.2 Phantom powered electret microphone circuit.

Figure 20.3 provides the necessary details for the transistor used in this circuit. Others can be substituted following the guidelines given in Chapter 17. As always, if an alternative transistor is to be employed, be sure to check the pinout for the particular part in question. Most crucially, notice here that the device used in this particular circuit is of the less common PNP type (with the arrow on the emitter pointing in, and the emitter itself operating at a positive potential relative to the collector).

Phantom powering is a commonly encountered method of providing power to an audio circuit such as a microphone or DI box. Notice in Figure 20.2 that no power supply connections are indicated, and the only external connections to the circuit are the three wires of a standard balanced audio connection (most typically implemented in the form of an XLR connector). In phantom powering the power is delivered to

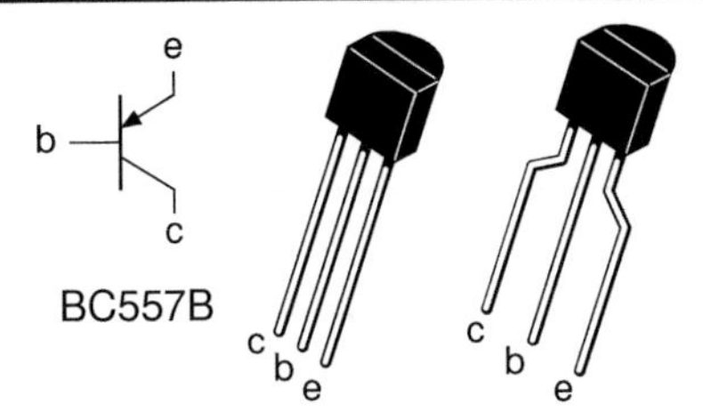

Figure 20.3 Pinout for the BC557B transistor used in the electret mic circuit given above. Note that this is a PNP device, and not the more commonly encountered NPN type. As usual it is always possible to try other transistors which look like they might be suitable replacements, but remember to confirm the specific pinout for any alternative device which you want to try.

the circuit over the same wires across which the audio emerges (Figure 20.4). Pins two and three both present +48V delivered through 6k8Ω, with pin one supplying the common ground path necessary in order to complete the circuit and allow current to flow. A phantom power supply can only deliver a fairly modest amount of current into the circuit being powered (limited by the 6k8 resistors in the supply), so care must be taken in the design of such circuits.

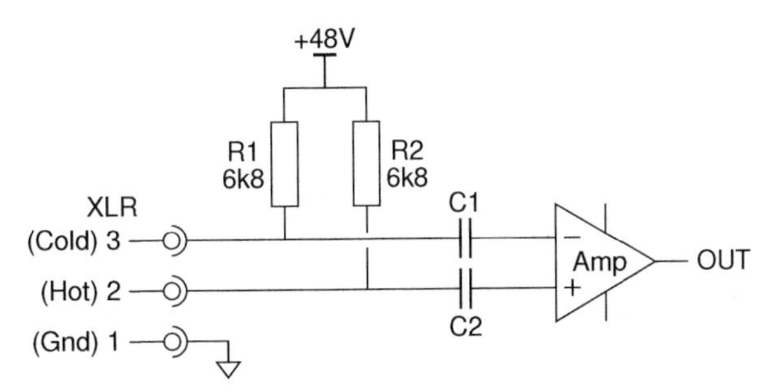

Figure 20.4 Standard phantom power supply configuration, as found in a phantom capable microphone input.

In drafting the schematic layout in Figure 20.2 the clarity of the circuit was better served by a neater layout, at the expense of the 'positive at the top, negative at the bottom' schematic diagram convention. Thus the ground connection, representing the most negative point in this circuit, is located in the middle of the diagram. This point attaches to pin one of the XLR.

The common connection point running along the bottom of the diagram is connected to this ground point through a 12V Zener diode (D1). This holds the collectors on the transistors, and all the other components connected here, at a potential of approximately +12V. The two emitters see the phantom power supply directly through the 6k8 resistors at the far end of the XLR. They are held at 48V minus whatever is dropped across these resistors. The microphone capsule itself drives the two bases through C1 and C2, providing the two anti-phase signals which ultimately drive the hot and cold outputs.

Learning by Doing 20.1

Basic Electret Mic Circuit

The circuit in Figure 20.2 represents an excellent project to spend a bit of time on, but in order to get going a little more quickly, and experiment with the basic workings of the electret microphone capsule, the simple circuit shown below allows you to get some sound out of one of these little capsules with minimal parts, and helps to illustrate the fundamental principles involved in using them.

In this circuit neither the value of R1 nor that of C1 is crucial. R1 controls the voltage seen by the mic. A value ten times as large would still be fine in this case. The job of C1 is to block the DC voltage from the battery while allowing the AC, audio signal to pass. Too small a value would strip all the low frequencies out of the audio. Also, C1's voltage rating must be sufficient to handle the voltages which it will be exposed to. See Chapter 13 for more details on these considerations.

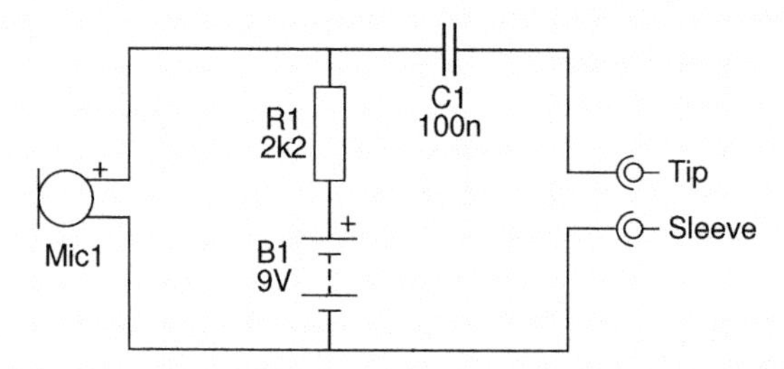

Step 1 (power the capsule): A typical capsule suitable for this application is the POM-2738P-C33-R from Pui Audio. This has a quoted standard operating voltage of $2V_{dc}$, with a maximum operating voltage of $10V_{dc}$, so the suggested 9V battery powering seems fair. The 2k2 resistor R1 will drop the voltage somewhat, so the voltage reaching the capsule should be well within range. It is always instructive to make measurements and perform calculations as you build circuits: connect Mic1, R1, and B1 into a circuit as shown, and measure the voltage across the capsule. Use this measurement to calculate the actual DC resistance of the capsule you are using. The impedance for this device is quoted as 2k2Ω, but its DC resistance will be substantially higher. Also calculate the current flowing. It shouldn't be much. I got just 0.33mA. The quoted maximum is 0.5mA for this device.

Step 2 (decouple the output): With the power connected, the capsule is live and generating a signal, but so far it has nowhere to go. Because of the DC voltage being used to power the device, it is important not to simply connect the terminals of the capsule into an amplifier in order to listen in. This would inject the DC voltage into the amp's input, which is probably not a good idea. Audio amplifiers are designed to expect an AC (i.e. audio) signal at their inputs. A given amp may be designed to cope with a DC offset on its input, but it is best not to test this possibility. The standard method for removing a DC bias from a signal is a DC blocking capacitor, in this case C1. Looking back to the standard phantom power configuration in Figure 20.4, the two capacitors between the +48V supply and the amp are performing exactly the same function. Audio outputs and inputs are very often 'capacitor coupled' like this, in order to accommodate differing DC offsets between one circuit and another.

Step 3 (connect to an amp): It is probably best not to plug this circuit into an amplifier with more than a watt or so of output power (at least not until everything is solidly held together and well tested). When working with a circuit such as this, the likelihood of loud pops on the output is high. A simple 386 based amplifier like that in Figure 18.8a would be an ideal candidate for use in an application like this. Indeed this is a pretty good rule of thumb in general – always fully test and characterise your circuits before considering connection to anything with more grunt than one of these little amplifiers. Your ears and your loudspeaker drivers will thank you.

Loudspeaker Drivers

Just as the microphone is commonly found at the start of the audio electronics signal pipeline, so the loudspeaker often represents the end of that chain. Loudspeakers come in a variety of types, but by far the most frequently encountered is the moving coil loudspeaker (Figure 20.5). This consists of a coil of wire suspended in a magnetic field and attached to the loudspeaker's cone. An AC signal in the coil generates a varying electromagnetic force, which causes the coil, and the attached cone, to move, producing sound.

Figure 20.5 A typical low power moving coil loudspeaker driver.

A moving coil loudspeaker has two terminals usually labelled plus and minus. These terminals connect to the ends of the coil within the device. There is no functional difference between the two terminals and often it makes no difference in which order they are connected to a signal source. The only time when the order becomes important is when two or more drivers are being used in conjunction. In this case it is important that all drivers use the same wiring convention, otherwise drivers will be operating in opposite polarities, one pushing as the other one pulls and vice versa. This can lead to acoustic cancellation, resulting in very 'thin' sounding audio. In its commonest wiring configuration a speaker's minus terminal is wired to the common circuit ground, while the plus terminal is driven by the amplifier signal, as in Figure 20.6.

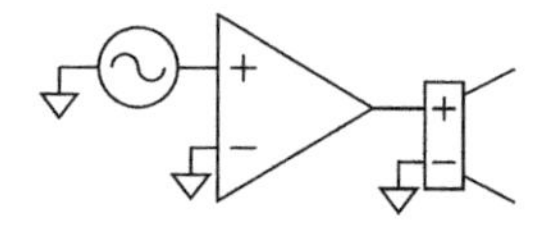

Figure 20.6 Basic loudspeaker wiring.

The two most basic specifications for any moving coil loudspeaker are its characteristic impedance and its power handling capability. By far the commonest impedance encountered is 8Ω, with 4Ω and 16Ω also being common. Most amplifiers will specify the minimum impedance load which should be attached to their output – four ohms is common but other limiting values are encountered also. Attaching a load with an impedance lower than the specified minimum will result in excessive distortion and may even cause damage to the amplifier.

The power handling capability of a loudspeaker is measured in watts. Exceeding this limit likewise incurs increased distortion and runs the risk of blowing the speaker, most commonly by overheating the voice-coil to the point where the wire burns through. The speaker shown in Figure 20.5 is a small (two and a half inch diameter), low power unit. As can be seen in the image it has an impedance of 8Ω, and is rated to handle just half a watt (0.5W). A small speaker like this works well with the 386 based amps presented here and elsewhere.

Coupling Capacitor

Often a capacitor will be placed between the amplifier output and the loudspeaker in a circuit design. The fundamental requirement which such a capacitor is intended to meet is that there be no DC component to the signal present across the loudspeaker's terminals. This is important for two reasons. Firstly, a DC component will result in the loudspeaker cone adopting an offset resting position, resulting in elevated distortion in the audio it produces. Secondly, a DC voltage will result in a corresponding DC current constantly flowing through the loudspeaker coil. This will result in an excessive heating effect, especially when no AC signal is present to keeping the coil moving back and forth, cooling it down. The inevitable result is a burned out speaker coil.

This observation leads to a perhaps unexpected conclusion for the matching of amplifiers and loudspeakers. It is obvious that a low power handling speaker should not be connected to a more powerful amplifier, as the amp is likely to blow the speaker when turned up. It may not be so obvious that the opposite arrangement – low power amp and high power speaker – also has the potential to blow the speaker. The amp, being lower power, is more likely to be overdriven and start clipping badly. Clipping sends repeated short bursts of high current through the loudspeaker coil, without any heat dissipating movement accompanying it, and so burnout can once again ensue.

When considering whether or not an output capacitor is required between amp and speaker, the quiescent voltage level at the output of the amp needs to be compared to the voltage level at the other speaker terminal. If they are the same, no capacitor is needed. Consider the two circuit configurations illustrated in Figure 20.7. In each case the minus terminal of the loudspeaker is connected to ground. Thus, when no signal is present at the input of the amplifier, the plus terminal of the loudspeaker must also sit at ground potential.

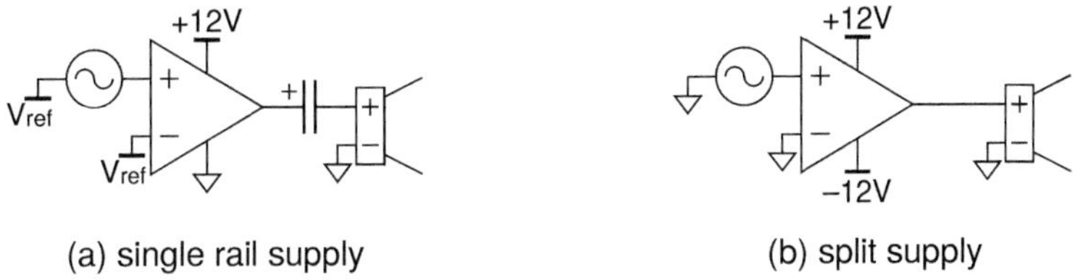

+12V
Vref
Vref
(a) single rail supply

+12V
−12V
(b) split supply

Figure 20.7 The role of an output coupling capacitor.

An audio amplifier needs to be able to amplify the positive and negative portions of an audio signal equally well. It can only produce outputs in the range of voltages available from its power supply, and therefore with no input signal the output must sit at a voltage half way between its positive and negative power rails. In Figure 20.7a a single rail supply is utilised. The midpoint in this case is +6V, and so this is where the output will sit. With the minus terminal of the speaker connected to ground (0V), clearly a DC blocking capacitor is needed between the amp and the plus terminal of the speaker. On the other hand, in Figure 20.7b a split supply is shown powering the amp. With its power rails at +12V and −12V, its output will sit at 0V, thus matching the ground potential on the minus speaker terminal, and so no capacitor is required.

Notice that the capacitor indicated in Figure 20.7a is shown as a polar type. The capacitor will act as a high pass filter, and since it is attached to a relatively low impedance in the form of the loudspeaker, it needs to be quite large in order not to block too much low frequency content from the audio signal. Large capacitors are most often polar, which is why this device is marked as such. The cutoff frequency of this high pass filter can be calculated using Eq. 13.13. A smaller value can be used to restrict lower frequencies, which might be desired when driving a small, low power handling driver such as the little half watt unit illustrated in Figure 20.5.

Bridge Mode Connection

There are two situations in particular in which the coupling capacitors discussed above are not needed: split rail and bridge mode operation. The split rail case has been addressed. Figure 20.8 illustrates the second situation, that of a bridge mode amplifier configuration. The requirement, as previously stated, is that no DC component exists across the terminals of the loudspeaker. The bridge mode configuration is unusual in that the minus terminal of the loudspeaker is not connected to ground. Instead the speaker floats around the quiescent output level of the amplifiers attached to the two terminals. Clearly these two amps must have matched outputs, so as to maintain the required equal quiescent levels, and avoid any unwanted DC current flow.

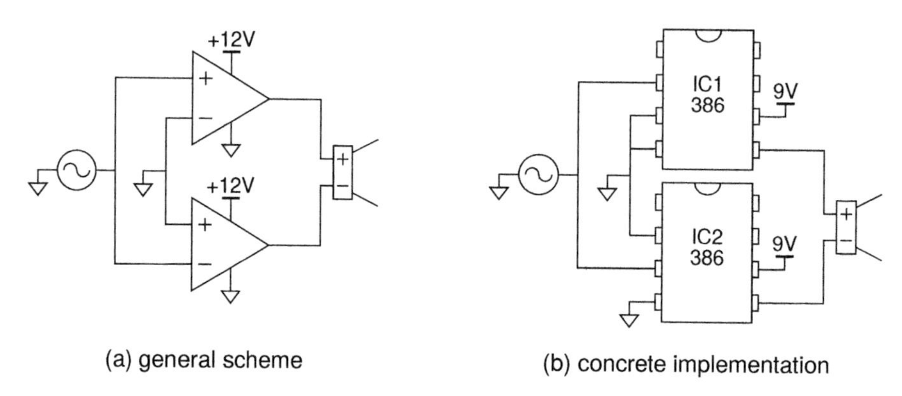

Figure 20.8 Driving a loudspeaker in bridge mode.

Figure 20.8a illustrates the basic bridge mode configuration: two differential amplifiers with the same signal fed to the noninverting input of one and the inverting input of the other. The other inputs are grounded, and the speaker is bridged across the two outputs (hence the name). In this case the amplifier powering scheme can be single or split rail, without the need for coupling capacitors in either case.

Figure 20.8b shows a simple concrete implementation of the bridge mode configuration using a pair of 386 low power audio amplifier ICs. Pins two and three are the inputs. The input signal is fed into pin two on IC1 and pin three on IC2, with the other input pins grounded. Pin five is the output on each chip, connected here to either terminal of the loudspeaker.

With absolutely no other components required apart from input, output, and power, this must be a contender for the simplest amp ever (alongside the bare-bones amp in Figure 18.8a, which uses just one 386, but does require an output capacitor in addition). In this very simple circuit pins one, seven, and eight are left unconnected. Obviously there are many enhancements which could be applied to this basic design, but it does operate quite effectively even as shown.

One further point must be given careful attention when examining a bridge mode configuration, that of amplifier loading. As previously stated, an amplifier will have a minimum load impedance, perhaps 4Ω. In the bridged configuration it is important to note that each of the two amplifiers involved sees only half of the load connected between them. Recall that load impedance is measured to ground, and note that there is a virtual earth point half way along the speaker's voice coil, with the terminal voltages seesawing up and down either side of it. Therefore, the speaker (or combination of speakers) connected across the amplifier outputs must have an impedance of at least twice the amplifiers' minimum load impedance.

Stabilisation Network

One other circuit element which is very often seen in conjunction with a standard moving coil loudspeaker is shown in Figure 20.9. This is what is usually referred to as a zobel network. In simple circuits, the network usually takes the form of a small resistor (usually 10Ω) and a capacitor (often around 47nF) in series. A lot of work and a lot of maths can go into modelling and designing ideal solutions, but the simple approach serves well for all but the most demanding applications.

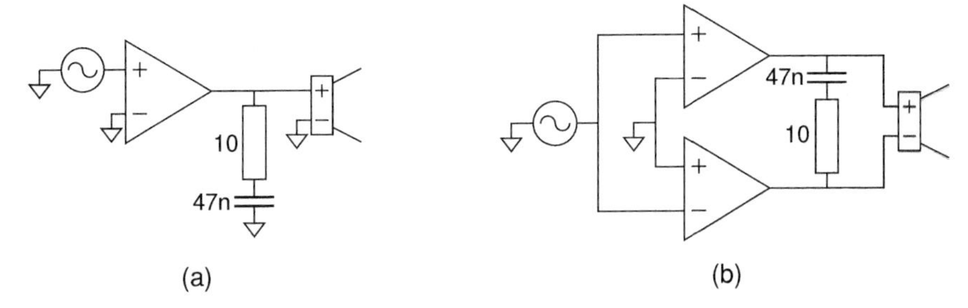

Figure 20.9 Stabilising a loudspeaker with a zobel network.

Electrically, a moving coil loudspeaker can be thought of as a resistor and an inductor in series. It is made from a coil of wire, so this should come as no surprise. Inductors have a frequency dependant impedance (Eq. 13.2). As such, if an amplifier has just a loudspeaker attached to its output, it will see a load which varies widely with frequency. A zobel network has an impedance which varies in the opposite direction – the speaker has inductance while the zobel network has capacitance.

The commonest moving coil loudspeakers have a quotes resistance of 8Ω, so the 10Ω resistor in the network matches this closely enough. The net result of the speaker and zobel in parallel is a load with a much more consistent impedance across the operating frequency range of the amplifier. This steadier loading tend to make an amplifier behave better.

Wiring Drivers in Series and Parallel

As mentioned above, loudspeakers have as one of their primary figures of merit, a characteristic impedance, the commonest being 8Ω, and likewise amplifiers have a minimum load impedance. It is a common situation to find multiple loudspeaker drivers wired together and connected to a single amplifier output (think of a guitar amp driving a two or four speaker cab). In such a case it is important to know how to determine the total impedance which a particular speaker combination will represent.

In actual fact it is very straightforward in most cases. The familiar resistor series and parallel rules are all that is required. Indeed, for almost all wiring configurations likely to be encountered the situation is even easier. It is usually the case (and usually the only advisable course) that only identical drivers are wired in combination like this (LF/HF combinations involving a crossover are a different matter, and are not covered here.)

Revising the series and parallel rules from Chapter 12, it quickly become apparent that two equal resistances in series result in a total resistance double the size, while two equal resistances in parallel results in a total half the size. So for instance, two 8Ω drivers in series looks like 16Ω, and two 8Ω drivers in parallel looks like 4Ω, see Figure 20.10.

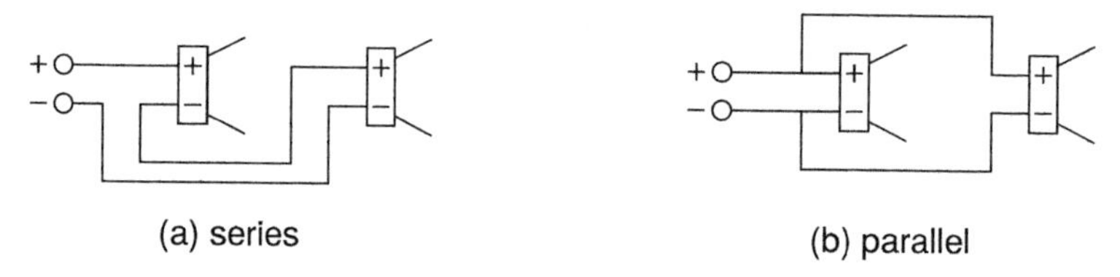

Figure 20.10 Wiring two loudspeakers in series (a) and in parallel (b). For loudspeakers of equal impedance Z_{LS}, two drivers in series: $Z_{series} = 2Z_{LS}$, and two drivers in parallel: $Z_{parallel} = Z_{LS}/2$.

Learning by Doing 20.2

Loudspeaker Wiring

Larger electric guitar cabinets often consist of four drivers connected to a single head amp. Below are shown two possible ways of wiring such a four driver setup. They are referred to as series-parallel and parallel-series wiring respectively. In the first, the pair of drivers on the left are wired in series, as are the pair on the right. These two pairs are then wired together in parallel: series-parallel.

In the second configuration, the top two are in parallel, as are the bottom two, and then these two parallel sets are connected together in series: parallel-series.

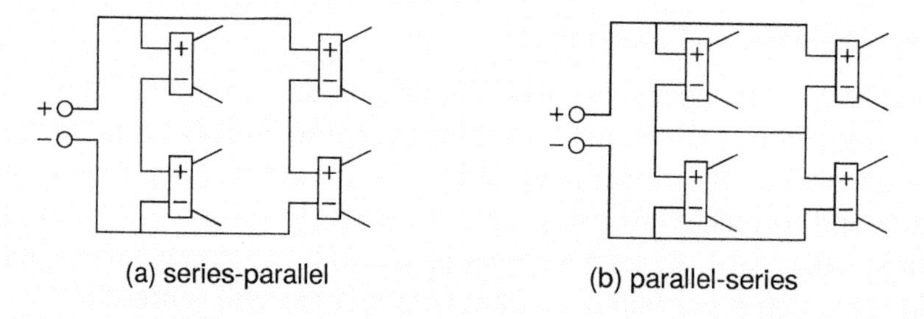

Assuming all drivers are 8Ω, use the series and parallel rules to determine the total impedance of each configuration.

Get four little eight ohm, half watt speakers like those shown in Figure 20.5. Use a multimeter to measure the voice coil DC resistance for each. It probably won't come out as exactly eight ohms, but it shouldn't be too far off.

Use crocodile clip leads to wire the drivers in various combinations of series and parallel. Calculate the expected impedance, and then make the measurement to see if it matches.

Build one of the various 386 amplifiers illustrated in the text here and in Chapter 11 and Chapter 18. Assess the sound achieved with various driver combinations. Don't connect a load impedance of less than 4Ω to the 386, and remember if you are using a bridged configuration the total load impedance should be kept at or above 8Ω because each 386 will only see half of this impedance.

The 386 is capable of driving even large guitar cabs remarkably well (for its size). Try connecting any larger loudspeakers or cabs available to compare the results. Remember that the really simple 386 circuits work well enough, but better performance can be achieved with a more elaborate circuit like the one described in Chapter 11.

Using a Speaker as a Dynamic Mic

The section on microphones earlier in this chapter focused on the commonly available electret condenser microphone capsule. Condenser microphones represent one of the two key microphone types found in audio work. The second type is the dynamic microphone. In many ways the dynamic mic is a more straightforward device than the condenser. Fundamentally it shares exactly the same working parts as the moving coil loudspeaker – in fact, so much so that it turns out to be a simple matter to use a moving coil loudspeaker as a microphone.

The key differences between the two are size and mass. Generally the detection element of a microphone (the diaphragm) is small and light so that sound can move it around easily. The equivalent part of a loudspeaker (the cone) tends to be much more bulky, so it can move a lot of air. So for a speaker to work effectively as a mic, it should be relatively light and the sound it is to be used to detect should be relatively energetic (i.e. loud and low frequency). Spend any time in a recording studio and you are bound to come across the NS-10 kick drum microphone or its commercialised variants. Basically its just a mid sized driver used as a microphone on loud, low frequency sources.

LEARNING BY DOING 20.3

LOUDSPEAKER AS A MICROPHONE

The subkick mic, as it is usually referred to, can generate excellent results. In this exercise we want to see how much action we can get out of a much more modest driver. Again we turn to the kind of small, cheap, low power speaker driver illustrated in Figure 20.5.

The generic amp in the diagram below can once again be any of the 386 utility amplifiers we have encountered, and all that is needed is to connect our little driver to its input. Either solder on a TS jack with a length of instrument cable, or even just zipper cable, or alternatively break out the clip leads, although this is likely to prove a little fiddly.

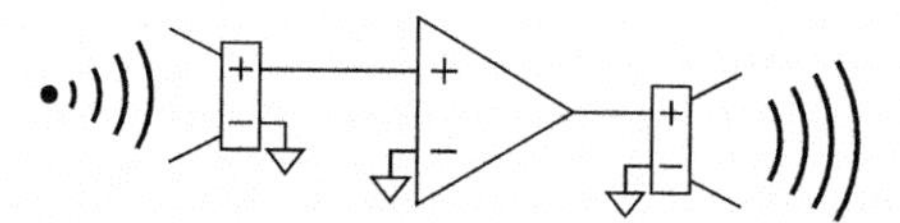

Talk into the speaker-mic, tap it, press it against anything that hums, buzzes, or otherwise vibrates, or up against the grill of another speaker playing music. And of course, there is acoustic feedback – move the speaker-mic around close to the output speaker, playing the feedback like a lo-fi, particularly annoying, theremin.

Coil Pickups

In common with dynamic microphones and moving coil loudspeakers, coil pickups operate on the principle of electromagnetic induction. In dynamic mics, a conductor moving in a magnetic field generates an electrical signal. In moving coil loudspeakers the process is reversed with an electrical signal in a conductor mounted in a magnetic field generating movement. Thus electromagnetic induction usually involves a conductor moving in proximity to a magnet.

Coil pickups represent something of a twist on this idea, in that the conductor and the magnet do not move relative to one another; both are firmly mounted in place within the pickup body, as illustrated in Figure 20.11. Instead a metal guitar string – which is not involved in the electrical circuit – vibrates in the magnetic field, causing a signal to be induced in the stationary coil of the pickup. As the metal string vibrates in the magnetic field, it deforms the field, effectively dragging it back and forth through the coil in the pickup, thus generating the requisite electrical signal. So rather than the conductor moving in the magnetic field, as is the more familiar incarnation of electromagnetic induction, the magnetic field actually moves through the conductor. The effect is however the same; a signal is induced in the conductor.

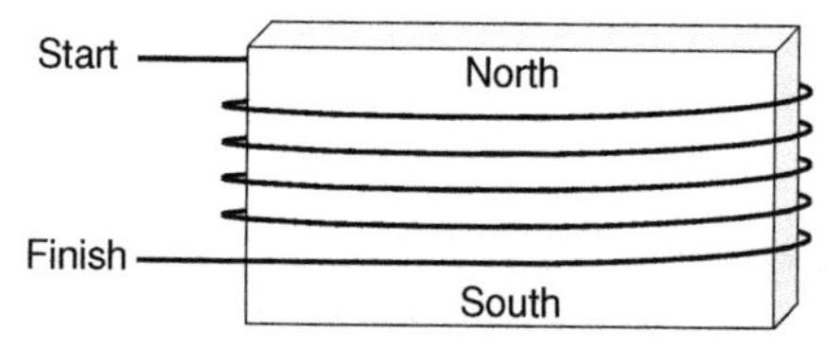

Figure 20.11 Structure of a standard single-coil electric guitar pickup.

Coil pickups come in two basic forms, single coil and humbucker pickups, as illustrated in Figure 20.12. The biggest problem with the basic single coil pickup is that a coil of wire can work like an ariel, detecting electromagnetic radiation floating through the air. Single coil pickups are known for being quite noisy in this regard. The solution is the humbucker. The ariel action of the pickup, mopping up stray noise, is independent of the magnet in the pickup, whereas the wanted signal induced by the vibrating string is directly dependent on the magnet. This difference can be used in order to retain the wanted signal while rejecting the unwanted airborne interference.

(a) single coil pickup

(b) humbucker pickup

Figure 20.12 A single coil and a humbucker electric guitar pickup.

A standard single coil pickup can be manufactured in two basic configurations, sometimes referred to as 'south to strings' and 'north to strings', depending on the orientation in which the magnet has been inserted into the assembly. This distinction is illustrated in Figure 20.13. The methods of constructing a pickup vary from manufacturer to manufacturer, and from one pickup model to another, but often it is the case that pickups are constructed such that it is relatively easy to remove and rotate the magnet, in order to reconfigure the pickup if required. Manufacturers will usually provide guidance to the process if such a procedure is possible with their particular devices.

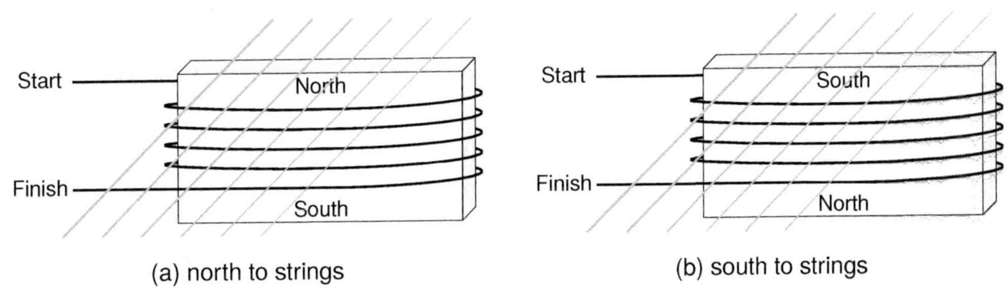

Figure 20.13 The magnet orientation can differ between different pickups.

Humbucker Pickups

To make a humbucker requires two single coil pickups whose magnets are mounted in opposite orientations. (In fact humbuckers often use a single magnet installed so as to direct its two poles one at each of the pickup windings.) The two pickups in a humbucker are placed side by side but are wired together out of phase, as in Figure 20.14. Each pickup generate both the wanted and unwanted signals described previously. The out-of-phase wiring means that the unwanted signal (which is not affected by the magnet orientations) cancels between the two pickups. The wanted signals however get flipped twice, once by the out-of-phase wiring and again by the reverse oriented magnets. This means that the wanted signals from the two pickups end up back in the same polarity and thus adds, actually giving a stronger signal.

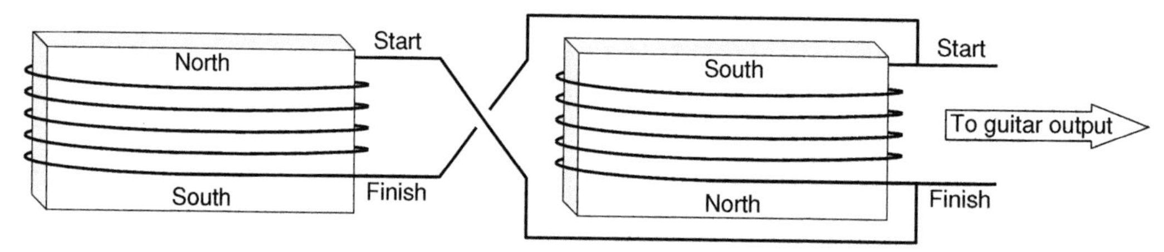

Figure 20.14 How to wire two single coil pickups in a humbucker configuration.

With two binary options (wiring phase and magnet orientation), there are four possible ways in which two single coil pickups can be wired together, and four corresponding types of behaviour. These four possible variations are laid out in Table 20.1. While many factors combine to make up the final tone emanating from an electric guitar, these basic wiring options can certainly prove a useful tool in tailoring the sound of an instrument.

Other considerations not discussed here include: series versus parallel pickup wiring, choice of magnet type (there are several), pickup positioning along the guitar body, and proximity of pickup to strings. Some pickups even have per-string magnet pole-pieces which can be individually adjusted to modify their response. And then there are many non pickup related factors, from the strings, to the guitar body, to the amp, all of which play a role – interesting topics but not for the current discussion.

LEARNING BY DOING 20.4

PICKUP WIRING

In order to take advantage of these wiring configurations when connecting multiple electric guitar pickups it is necessary to know three things about each pickup: the relative orientation of the magnets, the matched ends of each winding (in dual winding pickups), and the winding direction for each coil. Three simple tests can be performed to determine these for any given pickup. Get a selection of pickups together and perform the following procedures to characterise each one.

Table 20.1 Effect of different pickup combinations. All configurations are considered relative to a default in-phase, N-to-strings configuration

Configuration	In-phase		Out-of-phase	
	N-to-strings	S-to-strings	N-to-strings	S-to-strings
Signal				
Noise				
Result	Both add	Signal cancels, noise adds	Both cancel	Signal adds, noise cancels
Comments	Same as single coil	Worst of both worlds	Thin sound, can be useful	Standard humbucker

Magnet Orientation

Pickup magnets are mounted either 'south to strings' or 'north to strings'. There are various ways of determining the orientation in a particular case, including labelled bar magnets suspended by a string and purpose built tools that flip a tiny magnet to show either a black or a white indicator. Possibly the easiest and cheapest approach is to use a compass. A tiny button compasses works well. By holding the compass (or other free moving magnet) close to the top of the pickup, a reading can be made, as in the diagram below. It is not necessary to know which is north and which is south, just so long as they can be consistently differentiated.

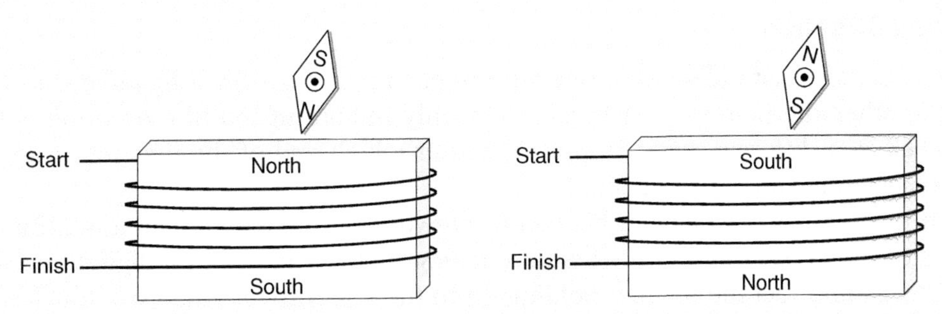

Using this test on a preassembled humbucker pickup should see the indicator quickly flip orientations as it is moved the small distance from in front of one of the two magnets (or pole pieces) to in front of the other. Most compasses will only rotate freely when held flat, so it is best to hold the pickup horizontally, with the two magnets one above the other. Then, holding the compass flat, bring it in from the side and move it up and down between the levels of the two magnets to observe their orientations.

Match Winding Ends

Many pickups have only two wires and so this step can be skipped. However in a pickup such as the one illustrated below (which has five wires) the matching connections must be sorted out first. Here there are two pairs for the two coils in this low profile dual coil pickup, and the fifth wire is a ground connection (effective grounding is another essential for a noise free electric guitar).

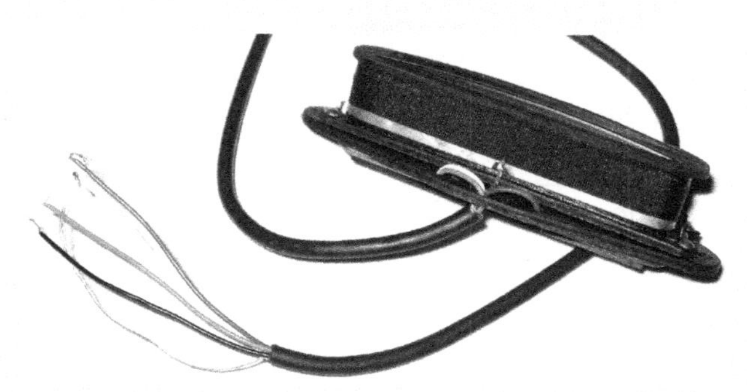

Finding the matching pairs is easy. Set your multimeter to the continuity tester range and test pairs of wires in succession. You should find two pairs of wires that give a beep, and all other combinations should give no reaction. Note the pairs which beep. Each pair represents the two ends of one coil. The remaining wire (if there are five wires, as here) is a ground wire.

In our case the five wires are: black, white, red, green, and bare. Testing the wires in this pickup shows the matching pairs to be black & white and red & green, and unsurprisingly the bare wire is the ground connection.

Winding Direction

Once pairs have been identified, one wire of each pair must be designated the start and the other the finish for each coil. It is only important that the ordering is the same between different coils. It doesn't matter which end of the first coil tested is designated start and which finish.

For this test you will need a big metal screwdriver (or similar) and a multimeter (this is one of those rare cases where an analog meter is best, but a digital one will do). You should feel the magnet holding on to the screwdriver when you touch it to the top of a pickup. If not, try a different metal tool until you find one that sticks solidly. The test setup is as shown below. As an illustration, the results obtained at each step of the tests will be given for the pickup pictured.

Step 1: Choose one wire from one pair and call it start (let's pick white). The other wire in the pair is finish (black in this case).

Step 2: Connect start (the white wire) to the red lead of your multimeter (connected to the meter's plus input) and finish (the black wire) to the black lead (connected to the meters minus input). Set the multimeter to a low DC voltage range, something like $200mV_{DC}$.

Step 3: Tap the screwdriver down onto the top of the pickup, and pull it back off again. The needle on the meter will kick in one direction when the screwdriver is tapped on, and in the other direction when it is taken off. For our example test, 'on' gives a negative kick and 'off' gives a positive kick of the needle. With a digital meter it can be a little more difficult to tell, but you're looking for whether the reading is going mainly negative or mainly positive.

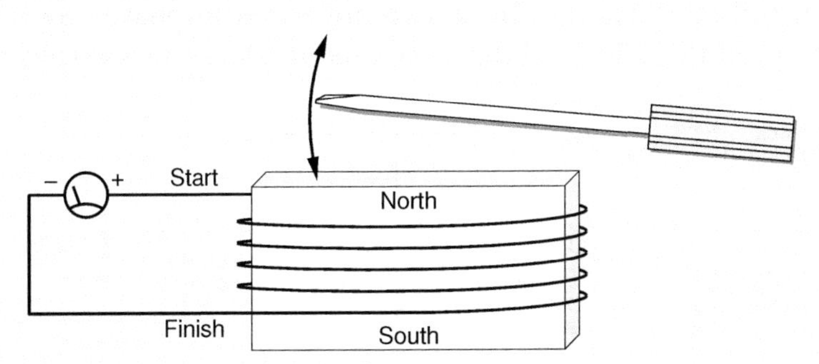

Step 4: Repeat Steps 2 and 3 for the second pair of wires and check your results against the table below.

Table 20.2: Identifying the ends of a coil pickup winding

Same magnet		Opposite magnet	
Same kick	**Opposite kick**	**Same kick**	**Opposite kick**
Ends match	Ends opposite	Ends opposite	Ends match

I connected red to the red test lead, and green to the black test lead, and got positive kick for 'on' and negative kick for 'off', the opposite of before. Because the magnet is also reversed in this second pickup (as is standard for a humbucker configuration) this means this coil is connected the same way as the first one (column four in the table – opposite magnet, opposite kick). Thus my second start wire is also connected to the red test lead. Here are my final results:

Coil 1: start = white, finish = black

Coil 2: start = red, finish = green

Magnets: opposite polarities

Therefore, in order to connect this pickup in a standard humbucker configuration, the white and green wires should be connected together on one side, and the black and red wires should be connected together on the other side, as indicated in Figure 20.14.

To round off this section on coil pickups, a simple electric guitar wiring diagram is presented (Figure 20.15). The two pickups indicated may be single coil variants, or they may represent complete humbucker pickups, or indeed other nonstandard wiring pickup pairs. If single pickups are assumed then it is certainly worth checking the magnet orientation of the pair. If they are opposite (or if one can easily be converted, as discussed above), then it might be advantageous to wire them out of phase to form a humbucker pair, even if they are not mounted right beside each other. The dot at the top of each coil indicates that these two are wired in phase. Swapping the two connections to one of the coils is all that is needed to place the two coils out of phase.

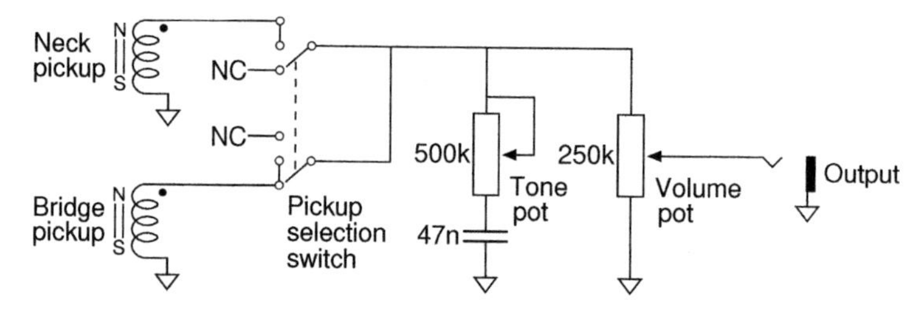

Figure 20.15 Typical wiring diagram for a basic two pickup electric guitar.

Pickup selection switches can be far more elaborate than the one shown here. In this case the three switch positions simply select one, the other, or both pickups in parallel. The symbol indicates an dpdt on-on-on device, and 'NC' stands for 'no connection'. A standard dp3t switch could also be used here, with jumper wires between the appropriate two pairs of throws.

The tone stack is a fairly standard arrangement for an electric guitar. With the pot turned up, the large resistance means little signal goes down that way, and the signal from the pickups passes un-affected. As the pot is turned down, the capacitor continues to block low frequencies from shunting to ground, but high frequencies can make their way through. Thus the tone control works by bleeding off more or less of the high frequencies from the incoming signal, depending on the position of the pot control.

Finally, the volume pot is absolutely standard, simply tapping off its output signal somewhere between full and zero, depending on the position of the pot knob, and the guitar output is provided via a standard TS jack.

Piezo Elements

Piezoelectric material has a very simple, very useful two way behaviour. If it is deformed (for instance bent) an electric charge appears on its surface, with opposite charges appearing on opposite faces. If the direction in which it is being bent is reversed, then so too does the sign of the charge on each face. Alternatively, if a charge is applied to its faces, then as a result it physically bends. Again, reversing the polarity of the charging signal reverses the direction of the bending which results.

Thus a vibration which is allowed to deform some piezoelectric material will produce an equivalent electrical signal between its two faces, and correspondingly, if an alternating signal, such as an audio signal, is connected across the faces of the material, then it will vibrate with the signal.

Thus piezo elements can be used as pickups or contact microphones, and they can be used as drivers or sounders. The piezoelectric material itself is a hard, brittle crystalline substance. In the piezo elements most commonly encountered, the material itself is adhered to a solid brass disc, in order to give it strength, and to provide a good electrical contact right across its back face.

The exposed top surface is covered in a slivered coating to form a second electrical contact, and typically two wires are soldered on, one to the brass disc and one to the silvered front surface, as shown in Figure 20.16a. Sometimes alternative electrical connection methods might be found. For instance, the buzz tester in most multimeters is implemented using a piezo element, and it is likely that the connections in such a device are made using carefully placed springs which contact the piezo disc when the multimeter case is put together. This technique avoids the need for soldering (which can be tricky with piezo discs, and easily broken).

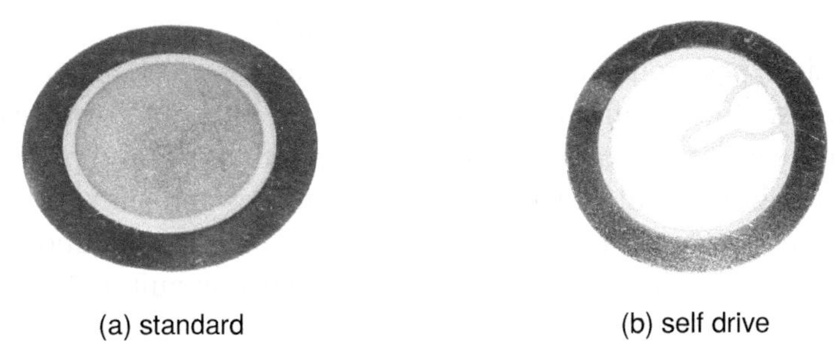

(a) standard (b) self drive

Figure 20.16 Two piezoelectric elements. A standard two terminal device and a self drive variant with a third 'sense' terminal.

As pickups, piezos are extremely easy to use; attach one to any amp input and you have a basic contact mic. Getting good quality sound out of them takes a little more effort. A piezo element presents an extremely high impedance, and so in order to maintain signal quality it is best to connect a piezo pickup to as high an input impedance as possible. Most typical audio inputs have input impedances no higher than a few tens of kΩs. Common guitar amps extend this up into the hundreds of kΩs range, which can be pretty good for a piezo, but even higher can be better. JFET buffers are commonly employed with piezos in order to get the most out of them. These are usually designed with an input impedance of one or two MΩs. The circuit in Learning by Doing 17.5 (p. 308) would be a good option for this application.

Piezos will not pickup sound travelling in the air; there just isn't enough energy in an acoustic signal to bend the brass disc and generate a signal. The piezo element must be attached firmly to a vibrating surface in order to achieve good sound pickup. Figure 20.17 illustrates the kind of device which can be found. The piezo element is enclosed in the plastic housing, with a tough rubber pad providing good mechanical linkage, and a sprung clamp for attaching the pickup to a surface. This one is most likely to be found clipped to the sound hole of an acoustic guitar.

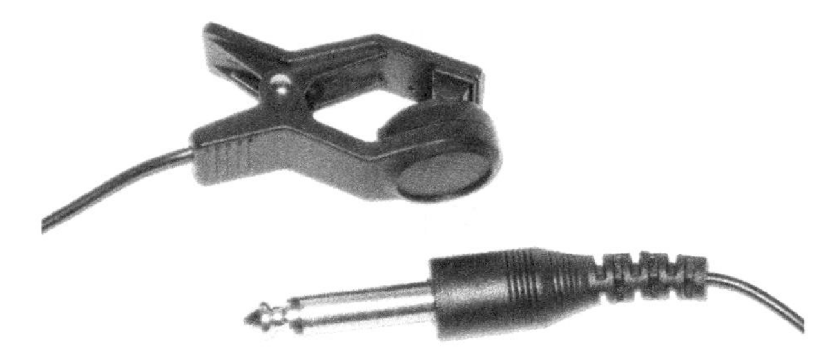

Figure 20.17 A piezo pickup with a spring clamp and a TS jack plug output.

LEARNING BY DOING 20.5

PIEZO AS A PICKUP

To do some basic experiments all that is needed is a piezo pickup such as the one in Figure 20.16a, a small amplifier and a matching jack plug, and some cabling or clip leads to attach the jack plug to the pickup. Scraping, tapping, and pressing the piezo to anything that vibrates will generate a sound. Springs and slinkies are a favourite source of interesting sounds. Homemade percussion instruments and general acoustic sound-makers like the tea chest bass and the cigar box guitar are often mic'd up using a piezo.

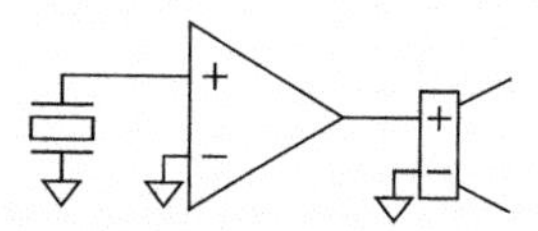

The sound from a raw piezo element can be quite tinny and harsh. Covering them in electrical tape can serve the dual purpose of taming the harsher rasp in their sound, and also insulating the electrical contacts to avoid pops and sound dropout due to inadvertent contacts to these exposed surfaces. Soldering on a good length of cable with an appropriate jack on the end makes the whole thing easier to use and a lot more versatile. And of course a good high impedance buffer can make a difference too, as discussed above. There is plenty of scope for expanding and experimenting with the capabilities of the basic piezo pickup.

Self Drive Piezo Elements

Figure 20.16b shown a less commonly encountered variant on the basic piezo element. In this type the top silvered contact area is separated into two unconnected zones – the main drive contact, and a smaller feedback region used to sense when the element has been deformed. These three terminal types are called self drive piezos, and are most commonly used to implement very simple oscillator sounders, as examined next in Learning by Doing 20.6. Regular piezos can also be used to reproduce a full audio signal (typically at a low fairly quality). Refer back to Learning by Doing 14.2 (p. 255) for an example of how to drive a piezo sounder with an audio signal.

LEARNING BY DOING 20.6

PIEZO AS A DRIVER

In the circuit shown below, a self drive piezo element is used in conjunction with a transistor to make a small oscillator sounder. The piezo both maintains the oscillations and generates the sound from the circuit. The basic operation is fairly straightforward.

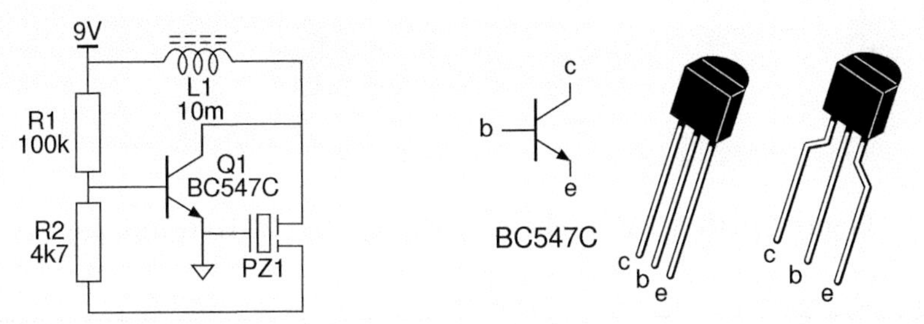

When power is applied, the main drive terminal connected to L1 is charged from the nine volts rail. This causes the piezo element to bend, but as it does so a positive charge is also induced on the smaller sense terminal connected to the bottom of R2. This positive voltage feeds through R2 and switches the transistor on. Switching Q1 on creates a low impedance path from the drive terminal to ground, and so the charge which was placed on the terminal dissipates and as a result the piezo element flattens again.

However, with the piezo flat once more, the sense terminal no longer has a charge on it. Since this is what initially switched the transistor on, Q1 once more switches off. This returns the circuit to its original state and so the whole process repeats.

The circuit can also be built with a third resistor in place of L1, but this inductor helps to drive the oscillator even harder (and hence louder) through the voltages it generates as it is energised and de-energised.

Sometimes this little oscillator can latch up and needs a little tap to get it going, but once running it can be very loud – be warned.

Reverb Pans

A spring reverb pan (Figure 20.18) is not strictly speaking a transducer; it takes as its input an electrical signal and produces as its output another electrical signal. However internally it actually consists of two transducer elements linked by one or more springs. The first transducer converts the incoming electrical signal into mechanical movement, which sends vibrations along the attached springs. These vibrations reflect back and forth along the springs producing the desired reverb effect. The movements within the springs also couple into the second transducer which is linked into the far end of the springs. This transducer operates in the reverse direction converting the spring's vibrations back into an electrical signal which appears at the output connection of the pan.

Figure 20.18 A TAD (Tube Amp Doctor) 8EB2C1B reverb pan.

Learning by Doing 20.7

Reverb Pan Drive and Recovery Circuit

While getting a signal out of a reverb pan is fairly straightforward, in order to get the best results it is necessary to follow the manufacturer's recommendations, especially as to the output and input impedances of the drive and recovery circuits used to operate the pan. As such, the circuit shown here is intended for use specifically with an 8EB2C1B reverb pan, although it is likely to work moderately well with others. For best results always look to follow manufacturers example application circuits. For these pans each part of the serial number tells you something:

8 – Type: 23.5cm (short), 3 spring
E – Input impedance: 800Ω
B – Output impedance: 2575Ω
2 – Decay: medium (1.75sec–3sec)
C – Connectors: input insulated, output grounded
1 – Locking device: none
B – Mounting: horizontal, open side down

At this point there should be nothing in the circuit which needs much explanation. Precise capacitor values are not generally crucial. Powering is split rail (notice the opamp power rails). For more headroom the power rail levels can be increased, depending on the components used. In particular the opamp is probably the limiting factor. The TL072 shown is spec'ed to run happily up to ±15V.

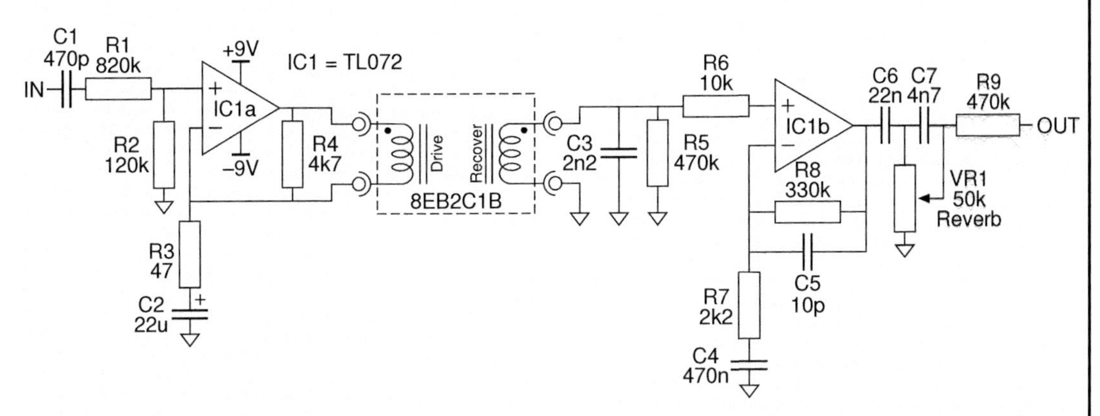

Build the circuit and try sending various signals through it. The electric guitar amp is where this circuit is most commonly deployed, but other sources work well too. This can make a very creditable general purpose reverb unit, and a fun circuit to experiment with further.

References

J. Borwick, editor. *Loudspeaker and Headphone Handbook*. Focal Press, 3rd edition, 2001.

N. Collins. *Handmade Electronic Music*. Routledge, 2nd edition, 2009.

R. Fliegler. *The Complete Guide to Guitar and Amp Maintenance*. Hal Leonard, 1998.

B. Hopkin. *Musical Instrument Design*. See Sharp Press, 1996.

W. Leach. Impedance compenstion networks for the lossy voice-coil inductance of loudspeaker drivers. *Journal of the Audio Engineering Society*, 52(4):358–365, 2004.

Appendices

A | Prefix Multipliers

When measuring the magnitude of electrical signals and the size of electronic components, very large and very small numbers are soon encountered. In order to make such quantities more manageable, standard multiplier prefixes are employed. The reader will surely be familiar with millimetres (mm) and kilograms (kg): a millimetre is one thousandth of a metre and a kilogram is a thousand grams. It is important to be familiar with the range of most commonly used multipliers. Those representing factors of 1,000 (multiplying and dividing by a thousand, a million, a billion etc.) are presented in Table A.1.

Table A.1 Multipliers (factors of 1,000)

Multiplier	Symbol	Prefix
10^{12}	T	tera-
10^{9}	G	giga-
10^{6}	M	mega-
10^{3}	k	kilo-
10^{0}	$= 1$	–
10^{-3}	m	milli-
10^{-6}	μ,u	micro-
10^{-9}	n	nano-
10^{-12}	p	pico-

Notice that the first three multipliers use capital letters, while the rest are all in lowercase. Misuses will often be encountered, especially capital K for kilo (e.g. KB instead of kB for kilobytes). It is wrong but not vitally important, except in the case of the prefixes m and M, as these two have different meanings, lowercase for milli- (divided by a thousand), and uppercase for mega- (multiplied by a million). Also note that the official symbol for micro-, one millionth is the lowercase Greek letter mu (μ) but the letter u can be used instead as indicated in the table, since Greek symbols are not always conveniently available when writing in a word processor.

As examples of the kinds of places where these multipliers can be encountered see the list below for each from biggest to smallest, with examples of where they might be found:

Tera – the capacity of large computer hard drives is measured in terabytes (e.g. 16TB)

Giga – the clock speed of a computer might be specified in gigahertz (e.g. 2.4GHz)

Mega – mid sized media files might be measured in megabytes (e.g. 22MB)

Kilo – distances in kilometres or frequencies in kilohertz are common (e.g. 12kHz)

Milli – millimetres of course for short distances, or milliwatts for power (e.g. 100mW)

Micro – microsecond, micrometre (aka micron) are both common (e.g. 500μm)

Nano and pico – less common, but used to measure capacitors (e.g. 22nF/33pF)

Factors to indicate multiplication and division by ten and a hundred are also in fairly common use. These are presented in Table A.2.

Table A.2 Multipliers (factors of 10 and 100)

Multiplier	Symbol	Prefix
10^2	h	hecto-
10	da	deca-
10^0	$=1$	–
10^{-1}	d	deci-
10^{-2}	c	centi-

Hecto – not many common usages. The hectare is 100 ares, 10,000 square metres

Deca – a multiple of ten, a decathlon has ten events, a decade is ten years etc.

Deci – the most important usage here is in the decibel (dB), one tenth of a bel

Centi – commonly used, one hundredth: centimetre, cent (in money and music) etc.

The symbol for a prefix is considered to be combined with the unit symbol to which it is attached, forming a new unit symbol, e.g. centimetre (cm), microfarad (μF) etc.

When writing decimal values, the prefix symbol is often used in place of the decimal point for clarity. A decimal point can be easily confused or misplaced in small or unclear writing or printing. So for instance a resistor labelled 4k7Ω has a value of four point seven kilohms or four thousand seven hundred ohms.

B | Quantities and Equations

This appendix brings together the most important reference material from throughout the book in one place. Important quantities are listed in Table B.1, along with the standard symbols used to represent them in equations, and the units in which they are measured.

Statements of useful rules and laws are provided in order to facilitate the straightforward analysis of simple electronic circuits and systems. The chapters where each group of equations is introduced are identified in the section headings. The index which follows these appendices can be used to guide the reader to more detailed information on the theory and the application of the individual rules and laws presented.

PRIMARY QUANTITIES

Table B.1 Primary quantities

Quantity		Units		Notes
Impedance	(Z)	Ohm	(Ω)	also Resistance (R) and Reactance (X)
Voltage	(V)	Volt	(V)	aka EMF, PD
Current	(I)	Amp	(A)	aka Ampere
Power	(P)	Watt	(W)	also Apparent power (VA)
Charge	(Q)	Coulomb	(C)	also Amp-hour (1Ah = 3,600C)
Frequency	(f)	Hertz	(Hz)	aka Cycles per second (cps)
Capacitance	(C)	Farad	(F)	
Inductance	(L)	Henry	(H)	plural Henries

Decibel Equations (see Chapter 5 – Signal Characteristics)

For power $\mathrm{dB} = 10\log\left(\frac{P_1}{P_2}\right)$ and $P_1 = P_2 \times 10^{\left(\frac{dB}{10}\right)}$

For voltage $\mathrm{dB} = 20\log\left(\frac{V_1}{V_2}\right)$ and $V_1 = V_2 \times 10^{\left(\frac{dB}{20}\right)}$

Reference levels $0dBu = 0.775V$ $+4dBu = 1.23V$

$0dBV = 1.0V$ $-10dBV = 0.316V$

Fundamental Laws (see Chapter 9 – Basic Circuit Analysis)

Ohm's law $V = I \times R$

Watt's law $P = I \times V \quad \left(= I^2 \times R = \frac{V^2}{R}\right)$

Kirchoff's current law (KCL) $I_{in} = I_{out}$ Currents into a point equal currents out

Kirchoff's voltage law (KVL) $V_o = 0$ Voltage drops around any loop equal zero

Resistor Equations (see Chapter 12 – Resistors)

Series rule $R_{series} = R_1 + R_2$

Parallel rule $R_{parallel} = \frac{R_1 \times R_2}{R_1 + R_2}$

Voltage divider rule $V_{out} = V_{in} \times \frac{R_2}{R_1 + R_2}$

$V_{out} = (V_a - V_b) \times \frac{R_2}{R_1 + R_2} + V_b$

RLC EQUATIONS (see Chapter 13 – Capacitors and Inductors)

Capacitors in series	(F)	$C_{series} = \dfrac{C_1 \times C_2}{C_1 + C_2}$
Capacitors in parallel	(F)	$C_{parallel} = C_1 + C_2$
Inductors in series	(H)	$L_{series} = L_1 + L_2$
Inductors in parallel	(H)	$L_{parallel} = \dfrac{L_1 \times L_2}{L_1 + L_2}$
Capacitor reactance	(Ω)	$X_C = \dfrac{1}{2\pi fC}$
Inductor reactance	(Ω)	$X_L = 2\pi fL$
RLC series rule	(Ω)	$\lvert Z_{series}\rvert = \sqrt{R^2 + (X_L - X_C)^2}$
LC series rule	(Ω)	$\lvert Z_{series}\rvert = \lvert X_L - X_C\rvert$
RLC parallel rule	(Ω)	$\lvert Z_{parallel}\rvert = \dfrac{1}{\sqrt{\left(\frac{1}{R}\right)^2 + \left(\frac{1}{X_L} - \frac{1}{X_C}\right)^2}}$
LC parallel rule	(Ω)	$\lvert Z_{parallel}\rvert = \left\lvert \dfrac{X_L \times X_C}{X_L - X_C} \right\rvert$
RC time constant	(sec)	$\tau = RC$
RL time constant	(sec)	$\tau = \dfrac{L}{R}$
RC cutoff frequency	(Hz)	$f_c = \dfrac{1}{2\pi RC} \quad \left(= \dfrac{1}{2\pi\tau}\right)$
RL cutoff frequency	(Hz)	$f_c = \dfrac{R}{2\pi L} \quad \left(= \dfrac{1}{2\pi\tau}\right)$
LC resonance frequency	(Hz)	$f_0 = \dfrac{1}{2\pi\sqrt{LC}}$
LC characteristic impedance	(Ω)	$Z_0 = \sqrt{\frac{L}{C}}$

TRANSFORMER EQUATIONS (see Chapter 14 – Transformers)

Turns ratio $k = \frac{N_{sec}}{N_{pri}}$

Power transfer $P_{sec} = P_{pri}$

Voltage transformation $V_{sec} = k \times V_{pri}$

Current transformation $I_{sec} = \frac{1}{k} \times I_{pri}$

Impedance transformation $Z_{sec} = k^2 \times Z_{pri}$

OPAMP EQUATIONS (see Chapter 18 – Integrated Circuits)

Open-loop transfer function

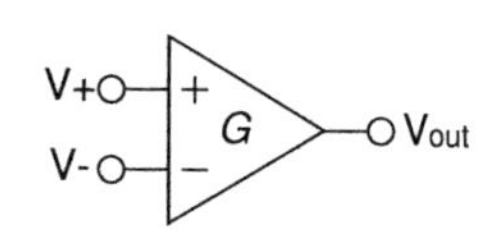

$V_{out} = G(V_+ - V_-)$

Voltage follower transfer function

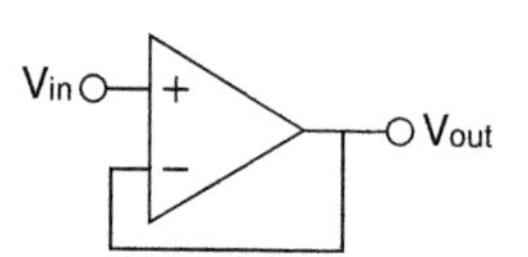

$V_{out} = V_{in}$

Inverting configuration

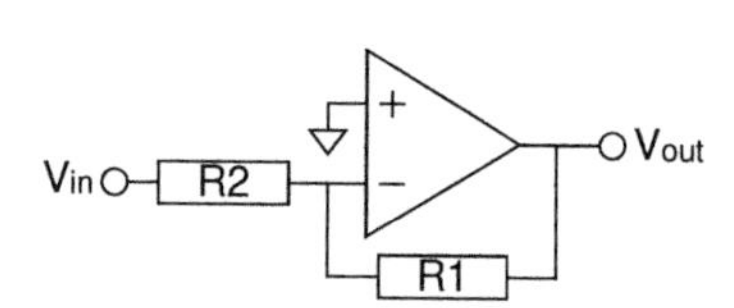

Gain equation: $g = -\frac{R_1}{R_2}$

Transfer function: $V_{out} = g \times V_{in}$

Noninverting configuration

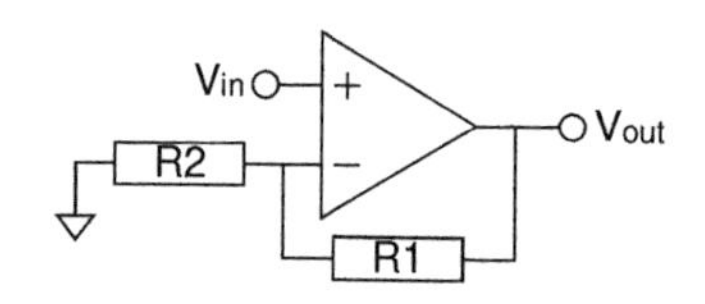

Gain equation: $g = 1 + \frac{R_1}{R_2}$

Transfer function: $V_{out} = g \times V_{in}$

C | Schematic Symbol Reference Chart

This chart brings together a large collection of circuit diagram symbols and conventions, used in the production of electronic circuit diagrams, including all the symbols used in this book, along with many others which are likely be encountered elsewhere. Some of the more important and widespread alternatives are referenced and described. Other variations can be encountered regularly. Usually their interpretations can easily be decerned by comparison to those illustrated here.

Table C.1 Schematic symbol reference chart

Name	Symbol	Description
Unconnected wires		Where two lines cross on a circuit diagram, but the vertical and horizontal wires are not intended to be connected together, a short break is often inserted in one line at the crossing point. An alternative convention of adding a loop to show one wire jumping over the other is also regularly encountered.
Connected wires		When two lines on a circuit diagram cross which are intended to show a common connection point between the vertical and horizontal wires, small offsets may be shown at the point of intersection in order to emphasise the fact that all wires are connected together. Alternatively a dot may be placed over the intersection point to indicate the same.
Ambiguous wires		Sometimes crossing lines are left undecorated. This can introduce uncertainty, as it may be intended to indicate either connected or unconnected wires, depending on what other convention has been adopted by the person drafting the circuit diagram. The intent is usually (but not always) obvious by examining the rest of the circuit. This style should be avoided.

Table C.1 (continued...)

Name	Symbol	Description
		Resistors
Resistor	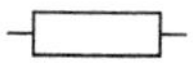	A fixed resistor is indicated by a rectangle with connection terminals at each end.
Resistor (alt.)		The zig zag symbol is another common way of indicating a fixed resistor. The zig zag style can also be used in place of the rectangle in all the other resistor types illustrated below. In this book the rectangular style is adopted.
Variable resistor	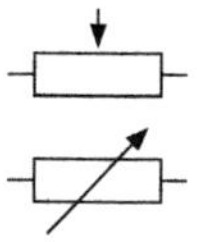	Also known as a pot or potentiometer (for the rotary kind), or a fader or linear fader (for the slider kind). Variable resistors have three terminals, two of which are fixed to either end, and one which moves between them. When only one of the two fixed terminals is used, the second symbol sometimes appears (see Chapter 12 for notes on this usage).
Trimmer resistor	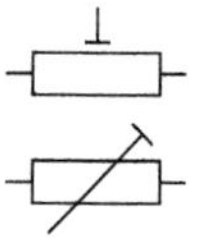	A variation on the variable resistor symbols sometimes appears where the arrow head is replaced by a bar. This indicates a preset or trimmer type component. These devices are much smaller than conventional pots, and are designed to be adjusted only during circuit calibration, not in normal circuit use.
		Capacitors, Inductors, and Transformers
Capacitor		As is the case with many standard symbols, the circuit diagram symbol for a capacitor suggests the physical construction of the device. Capacitors consist of two conductive plates, separated by a non conductive layer. this structure is clearly implied by the pair of parallel lines in the symbol. This symmetrical symbol indicates a non-polar capacitor, in contrast to the set of symbols which follow.

Table C.1 (continued...)

Name	Symbol	Description
Polar capacitor	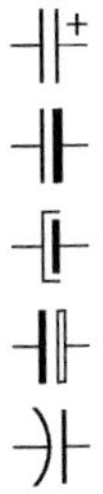	The plus sign in the first symbol here indicates a polar capacitor. In this case it is important which way round the component is connected. The positive lead goes on the side of the plus sign and the negative lead on the other. Alternative methods of indicating a polar device are also illustrated. These may or may not include a plus sign.
Variable capacitor		Variable capacitors are not all that common. Where they are encountered they typically have very small values. They can be used for tuning frequency dependant circuits.
Inductor (air core)	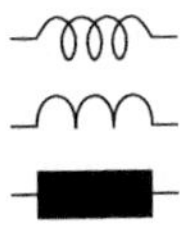	Again, the symbol is strongly suggestive of the structure of the component. An inductor is just a tightly wrapped coil of wire. In an air core inductor the coil has air or a non-ferromagnetic former in the middle.
Inductor (iron core)	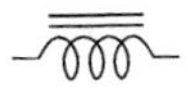	In an iron core inductor the coil is wrapped around a core made of solid iron or an iron laminate. This can improve its electrical characteristics.
Inductor (ferrite core)	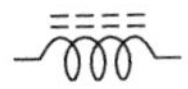	In this case the core is made out of a ferrite (iron) paste. Using this substance instead of solid iron can further improve some of the inductor's electrical characteristics.
Variable inductor		As in previous examples, the diagonal arrow indicates a variable value component. The inductance of a variable inductor can be altered. Variable inductors are rarely if ever found in standard audio electronics.
Transformer	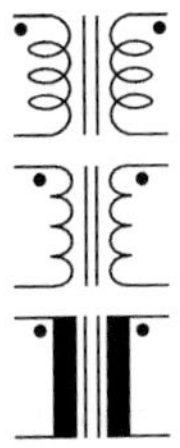	Transformers step voltages up or down. They are made of two coils of wire called the primary and the secondary. Once again, this is clearly reflected in the circuit symbol. The dot on the symbol indicates which end of a winding is designated the start.
Transformer (centre tap)	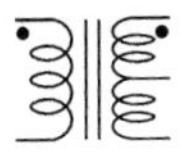	Often an additional connection is available from the middle of one or both of a transformer's windings. This is called a centre tap. Taps can also be taken off at other positions along a winding, providing multiple possible signal levels.

Table C.1 (continued...)

Name	Symbol	Description
Transformer (dual winding)		Occasionally a transformer will be manufactured with two separate coils on one or both sides. These can be wired together in different ways to achieve different step up or step down ratios, or can be used independently.
Diodes – the most common types are shown here. Many others exist.		
Diode		A standard diode allows current to flow in one direction and not the other. The symbol can be seen as an arrow showing the direction of flow, with the bar at the arrow tip blocking the reverse direction.
LED		A light emitting diode (LED) gives off light of a particular colour when it conducts current. The arrows emerging from the symbol illustrate this behaviour.
Laser diode		A laser diode produces coherent, columnated (i.e. laser) light, rather than the less controlled emissions from a standard LED.
Zener diode		A Zener diode is wired into a circuit in the opposite direction to a standard diode. The orientation should not be confused when connecting such a device. It is designed to provide a constant, well defined voltage across its terminals when reverse biased.
Schottky diode		A Schottky diode is very similar in function to a standard diode except that it has a lower switch on voltage and faster switching, but also a higher reverse bias leakage current.
Transistors come in a wide variety of types and subtypes. The variants shown here cover all the commonly encountered devices and a few more besides. All the symbols shown are also often seen with a surrounding circle. The two styles are interchangeable.		
Transistor (NPN BJT)	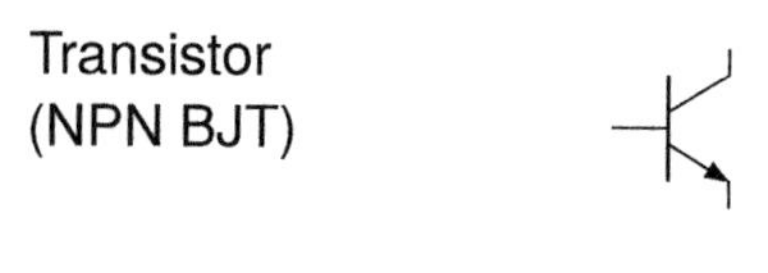	A bipolar junction transistor (BJT). The arrow heading out indicates that the polarity of the device is NPN. The three terminals are the collector at the top, the base to the side, and the emitter at the bottom, with the arrow.

Table C.1 (continued...)

Name	Symbol	Description
Transistor (PNP BJT)	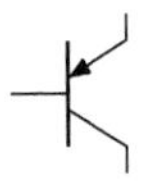	A bipolar junction transistor (BJT). The arrow heading in indicates that the polarity of the device is PNP. Here the three terminals are the emitter at the top (with the arrow), the base to the side, and the collector at the bottom.
Transistor (NPN Dar. BJT)	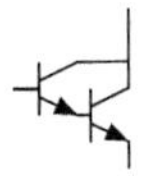	An NPN Darlington BJT is an NPN BJT with a very high gain. Outward arrows indicate the NPN type. Again, collector at top, base to the side, and emitter at bottom.
Transistor (PNP Dar. BJT)	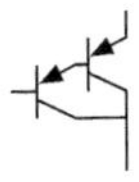	A PNP Darlington BJT is a PNP BJT with a very high gain. Inward arrows indicate the PNP type. Again, emitter at top, base to the side, and collector at bottom.
Transistor (n-ch. JFET)	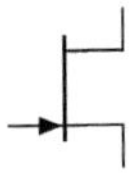 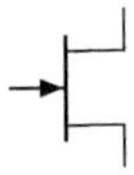	An n-channel junction field effect transistor (JFET). The inward arrow indicates the n-channel type. The three terminals are the drain at the top, the gate to the side, with the arrow, and the source at the bottom. The second variant of the symbol, with the gate entering at mid-level reflects the fact that some JFETs are manufactured symmetrical, and the collector and emitter connections may be reversed.
Transistor (p-ch. JFET)	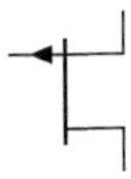 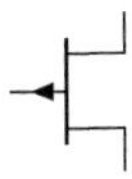	A p-channel junction field effect transistor (JFET). The outward arrow indicates the p-channel type. The three terminals are the source at the top, the gate to the side, and the drain at the bottom.
Transistor (n-ch. MOSFET enh.)	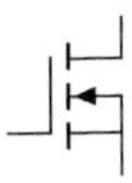	An enhancement mode (enh.), n-channel metal oxide semiconductor field effect transistor (MOSFET). The inward arrow indicates n-channel. The thick dashed line indicates enhancement mode (aka normally off). Terminal labelling is as for the n-channel JFET.
Transistor (p-ch. MOSFET enh.)	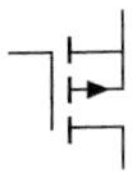	An enhancement mode (enh.), p-channel metal oxide semiconductor field effect transistor (MOSFET). The outward arrow indicates p-channel. The thick dashed line indicates enhancement mode (aka normally off). Terminal labelling is as for the p-channel JFET.

Table C.1 (continued...)

Name	Symbol	Description
Transistor (n-ch. MOSFET dep.)	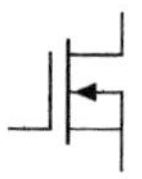	A depletion mode (dep.), n-channel metal oxide semiconductor field effect transistor (MOSFET). The inward arrow indicates n-channel. The thick solid line indicates depletion mode (aka normally on). Terminal labelling is as for the n-channel JFET.
Transistor (p-ch. MOSFET dep.)		A depletion mode (dep.), p-channel metal oxide semiconductor field effect transistor (MOSFET). The outward arrow indicates p-channel. The thick solid line indicates depletion mode (aka normally on). Terminal labelling is as for the p-channel JFET.
	Other Discrete Semiconductor Devices	
Photoresistor	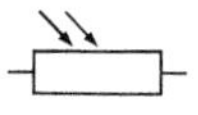	A photoresistor or light dependant resistor (LDR) is a resistor whose value changes depending on how much light is falling on it.
Thermistor	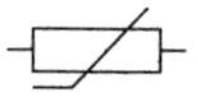	A thermistor is a resistor whose resistance changes with its temperature. Not often found in audio electronic, thermistors are commonly used in power electronics to detect overheating in a circuit or device.
Thermistor (PTC)	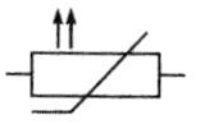	Thermistors come in two varieties depending on how the resistance changes with changing temperature. In positive temperature coefficient (PTC) devices the resistance rises as the temperature rises, and falls as the temperature falls. The PTC symbol has two arrows pointing in the same direction.
Thermistor (NTC)	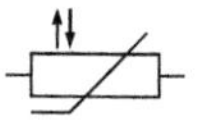	In a negative temperature coefficient (NTC) thermistor the resistance falls as the temperature rises, and rises as the temperature falls. The NTC symbol has two arrows pointing in opposite directions.
Varistor	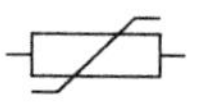	A varistor, MOV (metal oxide varistor), or VDR (voltage dependant resistor) (sometimes more generically referred to as a TVS or transient voltage suppressor) is a device which exhibits a very high resistance until a specified threshold voltage is applied, at which point the resistance falls sharply.

Table C.1 (continued...)

Name	Symbol	Description
Photodiode		A photodiode is a diode in which the semiconductor junction is exposed to allow light to fall upon it. When illuminated, photodiodes generate electricity. This component is the basis of the solar cell used in solar electricity generation.
Phototransistor		A phototransistor is a transistor which is controlled by light falling on the semiconductor junction, rather than the usual control signal applied to the base terminal.

The DIAC, thyristor, and TRIAC are related semiconductor devices which are often used in combination. One of their commonest applications is in the control circuitry used to implement adjustable lighting dimmers.

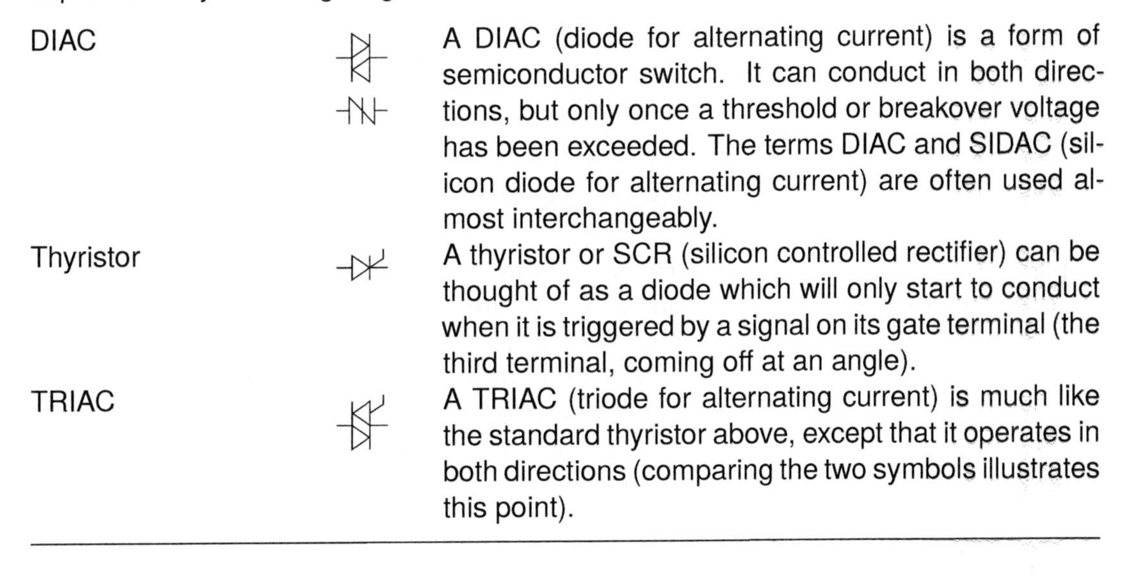

Name	Symbol	Description
DIAC		A DIAC (diode for alternating current) is a form of semiconductor switch. It can conduct in both directions, but only once a threshold or breakover voltage has been exceeded. The terms DIAC and SIDAC (silicon diode for alternating current) are often used almost interchangeably.
Thyristor		A thyristor or SCR (silicon controlled rectifier) can be thought of as a diode which will only start to conduct when it is triggered by a signal on its gate terminal (the third terminal, coming off at an angle).
TRIAC		A TRIAC (triode for alternating current) is much like the standard thyristor above, except that it operates in both directions (comparing the two symbols illustrates this point).

Table C.1 (continued...)

Name	Symbol	Description
		Integrated Circuits (ICs)
DIP IC	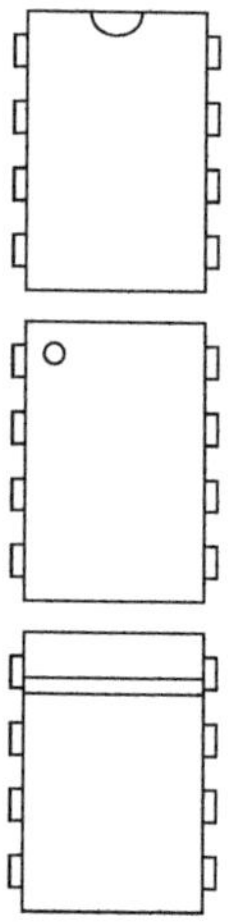	The commonest integrated circuits (ICs) come in dual inline packages (DIPs), most often having 8, 14, or 16 legs. The symbol shown is for an eight pin device but others will follow the same pattern. Legs are numbered counterclockwise from the leg at the top left. Any IC can be represented in this fashion, but some ICs also have special symbols; some of these are shown next. The special versions often omit some standard connections like the necessary power connections.
Opamp	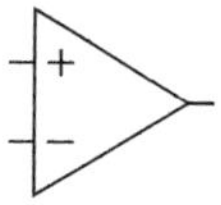	The operational amplifier (opamp) is probably the commonest IC used in audio circuits.
Opamp (power shown)	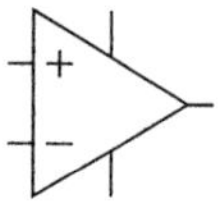	The two extra connections indicate power connections. These are always needed but are not always shown explicitly like this.
OTA	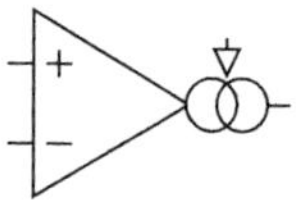	Operational transconductance amplifiers (OTAs) are especially used in synthesiser circuitry where they provide the voltage controlled elements in VCAs, VCOs, and VCFs.
Logic gates – the symbols which follow can be used to indicate various standard kinds of ICs in a circuit. Logic gates are not generally intended for use in audio circuits but some very inventive ways of using them have been devised.		
Buffer	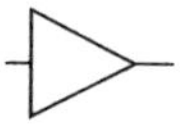	A buffer does nothing but isolate one circuit section from another.
Inverter	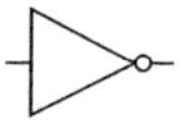	An inverter or NOT gate outputs the opposite of its input. If a low voltage goes in, a high voltage comes out, and vice versa.
AND	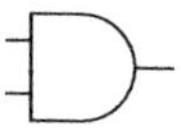	An AND gate's output only goes high when both inputs are high.

Table C.1 (continued...)

Name	Symbol	Description
OR		An OR gate's output goes high when one or both inputs go high.
XOR		XOR stands for exclusive OR. It's output goes high if one input is high and the other input is low.
NAND		NAND stands for NOT AND. It does the opposite of an AND gate. Its output only goes low when both inputs are high.
NOR		NOR (NOT OR) does the opposite of OR. Its output only goes high if both inputs are low.
XNOR		XNOR (NOT XOR) does the opposite of XOR. It's output goes high if both inputs are high or if both inputs are low.
NOT (Schmitt trigger) NAND (Schmitt trigger)		The Schmitt trigger version of a gate modifies the switching action. Normally switching simply happens around the mid voltage. This can cause intermediate or unstable output levels if the input stays around the midpoint. In order to avoid this hysteresis is applied to the switching.

Vacuum Tubes (also known as tubes, valves, and thermionic valves) were the precursor of semiconductor devices like diodes and transistors. They are rare now in general usage, but are still found in some audio electronics. The symbols below cover only two of what was originally a larger family of electronic components. In order to work the cathode must be heated. Symbols may or may not show the heater connections.

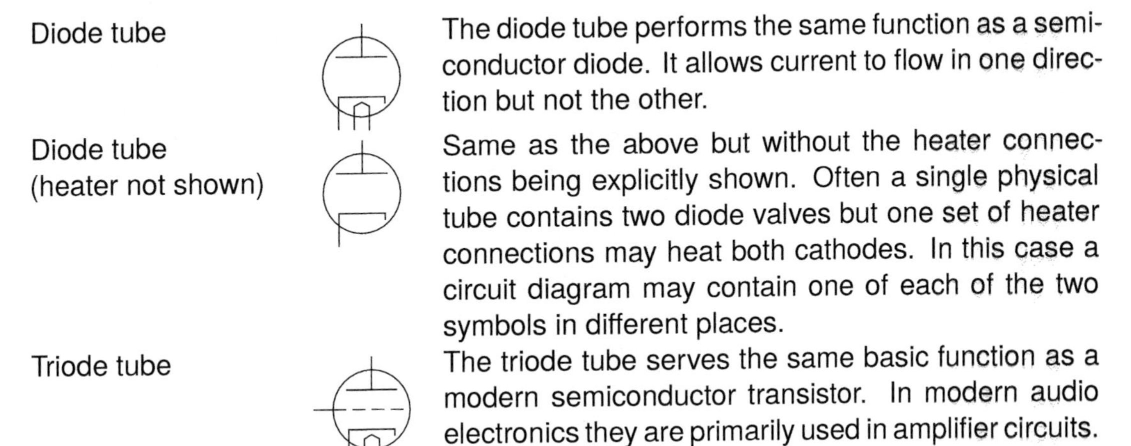

Name	Symbol	Description
Diode tube		The diode tube performs the same function as a semiconductor diode. It allows current to flow in one direction but not the other.
Diode tube (heater not shown)		Same as the above but without the heater connections being explicitly shown. Often a single physical tube contains two diode valves but one set of heater connections may heat both cathodes. In this case a circuit diagram may contain one of each of the two symbols in different places.
Triode tube		The triode tube serves the same basic function as a modern semiconductor transistor. In modern audio electronics they are primarily used in amplifier circuits.

Table C.1 (continued...)

Name	Symbol	Description
Triode tube (heater not shown)		Same as the above but without the heater connections being explicitly shown. Often a single physical tube contains two triode valves but one set of heater connections may heat both cathodes. In this case a circuit diagram may contain one of each of the two symbols in different places.
		Transducers
Microphone		A microphone turn acoustic sound into an electrical signal. The line on the left of the symbol is intended to represent the microphone's diaphragm. The two lines to the right are the connection points.
Loudspeaker		Loudspeakers turn electrical signals into acoustic sounds. The splayed lines on the right represent the cone of the loudspeaker. The two lines to the left are the connection points.
Piezo element		Piezo elements are bidirectional. They can turn vibrations into electrical signals and electrical signals into vibrations. As such they can be used as both pickups and drivers. The three terminal version is called a self drive element. It can be used to make a very simple buzzer circuit.
Coil pickup	N S	Coil pickups (aka electric guitar pickups) consist of a coil of wire around a magnet. The dot indicates which end of the coil is the start. This information is used when determining the wiring polarity of the pickup.
Reverb pan	Drive Recover	Reverb pan, reverb tank, or spring reverb adds reverb to a signal by transmitting it along a set of carefully mounted springs.
Optocoupler		Optocouplers (aka opto-isolators) come in a number of different forms, but usually involve an LED shining on a light sensitive device of some kind. In the two variants shown the sensors are a photoresistor and a phototransistor respectively. Optocouplers provide electrical isolation between their two sides. They can also add interesting distortion to an audio signal.

Table C.1 (continued...)

Name	Symbol	Description
		Switches
Switch (spst)		A standard single pole single throw (spst) switch provides the most basic on-off functionality.
Switch (spst NO)	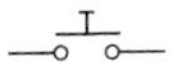	A single pole single throw normally open (spst NO) switch is a momentary switch which provides a contact when actuated and breaks the connection again once released.
Switch (spst NC)	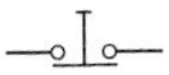	A single pole single throw normally closed (spst NC) switch is a momentary switch which normally provides a connection. The connection is broken when the switch is actuated and reconnects again once released.
Switch (spdt)	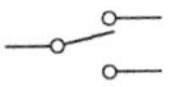	A standard single pole dual throw (spdt) switch provides for the connection to either of two different places.
Switch (spdt on-off-on)	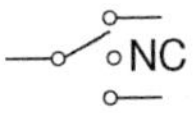 NC	An spdt on-off-on switch has three positions but only the outer two provide an electrical connection. NC = no connection.
Switch (dpst)	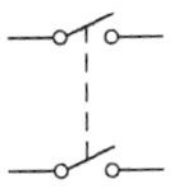	A standard dual pole single throw (dpst) switch is just two spst switches which move together. The two switches may be in totally different parts of a circuit. The dotted line between them indicates they are part of a single device. Sometimes labels are used instead of a dotted line; the two parts will carry labels of the form SW1a and SW1b for instance.
Switch (dpdt)	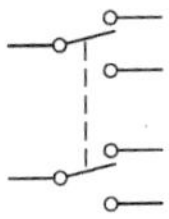	A standard dual pole dual throw (dpdt) switch is just two spdt switches which move together. The two switches may be in totally different parts of a circuit. The dotted line between them indicates they are part of a single device.
Switch (sp5t)	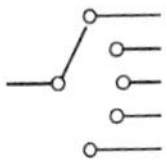	Higher numbers of throws are often laid out in an arc or circle. Here a five throw device is shown.
Switch (spst relay)	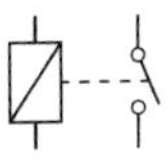	A relay is a switch which is actuated by an electrical signal rather than by being moved manually.

Table C.1 (continued...)

Name	Symbol	Description
		Miscellaneous Symbols
Gas discharge tube	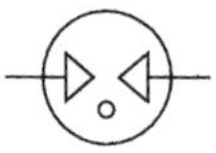	A gas discharge tube (GDT) can be encountered in a number of different variations and applied to various applications. The form most likely to be encountered in audio electronics acts as a TVS (transient voltage suppression) device, similar to the action of the MOV described earlier. They can help to protect circuitry from the damaging effects of lightning strikes and other high level, short lived voltage transients. When a transient of sufficient energy arrives, the gas in the tube becomes conductive (acting much like the gap in a spark plug), and the energy in the transient is shunted to ground, hopefully protecting the electronics which follows.
Incandescent bulb	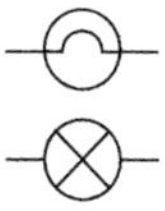	An incandescent bulb (or filament bulb) produces light when an electric current passes through its filament. Sometimes these devices are used as control elements in circuits. As the wire in the bulb heats up, its resistance increases, acting as a kind of PTC thermistor.
Fuse	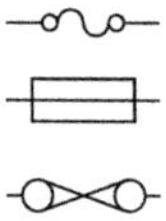	A fuse is a safety device in a circuit, usually associated with the circuit's power supply or input protection circuitry.
Cell		This symbol represents a simple cell such as a 1.5V battery like an AA or an AAA.
Battery	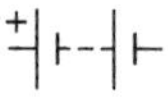	A battery providing anything more than 1.5V will usually be represented using something like this symbol.
Power rail		A general power rail can be represented with a symbol such as this. The voltage will usually be written above the line.
DC source	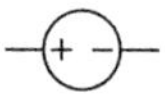	This symbol can represent either a battery or a DC power supply.
AC source	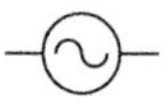	An AC signal source (which may represent either an audio signal or an AC power supply) can be represented like this.

Table C.1 (continued...)

Name	Symbol	Description
Current source	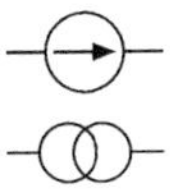	A current source in a circuit can often be represented by either of these two symbols. Notice that the OTA symbol introduced above incorporates the second of these two symbols on its output, indicating the fact that an OTA's output acts as a current source.
Meter	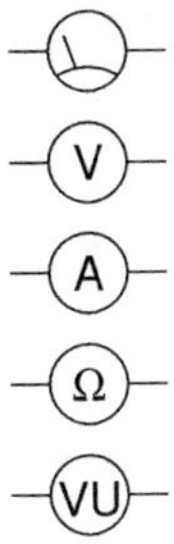 V, A, Ω, VU	A meter turns electrical signals into movements of a needle or numbers on a digital display. Each of these symbols can represent either a moving coil meter or a digital display meter. An annotation might indicate what the meter is designed to measure (shown here: V for volts, A for amps, Ω for ohms, and VU for volume units (a common audio scale)).
Terminal		A general connection point in a circuit. This might be an input or an output point in the circuit, or it might indicate a test point intended for testing and calibrating a circuit.
Interconnect		A general interconnection point. This symbol represents a single plug and socket pair, used to indicate a connection between two elements in a circuit diagram.
Socket	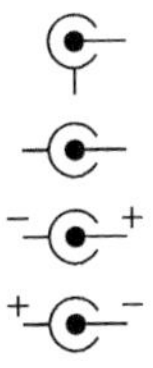 − +, + −	A generic two contact socket. Often used to show a power connector. Plus and minus signs may be included to indicate positive and negative terminals. Both tip positive and tip negative power supplies are common. In order to avoid damaging a circuit it is important that the wrong type is never connected to a circuit's power socket.
TS jack (unswitched)		A two terminal (Tip-Sleeve) unswitched jack or minijack socket, or similar.
TRS jack (unswitched)		A three terminal (Tip-Ring-Sleeve) unswitched jack or minijack socket, or similar.
TS jack (switched)		A two terminal (Tip-Sleeve) switched jack or minijack socket, or similar.
TRS jack (switched)		A three terminal (Tip-Ring-Sleeve) switched jack or minijack socket, or similar.

Table C.1 (continued...)

Name	**Symbol**	**Description**
Ground symbols – the four symbols below all represent connections to a circuit's ground. They all have slightly different meanings and different applications, as noted below, but they are often used pretty much interchangeably.		
Signal ground		Signal ground (or audio ground) represents the zero volts reference for the signal circuitry in a circuit.
Chassis ground		A point at which a circuit's ground plane is connected to the device's metal chassis.
Earth ground	⏚	A direct connection to the mains power earth pin or otherwise connected directly to the earth. Also just called ground.
Ground bar	⊥	A general ground connection might also be represented by this symbol. This can also represent the negative power rail in a split or dual power supply.

D | Circuit Catalogue

This appendix gathers together all the major circuits presented throughout the book and collects them here along with example breadboard and stripboard layouts for each. As has been stressed previously, there is no one right answer when it comes to drafting a circuit layout, and the reader is strongly encouraged to have a go at laying out these circuits themselves.

Circuit Index

Passive DI Box – see p. 253, Chapter 14 – Transformers

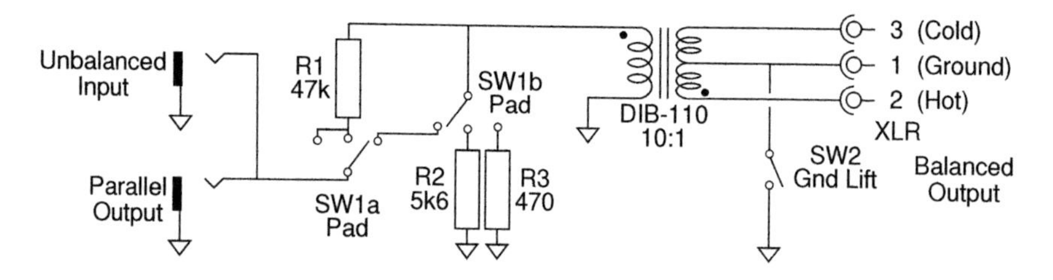

Single Transistor Fuzz – see p. 302, Chapter 17 – Transistors

9V
R1 100k
C2 220n
D1 1N4148
OUT
C1 4u7
IN
Q1 2N3904

Simple BJT Booster – see p. 302, Chapter 17 – Transistors

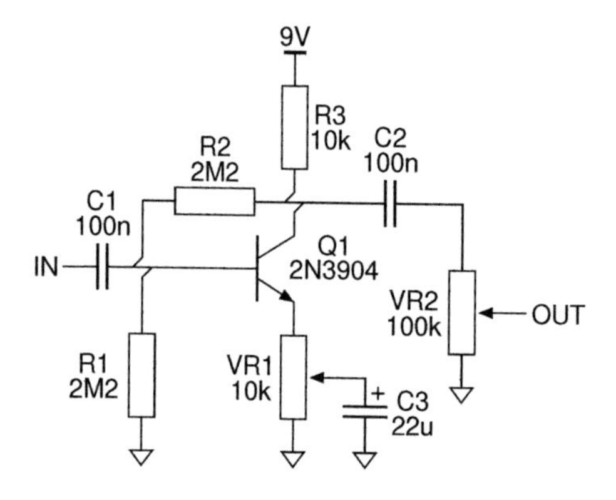

JFET Buffer – see p. 308, Chapter 17 – Transistors

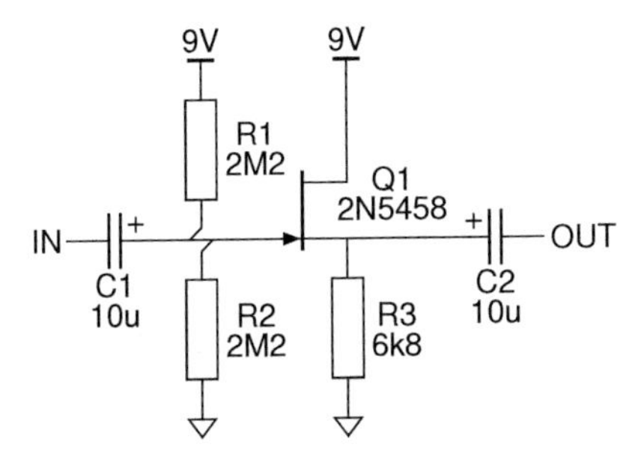

Opamp Preamp – see p. 323, Chapter 18 – Integrated Circuits

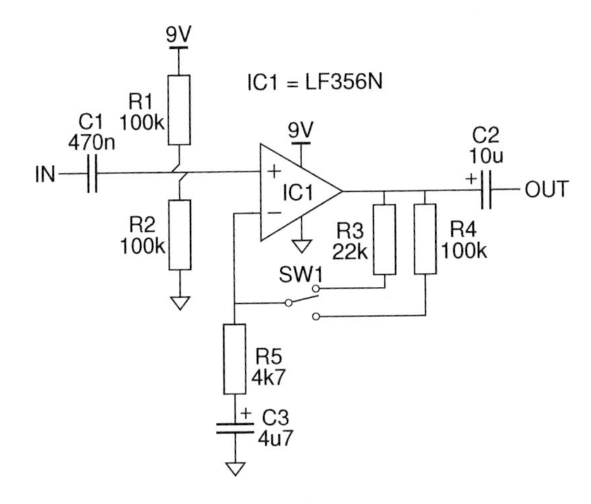

Regulated Power Supply – see p. 325, Chapter 18 – Integrated Circuits

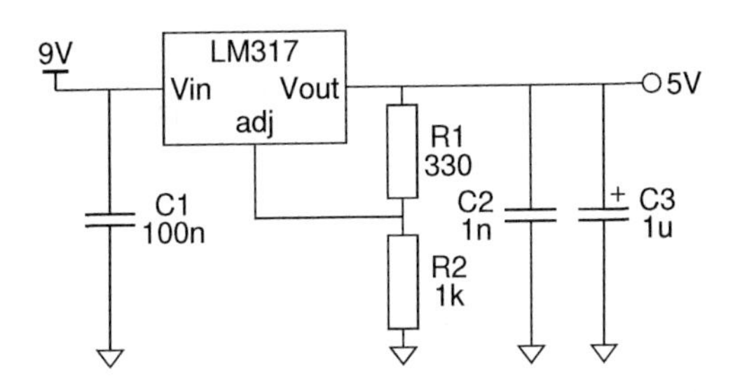

4093 Oscillator – see p. 330, Chapter 18 – Integrated Circuits

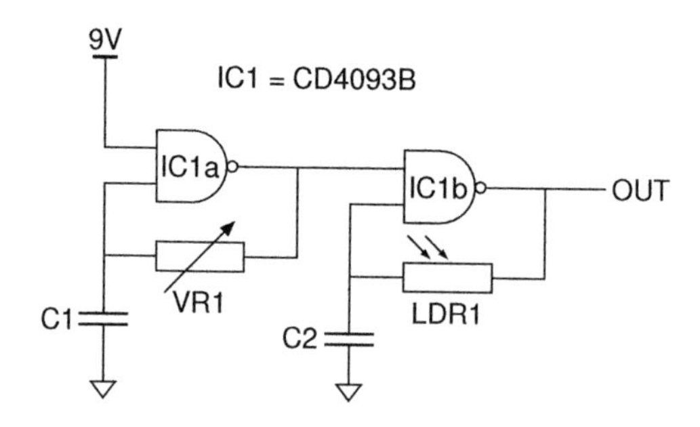

Low Voltage Tube Booster – see p. 342, Chapter 19 – Vacuum Tubes

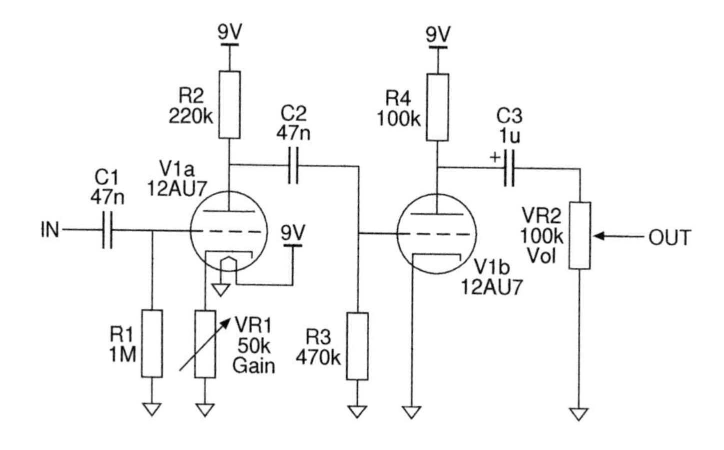

Phantom Powered Electret Microphone – see p. 348, Chapter 20 – Audio Transducers

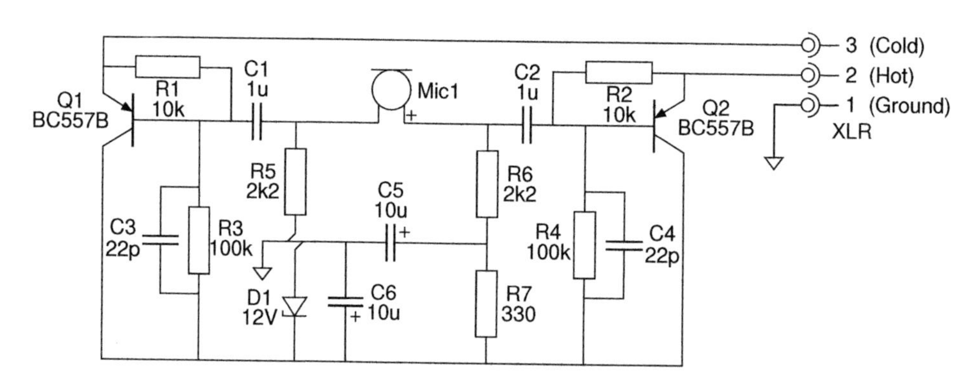

Spring Reverb Drive and Recovery – see p. 366, Chapter 20 – Audio Transducers

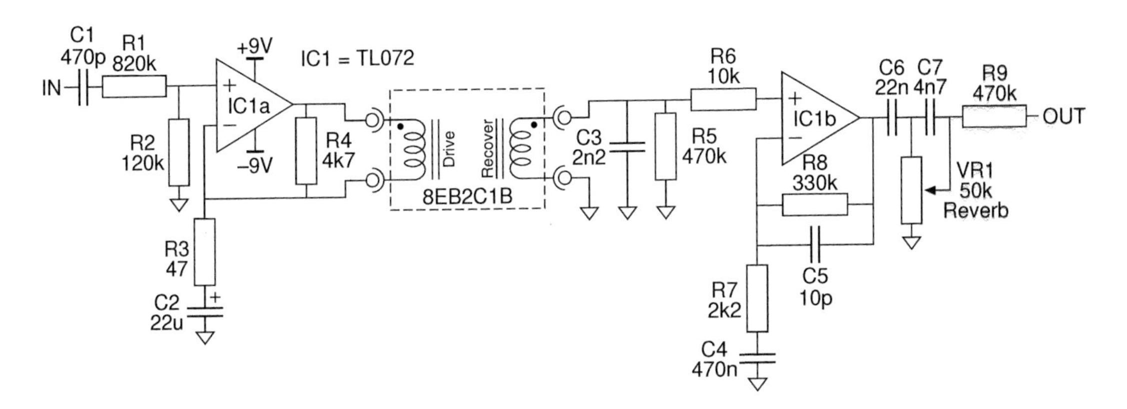

Piezo Oscillator – see p. 365, Chapter 20 – Audio Transducers

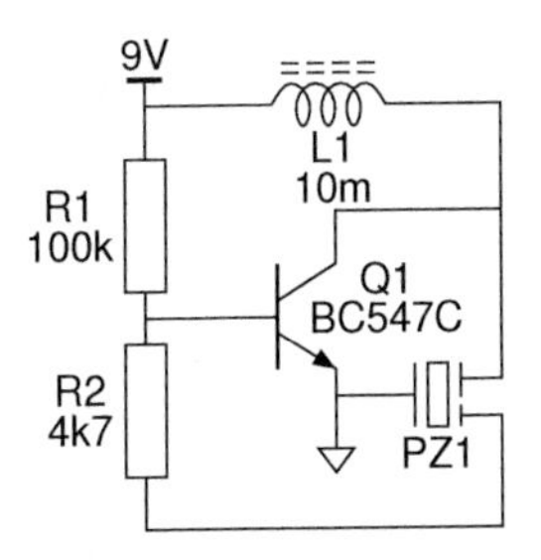

Low Power Audio Amplifier – see p. 163, Chapter 11 – Projects

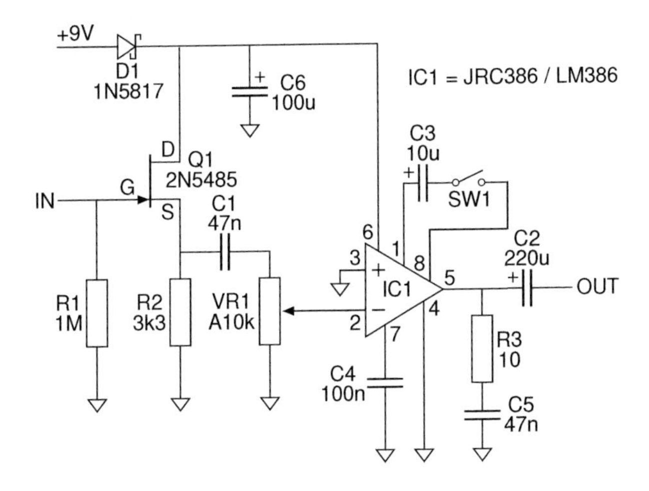

Stepped-Tone Generator – see p. 174, Chapter 11 – Projects

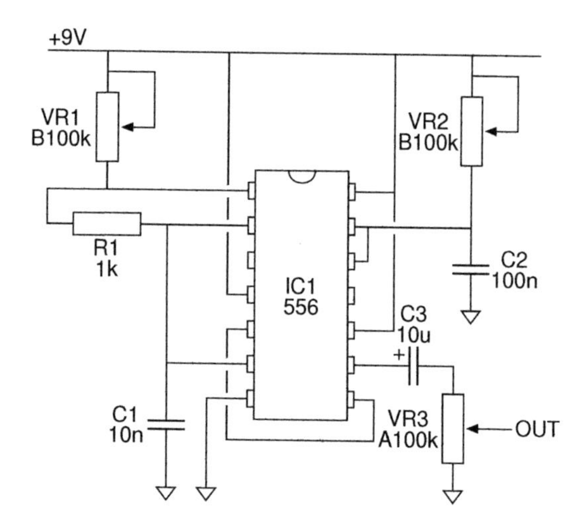

Modulation Effect – see p. 180, Chapter 11 – Projects

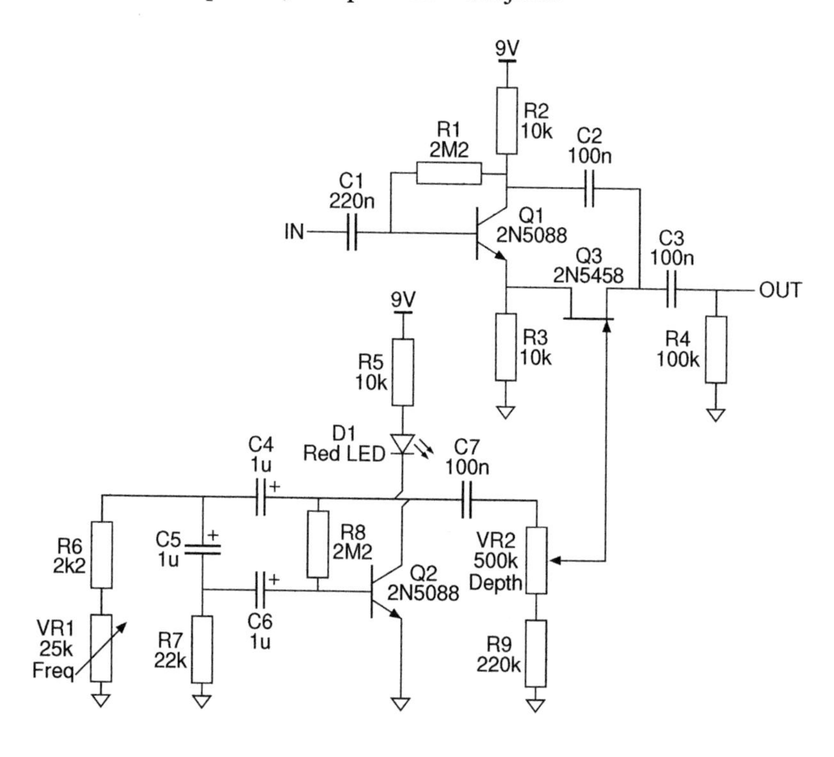

Noise Gate – see p. 115, Chapter 7 – Circuit Diagrams

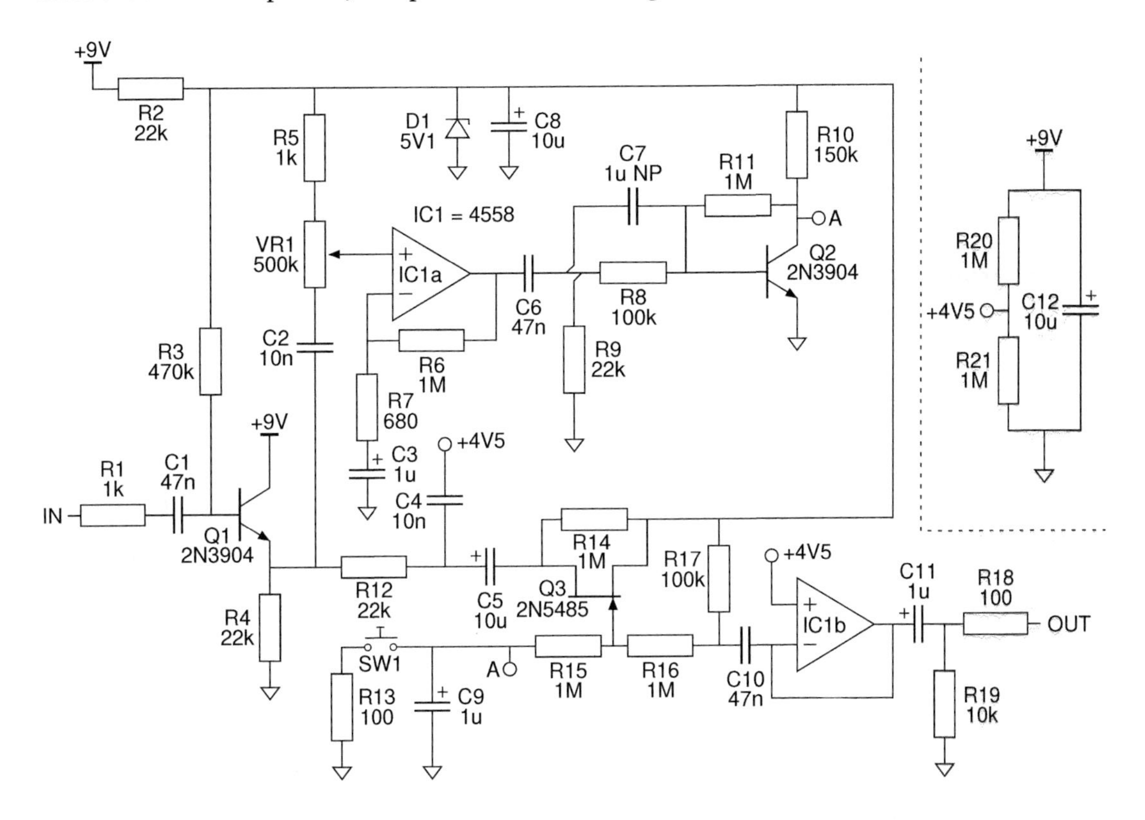

Breadboard Layouts

Passive DI Box – a breadboard layout of this circuit would consist of three resistors and a load of fly leads, and would not seem entirely worthwhile. The specified transformer has a suitable footprint for plugging in, but its legs are unlikely to fit into the breadboard holes and so it too would need fly leads soldered on. A stripboard layout, although also very simple, is furnished in the next section.

Single Transistor Fuzz

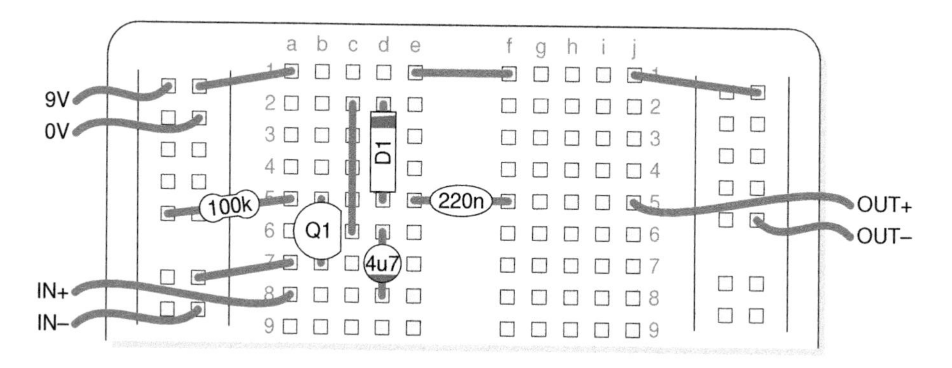

Simple BJT Booster

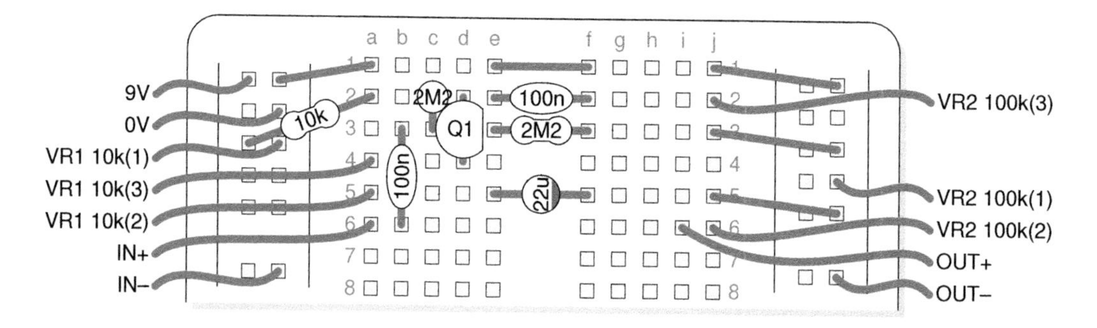

JFET Buffer

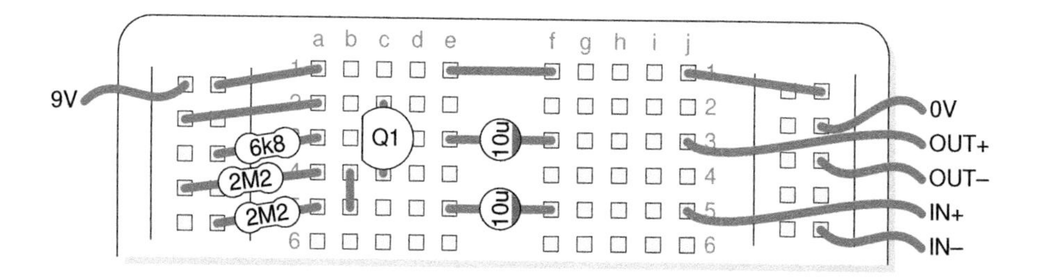

Opamp Preamp

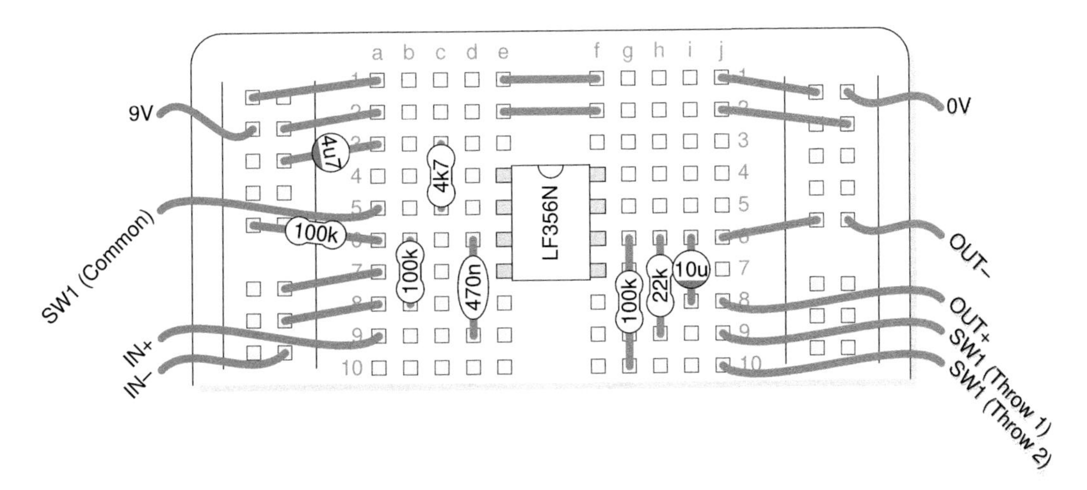

Regulated Power Supply

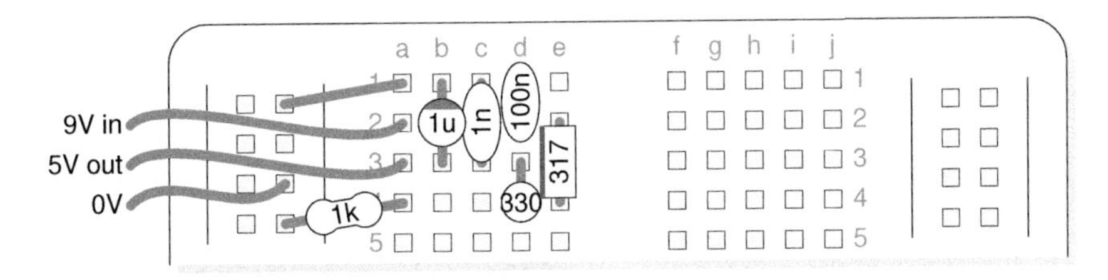

4093 Oscillator

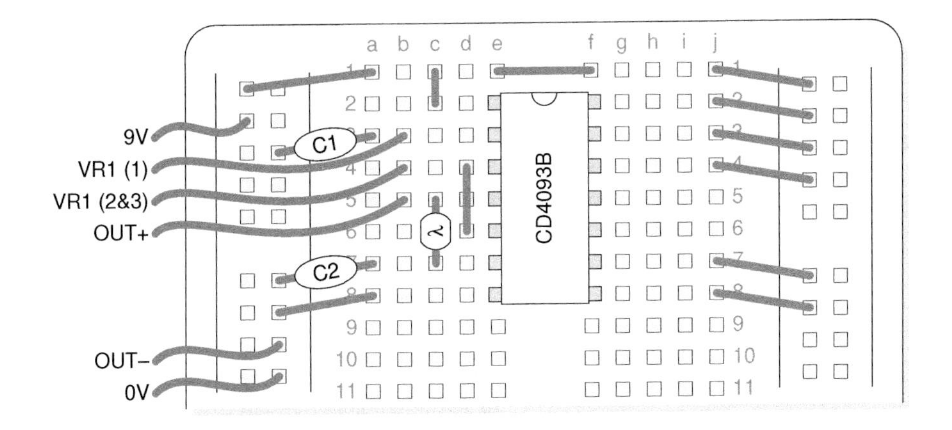

Low Voltage Tube Booster

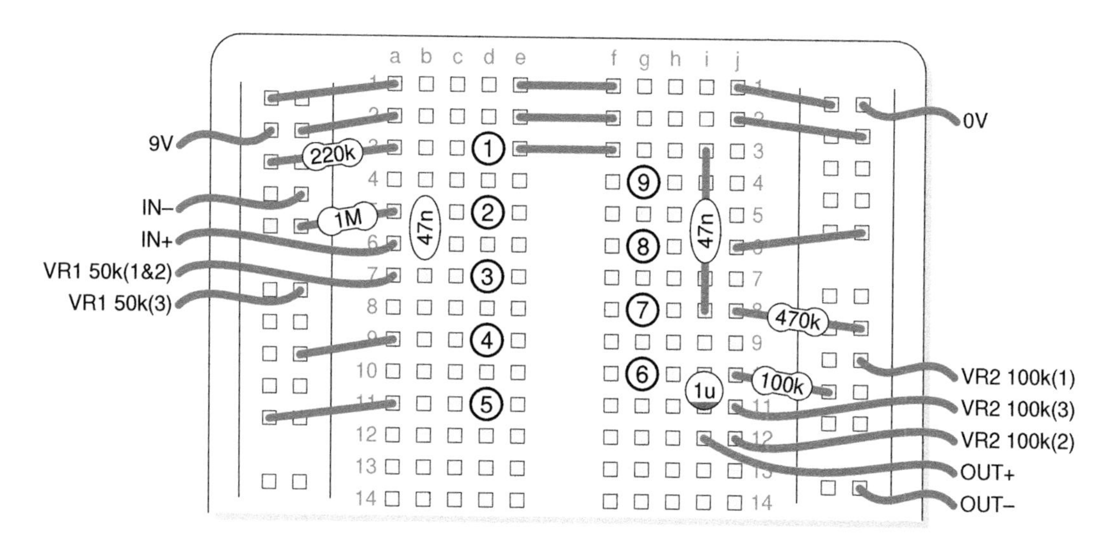

Phantom Powered Electret Microphone

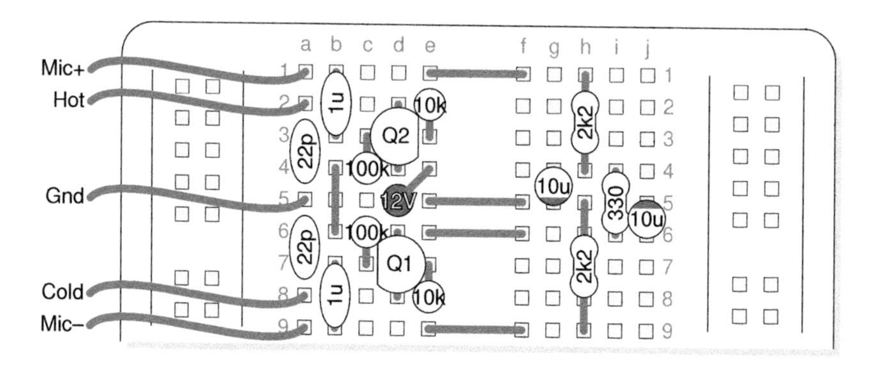

Piezo Oscillator

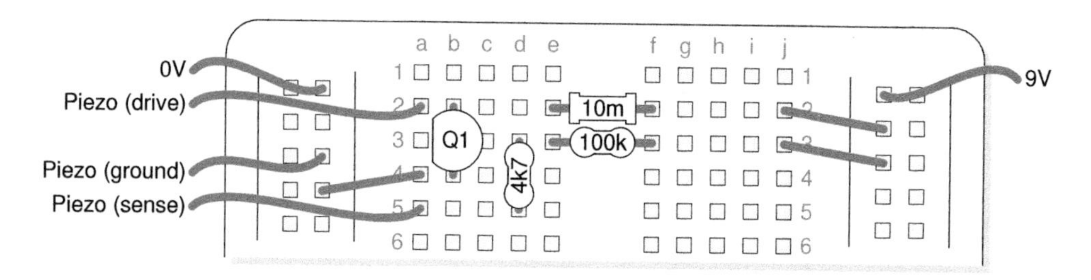

Spring Reverb Drive and Recovery

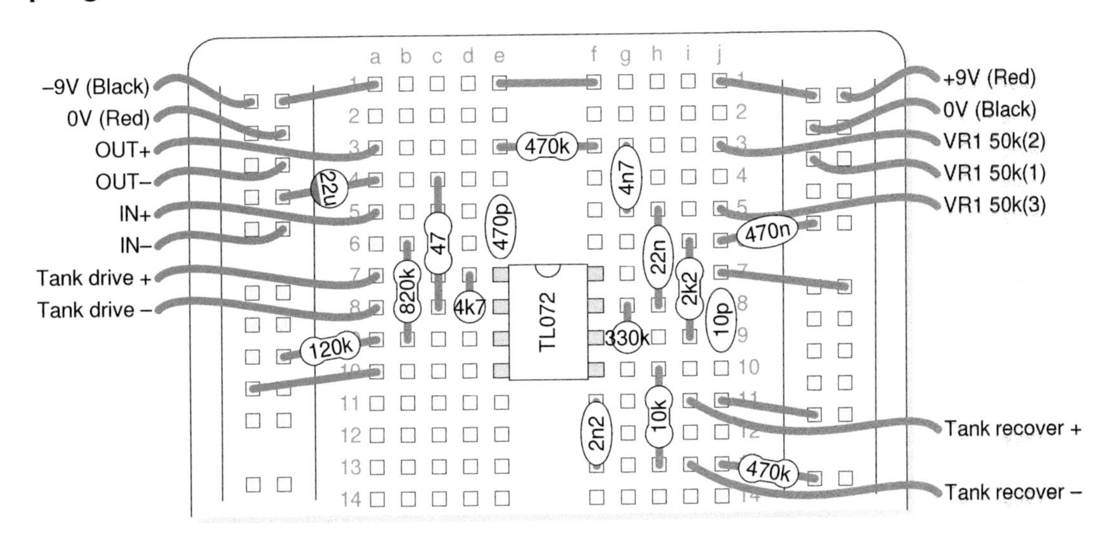

Low Power Audio Amplifier

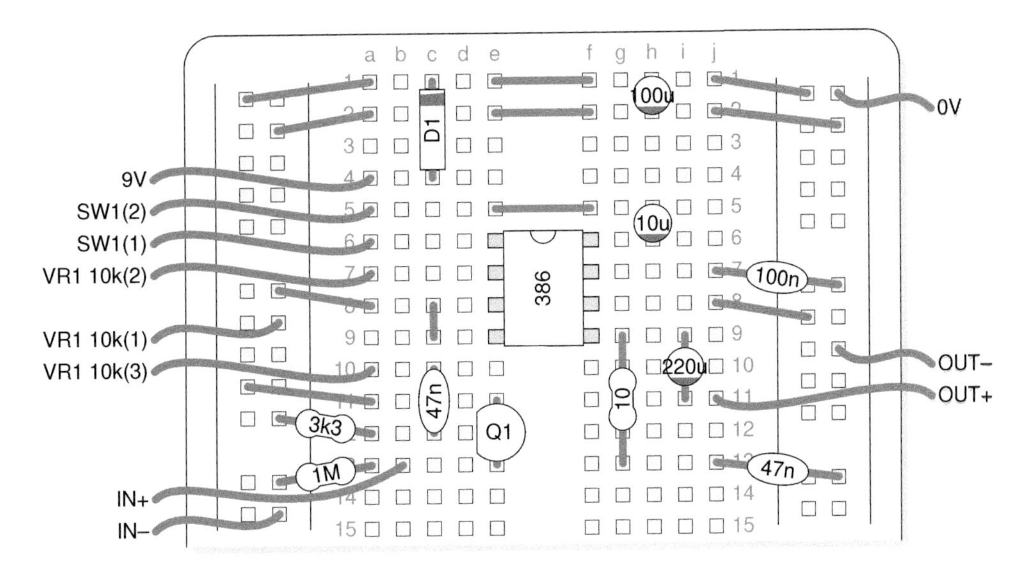

Stepped-Tone Generator

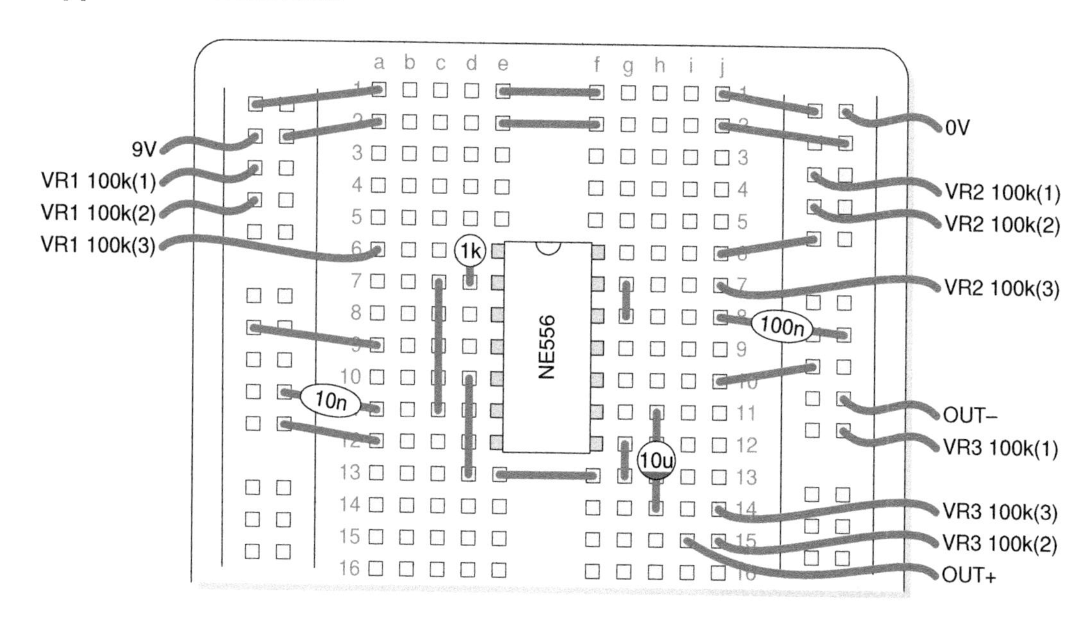

Modulation Effect

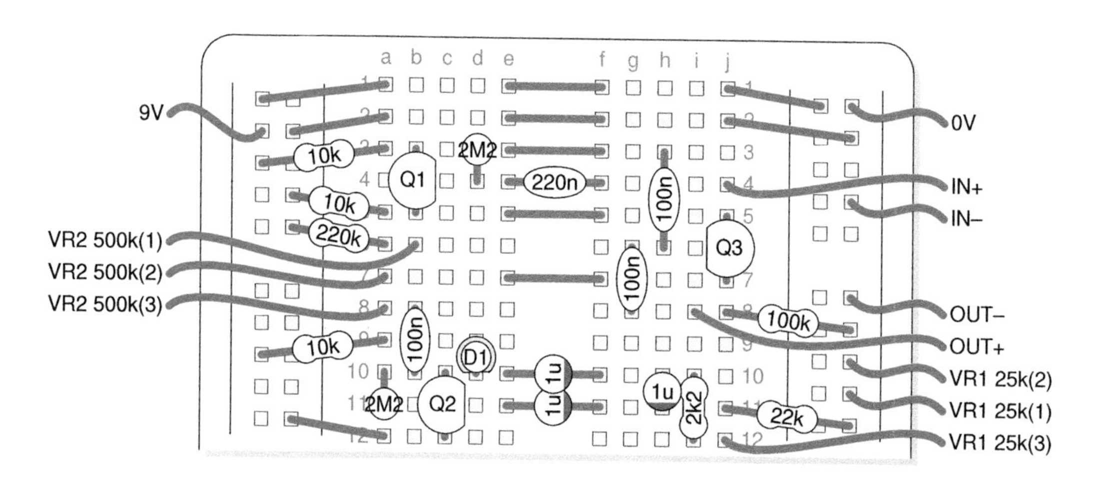

Noise Gate

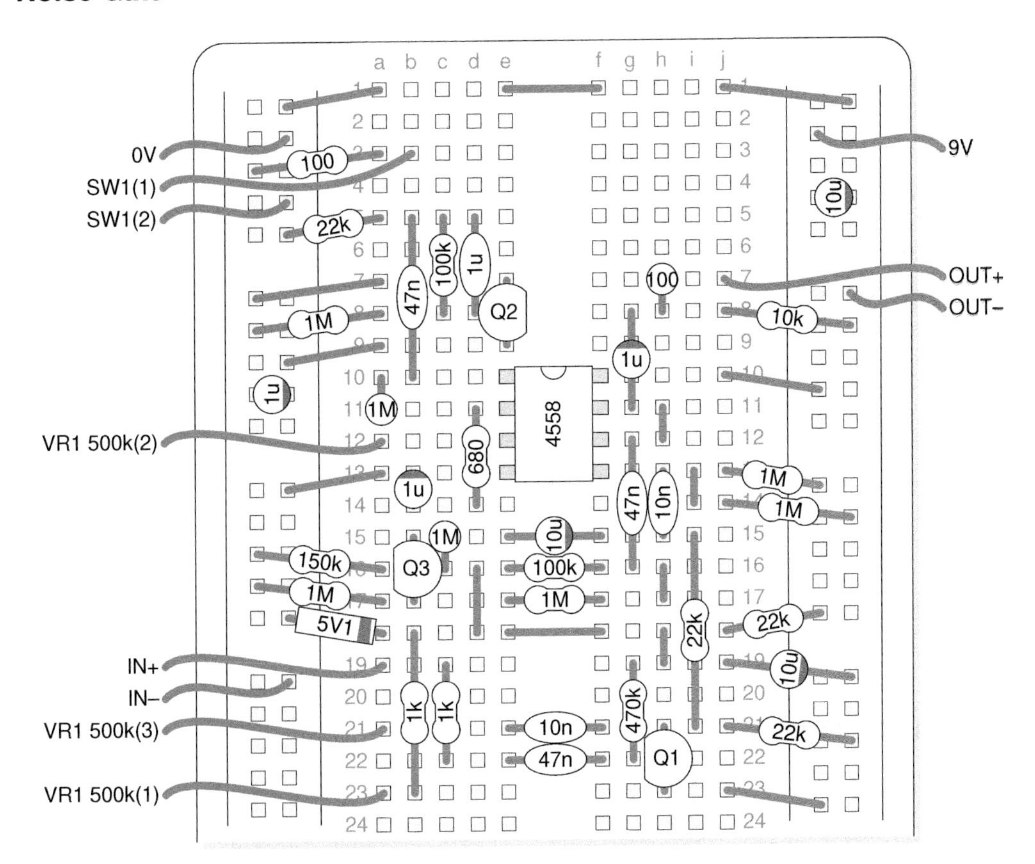

a b c d e f g h i j
0V
SW1(1)
SW1(2)
VR1 500k(2)
IN+
IN–
VR1 500k(3)
VR1 500k(1)
9V
OUT+
OUT–
100
22k
1M
1u
47n
100k
Q2
680
4558
Q3
150k
5V1
1k
10u
10n
470k
Q1
10k
22k

Stripboard Layouts

Passive DI Box – the five stripboard holes where the transformer legs are inserted will need to be slightly enlarged. A quick touch with a rotary tool will do the trick.

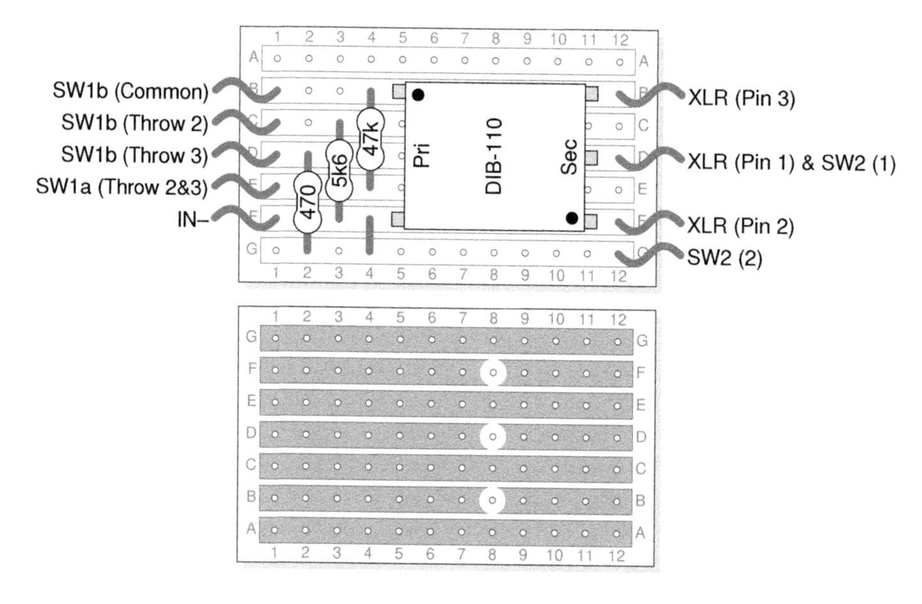

Single Transistor Fuzz

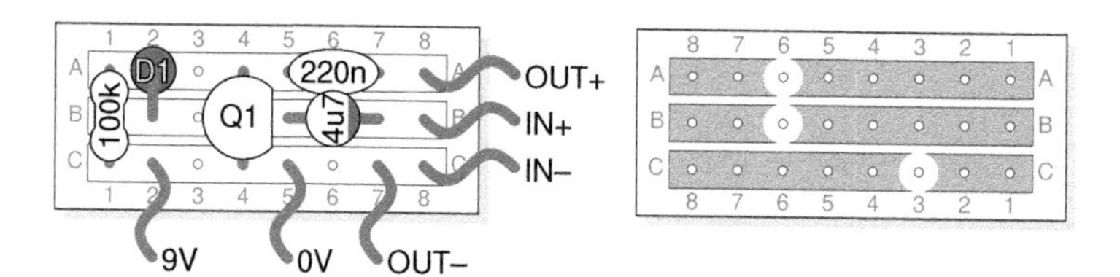

Simple BJT Booster

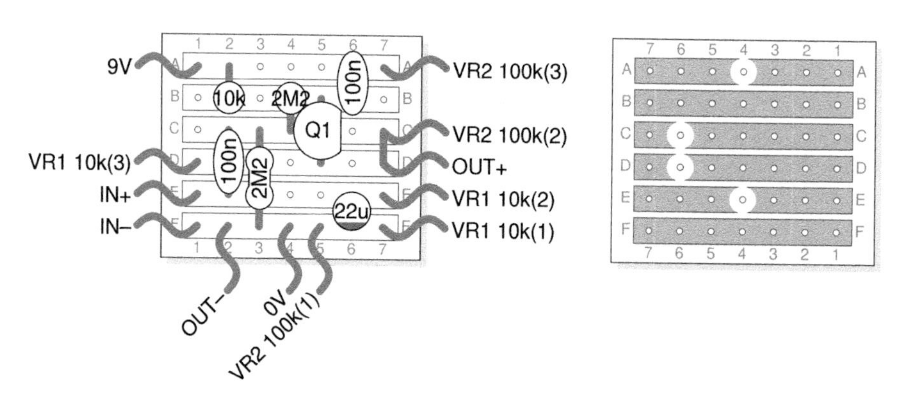

JFET Buffer

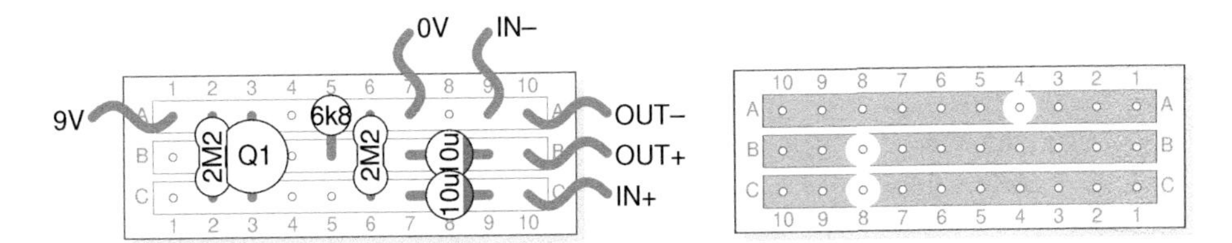

Opamp Preamp

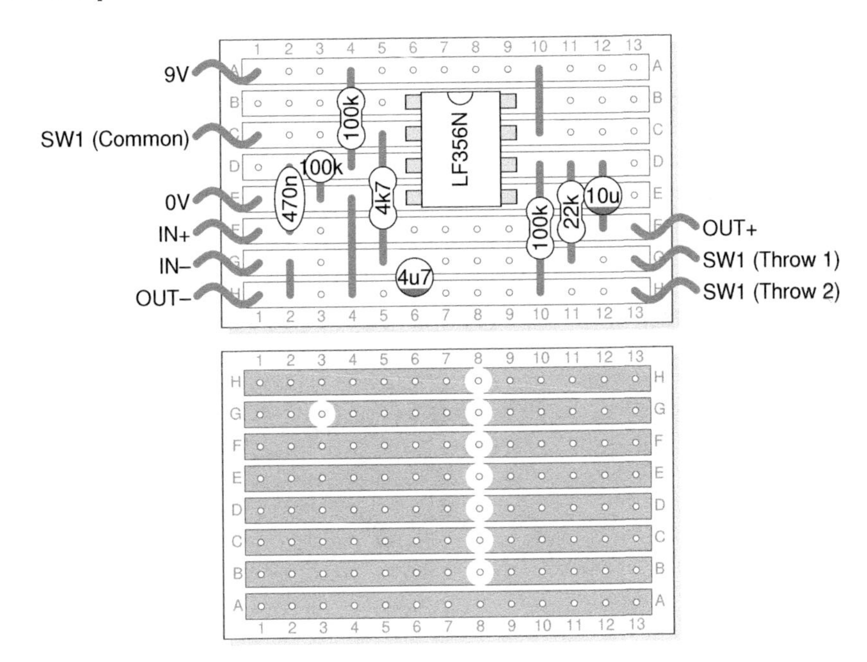

Regulated Power Supply

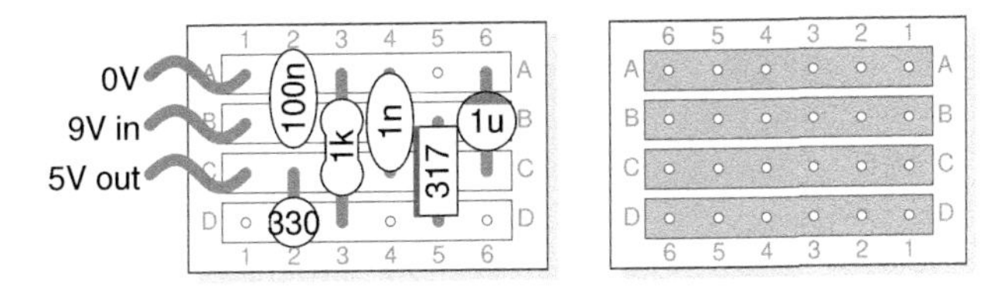

4093 Oscillator

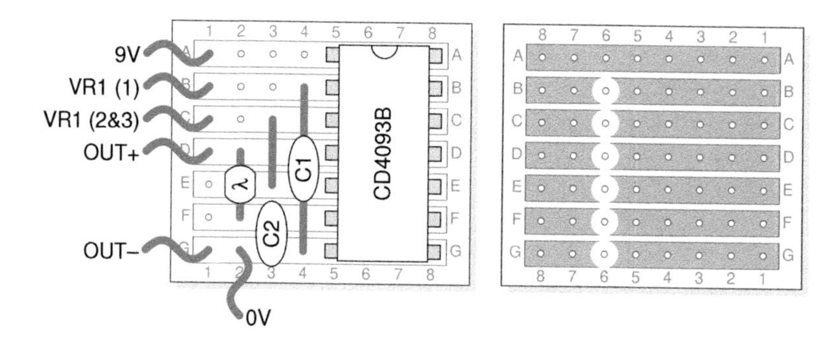

Low Voltage Tube Booster

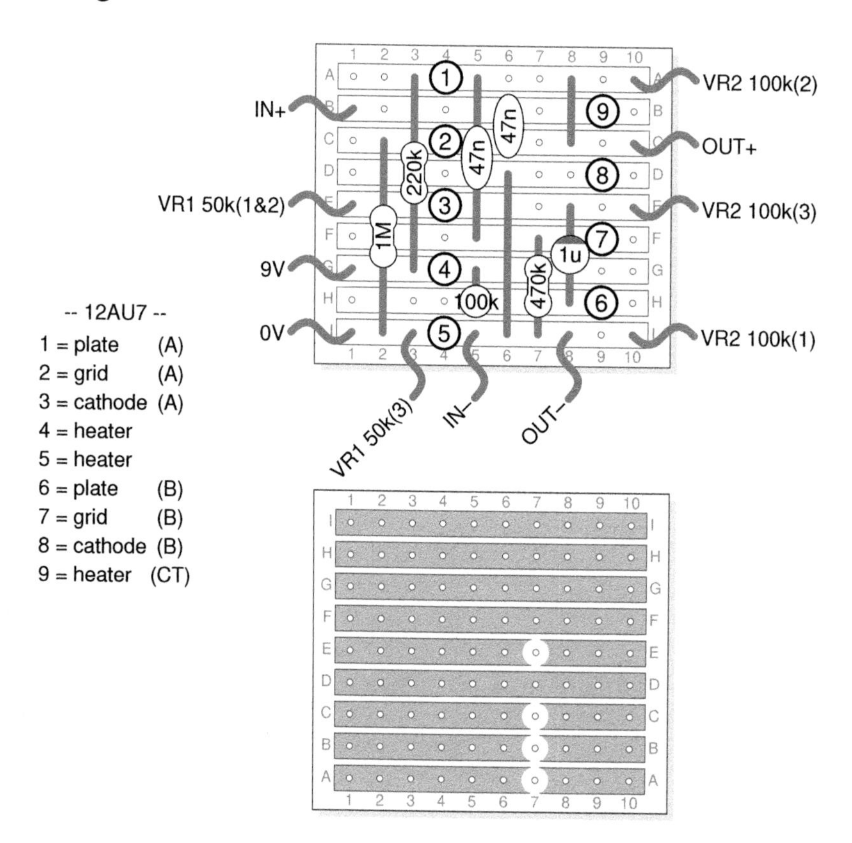

Phantom Powered Electret Microphone

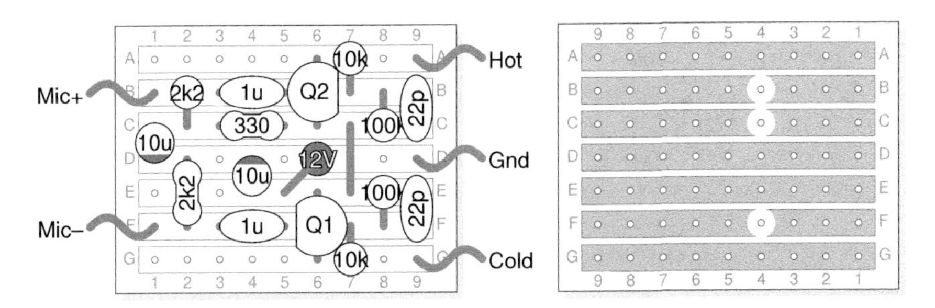

Spring Reverb Drive and Recovery

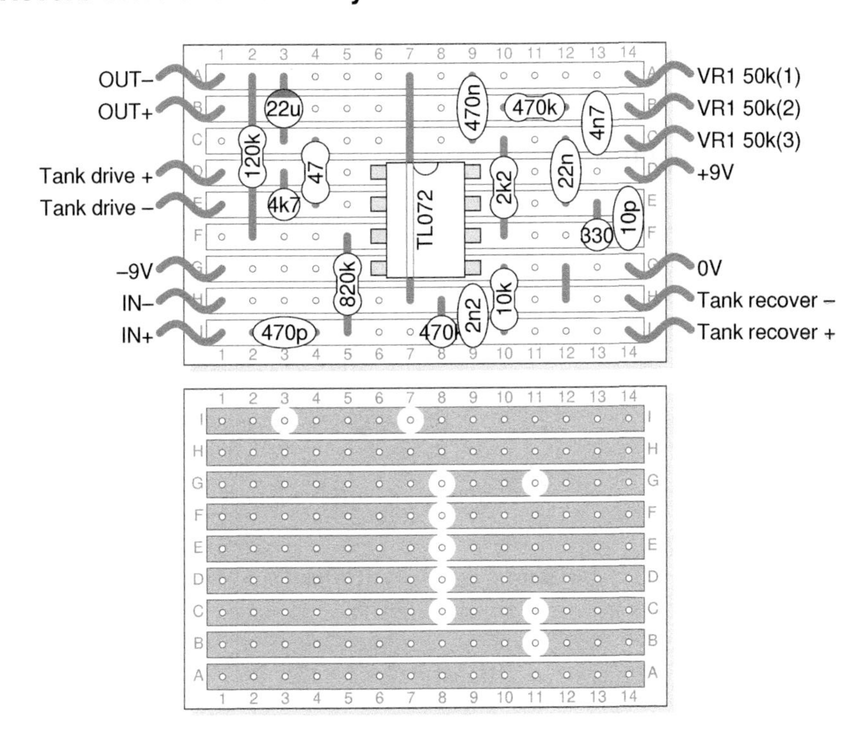

Piezo Oscillator

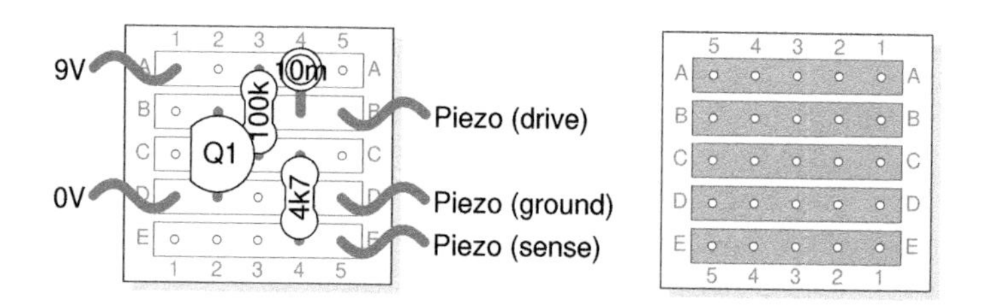

Low Power Audio Amplifier

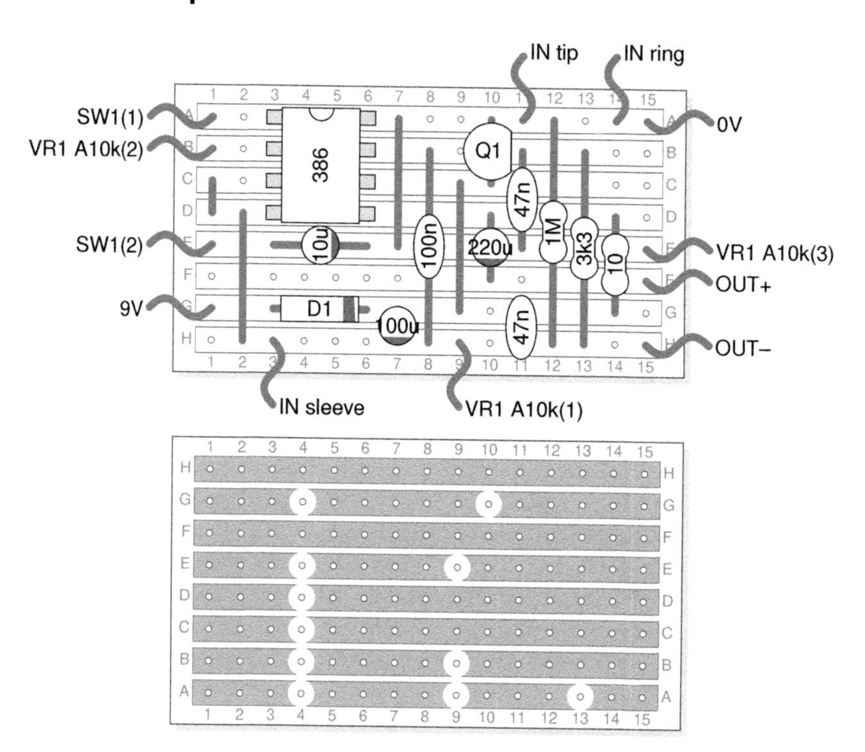

Stepped-Tone Generator

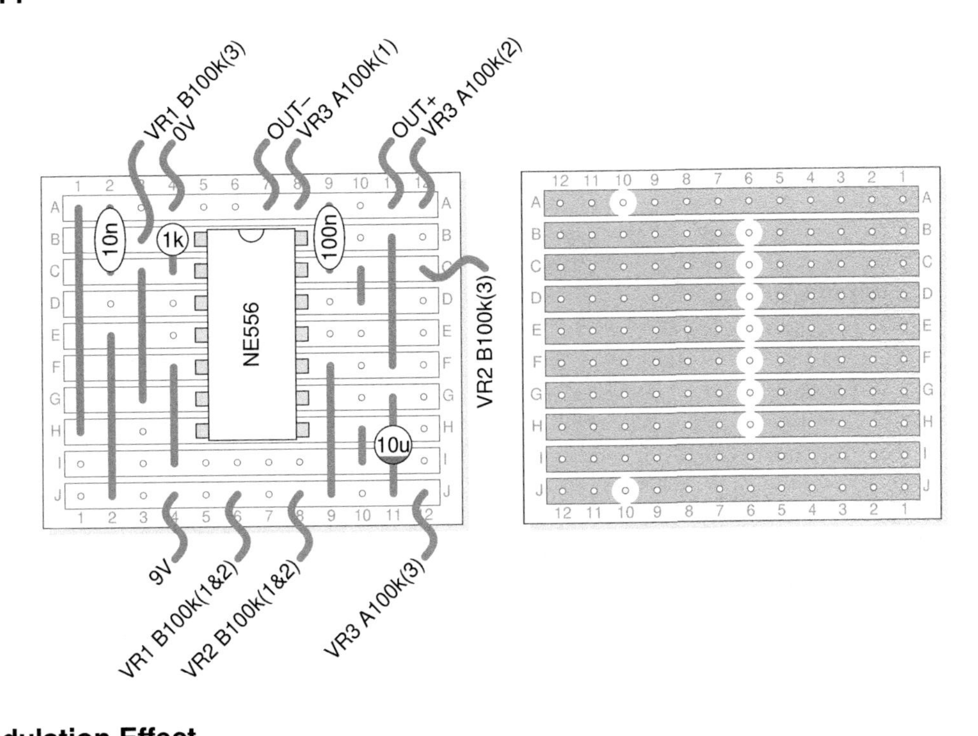

Modulation Effect

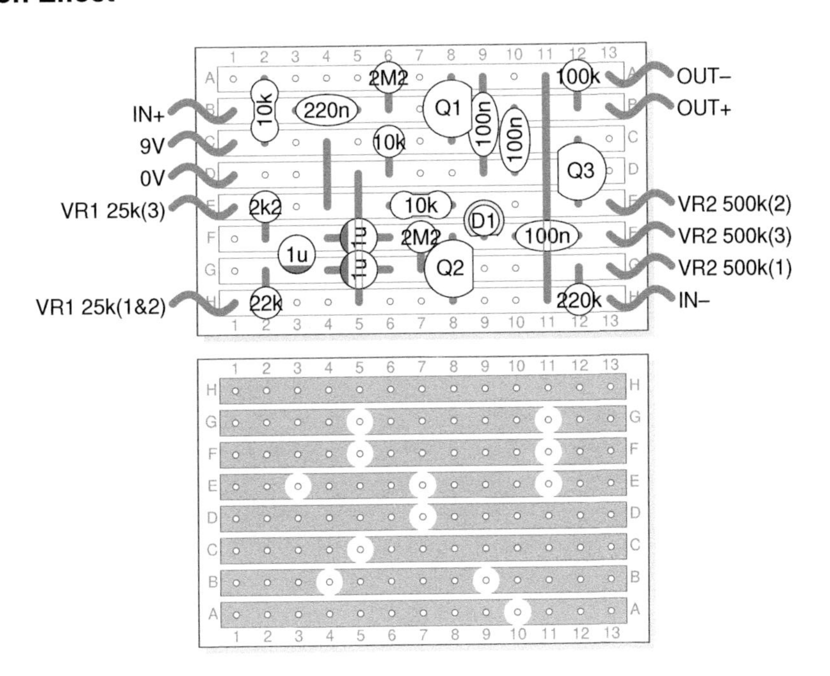

Noise Gate

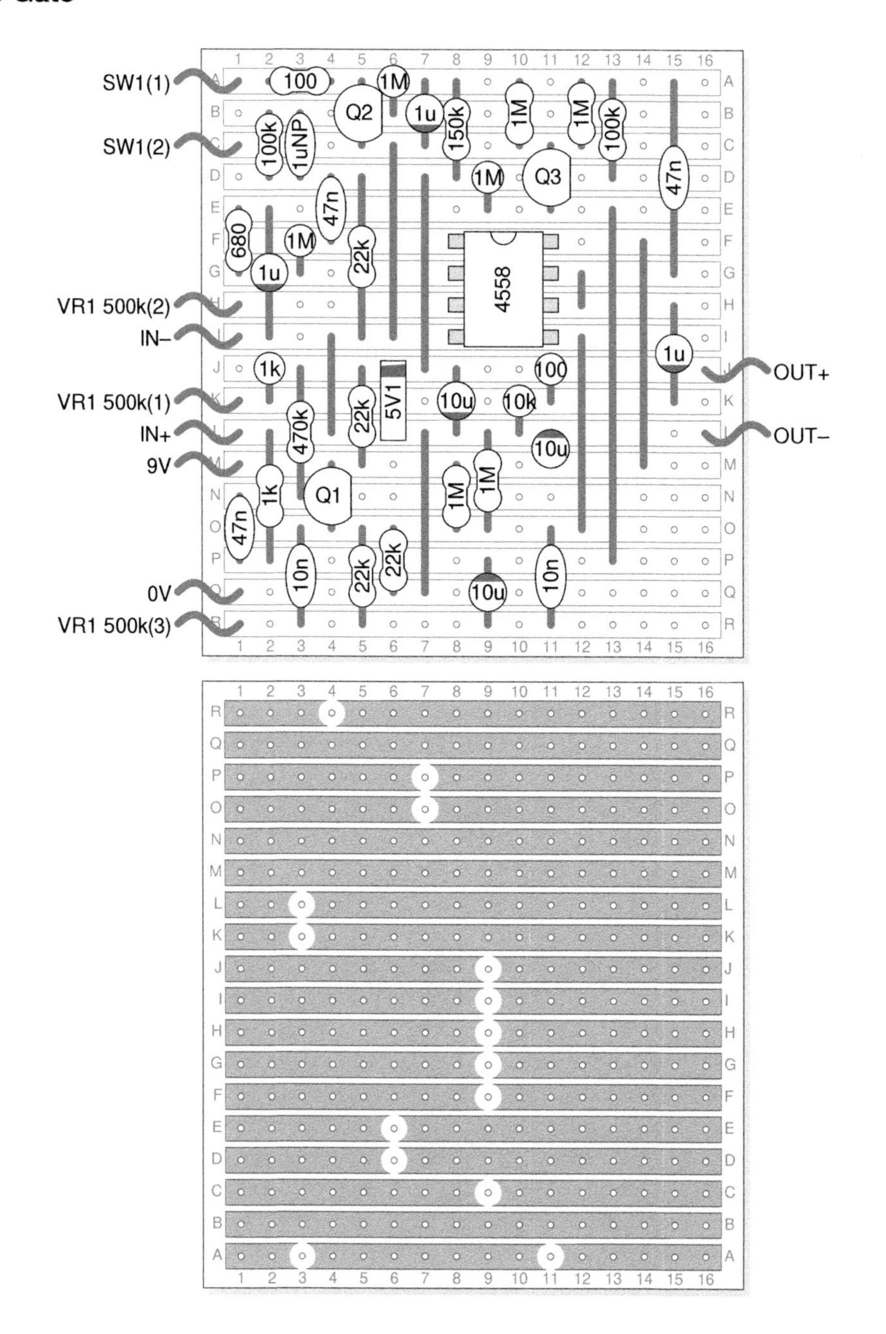
SW1(1)
SW1(2)
VR1 500k(2)
IN–
VR1 500k(1)
IN+
9V
0V
VR1 500k(3)
OUT+
OUT–
Q1
Q2
Q3
4558
5V1

Bibliography

Agilent. *Application Note 150: Spectrum Analysis Basics*. Agilent Technologies, 2004.

C. Anderton. *Electronic Projects for Musicians*. Amsco Publications, 1992.

D. Ashby, editor. *Circuit Design: Know It All*. Newnes, 2008.

G. Ballou, editor. *Handbook for Sound Engineers*. Focal Press, 4th edition, 2008a.

G. Ballou. Resistors, capacitors, and inductors. In G. Ballou, editor, *Handbook for Sound Engineers*, ch. 10, pp. 241–272. Focal Press, 4th edition, 2008b.

G. Ballou, editor. *A Sound Engineers Guide to Audio Test and Measurement*. Focal Press, 2009.

P. Baxandall. Negative-feedback tone control. *Wireless World*, 58(10):402–405, 1952.

P. Baxandall. Audio gain controls: A survey of the methods used to achieve acceptable control of gain in audio amplifiers. *Wireless World*, 86(1537):57–62, 1980a.

P. Baxandall. Audio gain controls 2: Obtaining equal gains in the two channels of a stereo pair. *Wireless World*, 86(1538):79–83, 1980b.

J. Bird. *Electrical and Electronic Principles and Technology*. Newnes, 3rd edition, 2007.

D. Bohn. *RaneNote 149: Interfacing AES3 & S/PDIF*. Rane Corporation, 2009.

J. Borwick, editor. *Loudspeaker and Headphone Handbook*. Focal Press, 3rd edition, 2001.

N. Boscorelli. *The Stomp Box Cookbook*. Guitar Project Books, 2nd edition, 1999.

C. Bowick. *RF Circuit Design*. Newnes, 2nd edition, 2008.

G. Briggs and H. Garner. *Amplifiers: The Why and How of Good Amplification*. Warfedale Wireless Works, 1952.

K. Brindley. *Starting Electronics*. Newnes, 4th edition, 2011.

E. Brixen. *Audio Metering: Measurements, Standards and Practice*. Focal Press, 2nd edition, 2011.

P. Buick and V. Lennard. *Music Technology Reference Book*. PC Publishing, 1995.

F. Camastra and A. Vinciarelli. Audio acquisition, representation and storage. In *Machine Learning for Audio, Image and Video Analysis*, ch. 2, pp. 13–55. Springer, 2015.

A. Carlson and D. Gisser. *Electrical Engineering Concepts and Applications*. Addison Wesley, 2nd edition, 1990.

G. Casella and R. Berger. *Statistical Inference*. Duxbury, 2nd edition, 2002.

G. Clayton and S. Winder. *Operational Amplifiers*. Newnes, 5th edition, 2003.

N. Collins. *Handmade Electronic Music*. Routledge, 2nd edition, 2009.

C. Coombs. *Printed Circuits Handbook*. McGraw-Hill, 6th edition, 2008.

P. Copeland. *Manual of Analogue Sound Restoration Techniques*. The British Library, 2008.

B. Cordell. *Designing Audio Power Amplifiers*. McGraw-Hill, 2011.

N. Crowhurst. *Audio Measurements*. Gernsback Library Inc., 1958.

N. Crowhurst. *Basic Audio, Vol. 1–3*. Rider, 1959.

H. Davidson. *Troubleshooting and Repairing Audio Equipment*. McGraw-Hill, 2nd edition, 1993.

H. Davidson. *Troubleshooting and Repairing Consumer Electronics Without a Schematic*. McGraw-Hill, 3rd edition, 2004.

G. Davis and R. Jones. *The Sound Reinforcement Handbook*. Hal Leonard, 2nd edition, 1989.

B. Duncan. *High Performance Audio Power Amplifiers*. Newnes, 1996.

J. Dunn. *Measurement Techniques for Digital Audio*. Audio Precision, 2004.

Elektor. Tup-tun-dug-dus. *Elektor*, 1(1):9–11, 1974.

A. Evans et al. *Designing With Field-Effect Transistors*. McGraw-Hill, 1981.

E. Evans, editor. *Field Effect Transistors*. Mullard, 1972.

F. Everest. *Critical Listening Skills for Audio Professionals*. Course Technology, 2007.

F. Everest and K. Pohlmann. *Master Handbook of Acoustics*. McGraw-Hill, 5th edition, 2009.

Fairchild. *AN-6609: Selecting the Best JFET for Your Application*. Fairchild Semiconductor Corporation, 1977.

J. Falin. *Reverse Current/Battery Protection Circuits*. Texas Instruments, 2003.

R. Fliegler. *The Complete Guide to Guitar and Amp Maintenance*. Hal Leonard, 1998.

T. Floyd. *Electronic Devices: Conventional Current Version*. Prentice Hall, 9th edition, 2012.

M. Geier. *How to Diagnose and Fix Everything Electronic*. McGraw-Hill, 2011.

S. Gelfand. *Hearing: An Introduction to Psychological and Physiological Acoustics*. Informa Healthcare, 5th edition, 2010.

R. Ghazala. *Circuit Bending: Build Your Own Alien Instruments*. Wiley, 2005.

A. Hackmann. *Electronics: Concepts, Labs, and Projects*. Hal Leonard, 2014.

L. Harrison. *Current Sources and Voltage References*. Newnes, 2005.

E. Heller. *Why You Hear What You Hear*. Princeton University Press, 2013.

J. Hood. *Valve and Transistor Audio Amplifiers*. Newnes, 1997.

B. Hopkin. *Musical Instrument Design*. See Sharp Press, 1996.

P. Horowitz and W. Hill. *The Art of Electronics*. Cambridge University Press, 3rd edition, 2015.

D. Howard and J. Angus. *Acoustics and Psychoacoustics*. Focal Press, 4th edition, 2009.

D. Hunter. *The Guitar Amp Handbook*. Backbeat Books, 2005.

IET. *Requirements for Electrical Installations BS 7671:2018 (IET Wiring Regulations)*. Institution of Engineering and Technology, 18th edition, 2018.

R. Jaeger and T. Blalock. *Microelectronic Circuit Design*. McGraw-Hill, 4th edition, 2011.

M. Jones. *Valve Amplifiers*. Newnes, 3rd edition, 2003.

M. Jones. *Building Valve Amplifiers*. Newnes, 2004.

W. Jung, editor. *Op Amp Applications Handbook*. Newnes, 2005.

Keithley. *Low Level Measurements Handbook: Precision DC Current, Voltage, and Resistance Measurements*. Keithley, 7th edition, 2016.

Keysight. *Impedance Measurement Handbook: A Guide to Measurement Technology and Techniques*. Keysight, 6th edition, 2016.

W. Leach. Fundamentals of low-noise analog circuit design. *Proceedings of the IEEE*, 82 (10):1514–1538, 1994.

W. Leach. Impedance compenstion networks for the lossy voice-coil inductance of loudspeaker drivers. *Journal of the Audio Engineering Society*, 52(4):358–365, 2004.

D. Levitin. *This is Your Brain on Music: The Science of a Human Obsession*. Dutton, 2006.

T. Linsley. *Electronic Servicing and Repairs*. Newnes, 3rd edition, 2000.

A. Malvino and D. Bates. *Electronic Principles*. McGraw-Hill, 8th edition, 2016.

R. Mancini, editor. *Op Amps for Everyone*. Texas Instruments, 2002.

M. Mandal and A. Asif. *Continuous and Discrete Time Signals and Systems*. Cambridge University Press, 2007.

C. Maxfield, editor. *Electrical Engineering: Know It All*. Newnes, 2008.

Maxim. *AN 636: Reverse Current Circuitry Protection*. Maxim Integrated Products, 2001.

B. Metzler. *Audio Measurement Handbook*. Audio Precision, 2nd edition, 2005.

E. Meyer and D. Moran. Audibility of a cd-standard a/d/a loop inserted into high-resolution audio playback. *Journal of the Audio Engineering Society*, 55(9):775–779, 2007.

F. Mims. *The Forrest Mims Engineer's Notebook*. LLH Technology Publishing, 1992a.

F. Mims. *Engineer's Notebook II: Integrated Circuit Applications*. LLH Technology Publishing, 1992b.

F. Mims. *Engineer's Mini-Notebook: 555 Timer IC Circuits*. Radio Shack, 3rd edition, 1996a.

F. Mims. *Engineer's Mini-Notebook: Op Amp IC Circuits*. Radio Shack, 2nd edition, 1996b.

F. Mims. *Getting Started in Electronics*. Master Publishing Inc., 2003.

C. Motchenbacher and J. Connelly. *Mullard Technical Handbook, Book 1 (Part 1): Semiconductor Devices, Transistors AC127 to BF451*. Mullard, 1974.

C. Motchenbacher and J. Connelly. *Low-Noise Electronic System Design*. Wiley, 1993.

N. Muncy. Noise susceptibility in analog and digital signal processing systems. *Journal of the Audio Engineering Society*, 43(6):435–453, 1995.

T. Needham. *Visual Complex Analysis*. Oxford University Press, 2000.

Y. Netzer. The design of low-noise amplifiers. *Proceedings of the IEEE*, 69(6):728–741, 1981.

S. Niewiadomski. *Filter Handbook – A Practical Design Guide*. Newnes, 1989.

S. Niewiadomski. Sharper by design. *Practical Wireless*, 80(9):34–35, 2004.

H. Ott. *Noise Reduction Techniques in Electronic Systems*. Wiley, 1988.

R. Penfold. *Computer Music Projects*. Bernard Babani, 1985.

R. Penfold. *More Advanced Electronic Music Projects*. Bernard Babani, 1986.

R. Penfold. *Electronic Projects for Guitar*. PC Publishing, 1992.

R. Penfold. *Practical Electronic Musical Effects Units*. Bernard Babani, 1994a.

R. Penfold. *Music Projects*. Newnes, 1994b.

A. Pittman. *The Tube Amp Book*. Backbeat Books, 2004.

W. Press, S. Teukolsky, W. Vetterling, and B. Flannery. *Numerical Recipes in C: The Art of Scientific Computing*. Cambridge University Press, 2nd edition, 1997.

RCA. *Receiving Tube Manual*. Radio Corporation of America, 1950.

M. Russ. *Sound Synthesis and Sampling*. Focal Press, 2nd edition, 2004.

P. Scherz and S. Monk. *Practical Electronics for Inventors*. McGraw-Hill, 3rd edition, 2013.

M. Schultz. *Grob's Basic Electronics*. McGraw-Hill, 11th edition, 2011.

A. Sedra and K. Smith. *Microelectronic Circuits*. HRW, 2nd edition, 1987.

A. Sedra and K. Smith. *Microelectronic Circuits*. Oxford University Press, 7th edition, 2014.

D. Self, editor. *Audio Engineering Explained*. Focal Press, 2010a.

D. Self. *Audio Power Amplifier Design*. Focal Press, 6th edition, 2010b.

D. Self. *Small Signal Audio Design*. Focal Press, 2nd edition, 2015.

D. Self. *Self on Audio*. Focal Press, 3rd edition, 2016.

W. Sethares. *Tuning, Timbre, Spectrum, Scale*. Springer, 2nd edition, 2005.

R. Severns, editor. *MOSPOWER Applications Handbook*. Siliconix Incorporated, 1984.

R. Shea, editor. *Transistor Audio Amplifiers*. Wiley, 1955.

I. Sinclair, editor. *Audio and Hi-Fi Handbook*. Newnes, 3rd edition, 1998.

I. Sinclair, editor. *Audio Engineering: Know It All*. Newnes, 2009.

I. Sinclair and J. Dunton. *Electronic and Electrical Servicing: Level 2*. Newnes, 2nd edition, 2007a.

I. Sinclair and J. Dunton. *Electronic and Electrical Servicing: Level 3*. Newnes, 2nd edition, 2007b.

I. Sinclair and J. Dunton. *Practical Electronics Handbook*. Newnes, 6th edition, 2007c.

G. Slone. *High-Power Audio Amplifier Construction Manual*. McGraw-Hill, 1999.

G. Slone. *The Audiophile's Project Sourcebook*. McGraw-Hill, 2002.

S. Smith. *The Scientist and Engineer's Guide to Digital Signal Processing*. California Technical Publishing, 2nd edition, 1999.

G. Sowter. Soft magnetic materials for audio transformers: History, production, and applications. *Journal of the Audio Engineering Society*, 35(10):760–777, 1987.

F. Stremler. *Introduction to Communication Systems*. Addison Wesley, 3rd edition, 1990.

T. Swike. *Guitar Electronics – Understanding Wiring*. Indy Ebooks, 2007.

E. Taylor. *The AB Guide to Music Theory – Part I*. ABRSM, 1989.

E. Taylor. *The AB Guide to Music Theory – Part II*. ABRSM, 1991.

Tektronix. *Fundamentals of Floating Measurements and Isolated Input Oscilloscopes*. Tektronix, 2011.

D. Thompson. *Understanding Audio*. Berklee Press, 2005.

TI. *The TTL Data Book*. Texas Instruments, 5th edition, 1982.

R. Tomer. *Getting the Most out of Vacuum Tubes*. Sams, 1960.

B. Trump. *The Signal: A Compendium of Blog Posts on Op Amp Design Topics*. Texas Instruments, 2017.

G. Valley, H. Wallman, and H. Wenetsky, editors. *Vacuum Tube Amplifiers*. McGraw-Hill, 1948.

B. Vogel. *How to Gain Gain: A Reference Book on Triodes in Audio Pre-Amps*. Springer, 2008a.

B. Vogel. *The Sound of Silence – Lowest-Noise RIAA Phono-Amps: Designer's Guide*. Springer, 2008b.

J. Walker. *A Primer on Wavelets and Their Scientific Applications*. CRC Press, 1999.

G. Weber. *Tube Amp Talk for the Guitarist and Tech*. Kendrick Books, 1997.

B. Whitlock. *AN-002: Answers to Common Questions About Audio Transformers*. Jensen Transformers, 1995.

B. Whitlock. Audio transformer basics. In G. Ballou, editor, *Handbook for Sound Engineers*, ch. 11, pp. 273–307. Focal Press, 4th edition, 2008.

R. Widlar et al. *Linear Applications Handbook*. National Semiconductor, 1994.

A. Williams and F. Taylor. *Electronic Filter Design Handbook*. McGraw-Hill, 4th edition, 2006.

S. Williams. Tubby's dub style: The live art of record production. In S. Firth and S. Zagorski-Thomas, editors, *The Art of Record Production: An Introductory Reader for a New Academic Field*, ch. 15, pp. 235–246. Routledge, 2012.

T. Williams. *The Circuit Designer's Companion*. Newnes, 2nd edition, 2005.

R. Wilson. *Make: Analog Synthesizers*. Maker Media, 2013.

E. Young. *Dictionary of Electronics*. Penguin, 1985.

H. Zumbahlen. *Linear Circuit Design Handbook*. Newnes, 2008.

Index

9780367359850